THE NEW NATURALIST

A SURVEY OF BRITISH NATURAL HISTORY

INHERITANCE AND NATURAL HISTORY

The aim of this series is to interest the general reader in the wild life of Britain by recapturing the inquiring spirit of the old naturalists. The Editors believe that the natural pride of the British public in the native fauna and flora, to which must be added concern for their conservation, is best fostered by maintaining a high standard of accuracy combined with clarity of exposition in presenting the results of modern scientific research.

THE NEW NATURALIST

INHERITANCE AND NATURAL HISTORY

R. J. BERRY
Professor of Genetics in the University of London

WITH 12 COLOUR PHOTOGRAPHS
19 PHOTOGRAPHS IN BLACK AND WHITE
AND 110 LINE DRAWINGS

COLLINS
ST JAMES'S PLACE, LONDON

William Collins Sons & Co Ltd
London · Glasgow · Sydney · Auckland
Toronto · Johannesburg

First published 1977

ISBN 0 00 219084 2
Filmset by Jolly and Barber Ltd., Rugby
Made and Printed in Great Britain by
William Collins Sons & Co Ltd Glasgow

CONTENTS

PLATES

(Plates 5–8 are in colour)

ACKNOWLEDGEMENTS

The originals of the colour plates are based on photographs by Professor E. B. Ford, F.R.S., Drs C. J. Cadbury, D. L. T. Conn, D. Horsley, J. R. Heal, D. R. Lees, and the author. Plate 8 (bottom half) originally appeared in the Annals of Eugenics, *10*, 1940. The black and white plates are based on photographs by Professor A. D. Bradshaw, Drs J. P. Baker, C. J. Cadbury, M. Daker, D. R. Lees, U. Mittwoch, H. N. Southern, Mr R. Nash, and Aerofilms, Ltd. The bottom photograph of plate 10 originally appeared in a British Ecological Society Symposium on the *Scientific Management of Animal and Plant Communities for Conservation*, published by Blackwells; the top of plate 11 in the Proceedings of the Royal Society, *175* B, 1970; and plate 12 (top) in *Ecological Genetics* by E. B. Ford, published by Chapman & Hall.

Text figures previously published include:

1, 22, 26, 96, 97, 110, from *Teach Yourself Genetics*, by R. J. Berry (English Universities Press, 1965). 9, from *The Process of Evolution* by P. R. Ehrlich & R. W. Holm (McGraw-Hill, 1963). 11, 13, 88, from Eugenics Review. 24, from *Symposium of Zoological Society*, 26, 1970 (Academic Press). 25, from *Critical Supplement to the Atlas of the British Flora* edited by F. H. Perring & P. D. Sell (Nelson, 1969). 33, 95, from Mammal Review. 19, 34, 43, 61, 62, from Heredity. 35, 38, from *The Evolution of Melanism* by H. B. D. Kettlewell (Oxford University Press, 1973). 39, from *Symposia of Society for Experimental Biology*, 7, 1953. 40, from *Human Diversity*, by K. Mather (Oliver & Boyd, 1964). 41, 73, 77, 79, 94, from Journal of Zoology. 45, from Proceedings of the Royal Society, ser. B. 51, from Science, New York (copyright 1973 by the American Association for the Advancement of Science). 53, 54, from *Farm Animals* by J. Hammond (Arnold, 1971). 55, from Science Progress (Blackwells). 56, from *Principles of Human Biochemical Genetics* by H. Harris (North-Holland, 1970). 60, from *The Fulmar* by J. Fisher (Collins, 1952). 41, 65, 70, 80, 102, from *Scientific Management of Animal and Plant Communities for Conservation*, edited by E. Duffey & A. S. Watt (Blackwells, 1971). 74, 75, from *Symposia of the Society for the Study of Human Biology*, 8, 1968. 82, from Hereditas, Lund. 84, from *Mollusca* by A. H. Cooke (Cambridge University Press, 1895). 89, from Nature in Wales. 91, from Field Studies, by permission of the Field Studies Council. 93, from Experimental Gerontology. 99, 100, 101, from *Ecological Genetics and Evolution*, edited by E. R. Creed (Blackwell, 1971). 103, from *The Domestication and Exploitation of Plants and Animals*, edited by P. J. Ucko & G. W. Dimbleby

(Duckworth, 1969). 104, from *Man's Natural History*, by J. S. Weiner (Weidenfeld & Nicolson, 1971). 105, from Journal of Biosocial Science, 106, from *Phenetic and Phylogenetics Classification*, edited by V. H. Heywood & J. McNeill (Systematics Association, 1964). 109, from *Genetic Effects of Radiations*, by C. E. Purdom (Newnes, 1963).

EDITORS' PREFACE

At first sight 'Inheritance and Natural History' may seem a surprising title for an addition to the New Naturalist series, which claims to deal primarily with the native fauna and flora of Britain. However, readers will soon find that Professor R. J. Berry's book fulfils all the requirements of the New Naturalist reader. It does not simply describe the conditions which exist today; it does much to explain how the animals and plants which survive in Britain have adapted to the changing conditions of our islands over the past centuries, and how that adaptation continues until the present day.

Many naturalists in the past have considered that Britain has what they call its 'native' fauna and flora, consisting of those organisms which were present when the land bridge with the continent of Europe was cut some 12,000 years ago. Some have even implied that these plants and animals are in some way 'superior' and more worthy of study than the so-called 'aliens' which have reached our shores, generally as the result of man's activities, during more recent years. It is, of course, impossible to divide these two groups, natives and aliens, in a completely logical manner. We can easily recognize recent immigrants such as the Grey Squirrel and the Colorado Potato Beetle, particularly when they are serious or potential pests, but many of us accept as British the rabbit, which was introduced nearly a thousand years ago. We also generally welcome birds and insects which invade Britain under their own power, particularly in the case of species which were previously resident and were exterminated by man. So we know that our wildlife is changing in composition; some species disappear and others are introduced. However, it is generally believed that the 'native' species are of particular interest as they are very similar if not identical to their ancestors who moved into our country after the last ice age. Many conservationists have striven to maintain their genetic purity, and have tried to prevent contamination from alien sources, for instance by 'boosting' the population of rare species by the introduction of reinforcements from flourishing colonies in other parts of the world.

Professor Berry shows that our fauna and flora has not remained unchanged over thousands, or even hundreds, or tens, of years. He describes, from his own meticulous studies of small mammals and other creatures, particularly on many of our smaller islands, and from a comprehensive analysis of the work of many other scientists and naturalists, how evolution and natural selection still operate and how rapidly populations change in their anatomy, physiology and behaviour. His explanations are based on the use of laboratory techniques for the biochemical study of changes in the cells of the animals, and on an understanding of modern genetics and the genetic code, but he also follows in the tradition of Charles Darwin and the great British naturalists in his ability to make his own observations in the field – and, more important, to recognize and interpret the significance of these observations.

Inheritance and genetics are not easy subjects, and although Professor Berry has a genius for clear exposition some New Naturalist readers may find parts of his text, particularly when mathematical notations are included, rather heavy going. Our advice is that, on first reading, they should enjoy the main descriptive sections and skip any paragraphs which they find difficult. They will probably return to these later, and find that comprehension comes more easily than they expected, when they see how the theoretical arguments illuminate the information derived from the studies of living populations of wildlife.

This absorbing book, is not, as Professor Berry insists, a text book of genetics. It is primarily intended to explain how the fauna and flora of Britain have evolved, and are still evolving. Nevertheless it should be of interest to naturalists and to students of nature in all parts of the world. Though many of the examples are taken from the British Isles, work in other countries is included when it is relevant. It is also clear that the same evolutionary process is operating in all parts of the world, and that Britain is perhaps serving as a laboratory, where the results obtained obviously have a much wider application.

AUTHOR'S PREFACE

THIS book has its origins in a letter to me from John Barrett (author of the *Pocket Guide to the Sea Shore*, and *Life on the Sea Shore*), whom I got to know well when he was Warden of Dale Fort Field Centre in Pembrokeshire and I was researching on mice on the nearby island of Skokholm (see Chapter 7). He wrote, 'The other day it occurred to me that nobody had written a book about genetics for everyman – genetics as they apply to the world around, gardeners, farmers, naturalists, racialists, wild populations, evolution; genetics that would lie behind almost all the *New Naturalist* books. I know there must be textbooks, probably galore, but these are inevitably somewhat or very technical and therefore closed to all those who have not had some training and experience in the language and techniques. I at once put this idea to Collins, suggesting that you might be persuaded to start from the beginning and explain genetics to the layman . . . Think about it – probably the notes you must have for your course of lectures to 1st year students are a substantial part of the book almost ready written'.

My first reaction to this was to revile John Barrett (and I am far from the first to do that), but having talked with Michael Walter of Collins, my enthusiasm began to grow for the idea of using examples largely from British natural history to explain why inherited properties are important for a full understanding of ecological problems, and how genetical variation is maintained in animal and plant populations. The great joy of this sort of book from the author's point of view is that it does not have to be as inclusive and balanced as a textbook; the chapters that follow involve a selective and therefore idiosyncratic choice of topics and examples. My main aim has been to show the ways in which inherited variation can help to explain the properties of natural populations. I have been concerned to describe and expound on British populations; for this reason some classical studies (like those of Dobzhansky on *Drosophila* species, Clarke and Sheppard on swallowtail butterflies, or Lack on the Galapagos island finches) receive only mention. I am unrepentant about this restriction I have chosen. British biologists have contributed disproportionately to the understanding of genetical processes in natural populations. It is doubtful if the reason for this is 'the fascination with birds and gardens, butterflies and snails which was characteristic of the prewar upper middle class from which so many British scientists came' (Lewontin, 1972). Sociologically and scientifically, such a statement betrays an ignorance of the British attitude to their countryside which has run from Neckham and Ray to Gilbert White and *Watership Down*. Moreover, in the context of the ideas of this book, it was two Oxford biologists, Charles Elton and E. B. Ford who inspired many throughout the world to begin thinking, and it is understandable that the fruits of their work should so far have been mainly in Britain.

Inheritance and Natural History is not my first-year lectures to university students, but it is related to a course my colleague, Dr J. S. Jones, and I have given for several years past to final year biology undergraduates at University College London. (To be honest, my 1st year lectures have already been published – as *Teach Yourself Genetics*, 2nd edition, 1972; published by English Universities Press.) I have left out most of the sums necessary for specialists grappling with quantitative problems. Readers wanting to follow up these topics will have to turn to one of the text-books listed in the references (such as Li, 1955; Falconer, 1960; Crow and Kimura, 1970; Mather and Jinks, 1971; or Cavalli-Sforza and Bodmer, 1971). Earnest students may like to know that two small books are 'required reading' before our University of London course. These are Wilson and Bossert's *Primer of Population Biology* (1971, Sinauer) and Sheppard's *Natural Selection and Heredity* (4th edition 1975, Hutchinson).

But I have not written this book with earnest students in mind. The people I am addressing are all those of us who wonder why people or organisms differ; who want to know why starlings are more speckled in Scotland than in England, or why pansies may be blue or yellow (I do not know myself, but the clue to the answers should be found in the following pages); in short, I have written for enquiring observers. I have tried to avoid jargon and the sort of 'in-writing' that some scientists love to cultivate to make sure that they can be understood only by the initiated. Wherever possible I have used the scientific name for a species only at its first mention. Professional biologists will recognize that I have sometimes dealt with controversial points in a dogmatic way. I hope that I have always given sufficient reason for my beliefs, as St Paul would have said. Certainly professionals have not yet appreciated to the full the implications of either the enormous amount of inherited variation found in virtually all natural animal or plant populations, or the repeated finding that natural selection is much stronger than used to be suspected by either naturalists or theoreticians.

Charles Darwin wrote in his *Autobiography*:

'I have no great quickness of apprehension or wit which is so remarkable in some clever men, for instance, Huxley. I am therefore a poor critic; a paper or book, when first read, generally excites my admiration, and it is only after considerable reflection that I perceive the weak points. My power to follow a long and purely abstract train of thought is very limited; and therefore I could never have succeeded with metaphysics or mathematics. My memory is extensive, yet hazy: it suffices to make me cautious by vaguely telling me that I have observed or read something opposed to the conclusion which I am drawing, or on the other hand in favour of it; and after a time I can generally recollect where to search for my authority. So poor in one sense is my memory, that I have never been able to remember for more than a few days a single date or line of poetry.

'On the favourable side of the balance, I think that I am superior to the common run of men in noticing things which easily escape attention, and in observing them carefully. My industry has been nearly as great as it could have been in the observation and collection of facts. What is far more important, my love of natural science has been steady and devout.'

I have no aspirations greater than these, and if I can help more people to achieve such a blend of perception and humility, I will have achieved a great deal.

I have been singularly fortunate in my teachers. I took my first degree in R. A. Fisher's department at Cambridge, then worked under Hans Grüneberg in a department where the other professors were J. B. S. Haldane and L. S. Penrose. All of these would have expressed sardonic displeasure at the lack of precision in the pages that follow. Nevertheless they must all accept some responsibility for moulding (or, perhaps, failing to mould) my thinking on genetical problems. More especially, I would like to pay tribute to Bernard Kettlewell who first taught me to understand and investigate genetical problems in natural populations, and to whom I shall always remain grateful for demonstrating so clearly that true science is vastly greater than the absurd reductionism that too often goes on in laboratories. To my teachers, colleagues, and students I offer this book in the hope that it will drive them to study natural history more intelligently, even if their motivation is only to prove how stupid I have been in some of my statements.

The whole text was read in manuscript by Dr Caroline Berry, Sir Timothy Hoare, and Dr Josephine Peters; Dr J. S. Jones read most of it; and parts have been seen by Professor A. D. Bradshaw, and Drs C. Bantock, A. Cook, D. Heath, H. B. D. Kettlewell, D. R. Lees and H. N. Southern. All these have commented with various degrees of asperity, and I am extremely grateful to them. My thanks are due also to Mr A. J. Lee for drawing the original figures, to all who have taken trouble to help me with the illustrations, especially Professor E. B. Ford, F.R.S., for permission to reproduce previously published figures and photographs (see *Acknowledgements*), to Miss Marion Roper for her care with typing and checking and to Dr Jane Hughes for help with the proofs.

CHAPTER I

THE STUDY OF DIFFERENCES: GENES AND NOT-GENES

GENETICS is about differences between individuals, and about the transmission of these differences from generation to generation. This means that it operates where history and biology meet. Not surprisingly such a meeting produces a host of 'might-have-beens'. For example, George III lost Britain's American colonies, and one reason for this was that he lacked control over himself for much of his reign. Now George III probably suffered from a rare hereditary disease called porphyria (Macalpine & Hunter, 1968). It was this that led Shelley to write that he was 'an old, mad, blind, despised and dying king'. Would George IV have ever come to the throne if it had been known that his father had an inherited disease? Would the recent history of Western civilization have been any different if George III's illness had been correctly diagnosed and treated?

Noah was an albino. The Book of Enoch, a second century BC Jewish document describes him as having 'flesh as white as snow and red as a rose; the hair of his head was white like wool, and long; and his eyes were beautiful'. Another early Jewish book (the book of Jubilee: one of the Dead Sea scrolls from Qurman) tells us that he married his first cousin (Sorsby, 1958). If Noah and his wife were responsible for the repopulation of the world after the Ark episode, a quarter of the world's population should be albinos. The fact that only one in 10,000 of us is albino gives information about the uneven distribution of genes and perhaps about natural selection.

These stories illustrate the recurring themes of this book:

1. The differences that exist between individuals in a population. These are so many that any individual is likely to be distinct and possibly unique. This is a generalization that applies to a great range of both animals and plants as well as humans.

2. The extent and mechanism for transmitting characteristics from one generation to another.

3. The factors which control the frequency of a trait in a local group.

These three themes deal with the basic questions that we ask about any situation: What? How? and Why? In practice answers to them involve describing the diversity that exists in a group, determining the way in which the traits in question are inherited, and then measuring the contribution that each trait makes to the life of its possessor. This is not the usual way of working for a field biologist. His concern is most commonly with the relative abundance of the different sorts (species) of animals and plants that make up a community, and how individuals of a species function and interact.

However, not all individuals of a species contribute equally to a community. To take a well-known example, moths which rely for their survival on camouflage have a high chance of living long enough to breed in industrial areas only if they are darker coloured than their relatives living in unpolluted countryside. Twice as many dark than light Peppered Moths (*Biston betularia*) released in a wood on the edge of polluted Birmingham were recaptured in following nights. Direct observation of the moths during the daytime showed birds (notably Hedge Sparrows (*Prunella modularis*), and Robins (*Erithacus rubecula*)) searching the trees where the insects were sitting, and taking a high proportion of the light, more conspicuous ones (Kettlewell, 1955*a*) (see pages 120–30). By comparing the relative success of dark and light forms of the same species, we may learn more about the biological pressures acting on the moths and their predators than by studying the ecology of the moths in its own right. The colour variation between individuals of the same species gives us a natural experiment which we must learn to recognize and interpret.

The spread of melanic forms in a hundred species of trunk-sitting moths in Britain since the industrial revolution in the middle of the nineteenth century is probably the best known example of genetical change. In man, of course, examples of variation and population differences abound – blood groups, eye colour, particular abilities (such as the musical talents of the Bachs or the literary ones of the Pakenhams), racial characteristics (skin colour, hair form, general physique, and so on), and many others. A few other examples come easily to mind: breed differences between domesticated animals and plants; insecticide resistance in house-flies, mosquitoes and aphids; warfarin resistance in rats and mice; antibiotic resistance in disease organisms. Obviously genetical variation must have occurred in the dim past to allow evolutionary divergence, but for most people genetics exists more as a matter of pride (or despondency) in their own families than as a fact of real biological importance.

Until the mid-1960s this would have been a justifiable attitude. Despite the enormous diversity that obviously existed in human populations, it was generally accepted that the general rule for animals and plants was uniformity. After all, a mouse was much the same wherever it was caught, and features such as a white belly or tail tip were apparently of little general significance. It was the insight of pioneers such as E. B. Ford in Britain and Theodosius Dobzhansky in the USA who combined ecological techniques with genetical ones which has led to our present understanding of genetical variation. A major contribution to positive thinking was Ford's (1940) emphasis on the significance of the occurrence together in a population of two or more distinct types (such as black and white moths, or blood groups in man). He defined as *polymorphism* 'the occurrence together in the same locality of two or more discontinuous forms of a species in such proportions that the rarest of them cannot be maintained by recurrent mutation'. We shall have to discuss in some detail the meaning of the terms described. For the moment it is sufficient to say that the existence of polymorphism implies a balance of forces of natural selection and has led to some valuable insight into population structure.

GENETICAL UNIFORMITY AND HETEROGENEITY

However, polymorphism was always believed to be the exception; genetical uniformity was the rule. We now know this to be incorrect: a much higher level of heterogeneity occurs in natural populations than used to be thought possible. It seems likely that we all inherit slightly different forms from our two parents of about one in thirteen of the 10,000 genes we receive from each, and that perhaps a quarter of these 10,000 genes will be represented by more than one form in a large population.

This change in understanding has come about largely through applying the simple techniques of electrophoresis (pages 24–5), which has provided an apparently objective way of measuring protein variation.

Proteins (or more strictly, the amino-acid chains of which they are made) are the primary products of genes (pages 34–5). The simplest alteration in a protein is when one amino-acid is substituted for another, and this will usually come from a change in a gene. Electrophoresis detects changes in the net electrical charge on a molecule, which means about a third of all possible amino-acid substitutions.

If we look at a range of proteins in a group of organisms, there are two estimates possible: the number of different versions of a protein that exist, and

TABLE 1. Frequency of enzyme variations in humans (after Harris, Hopkinson and Robson, 1973)

Enzyme	*Number of unrelated individuals tested*	*Number of different rare alleles*	*Number of heterozygotes per 100 individuals*	
			for rare alleles	*for common alleles*
Phosphoglucomutase-1	10,333	5	1.16	366
Phosphoglucomutase-2	10,333	3	0.68	—
Glutamate oxalate transaminase	1,195	0	—	33
Adenylate kinase	6,760	1	0.15	77
Nucleoside phosphorylase	1,542	2	1.30	—
Pyrophosphatase	2,190	0	—	—
Alkaline phosphatase	3,244	10	23.43	502
Peptidase-B	7,041	3	2.13	—
Triosephosphate isomerase	1,750	2	1.17	—
Superoxide dismutase	11,237	1	0.62	—
Inosine triphosphatase	641	0	—	—

the commonness of each. If a large number of individuals are tested, some variants turn out to be very rare, while others occur in more than one or two per cent of individuals (Table 1). The latter will fall within Ford's definition (*v.s.*) of polymorphic forms.

In 1966 Lewontin and Hubby working on a species of the Fruit Fly *Drosophila* in the United States and Harris working on man in England published estimates of the amounts of electrophoretic variation. Both studies revealed much more variation than had previously seemed possible. Since then, investigations of more species, both animal and plant, have shown that these first estimates were entirely typical (Table 2). As already noted, genetical heterogeneity rather than homogeneity is the rule.

TABLE 2. Frequencies of genic variants detected by electrophoresis (after Selander and Kaufman, 1973*b*)

	% of polymorphic loci	*Average % of loci in an individual which are heterozygous*
INVERTEBRATES		
Drosophila spp.	25–42	14.5
Land snails	—	20.7
Horseshore Crab	25	9.7
VERTEBRATES		
Fish	—	11.2
Lizards	—	5.8
House Sparrow	—	5.9
Rodents	23–29	5.6
Elephant Seal	17	3.0
Man	28	6.7

VARIATION: OLD AND NEW

Electrophoresis gives us unbiased estimates of the amount of inherited variation in the proteins tested. Previous estimates were always complicated by two problems, the difficulty of distinguishing traits determined by inheritance from those controlled by the environment, and of separating the action of one gene from that of others.

We can recognize the gene-environment problem wherever we turn. For example, snail shells are usually coiled dextrally (*i.e.* with the shell spiral turned in a right hand way), but sinistrals (left-handed shells) are found in most species. In the common Pond Snail (*Limnaea peregra*) the difference between left and right handed coiling is determined by a single gene; in the Roman Snail (*Helix pomatia*) the rare sinistrals seem to be the result of an embryological upset and not to be inherited.

The effect of the environment tends to be a more troublesome problem for botanists than zoologists. A dwarf plant may be the result of an undeveloped root system, an exposed situation, lack of an essential nutrient, disease – or inherited characteristics. Horticulturists deal with this situation experimentally by varying the conditions under which they grow their plants. If they are concerned with economic yield, they try to produce the optimum conditions for growth of their plant.

At the British Ecological Society's Research Station at Potterne in Wiltshire, Marsden-Jones & Turrill (1938) experimented for over 20 years to determine the effects of soil conditions on various native plants. They found considerable differences between species. For example the Knapweed (*Centaurea nemoralis*) showed little variation in widely different soils. On the other hand, the Broad-leaved Plantain (*Plantago major*) proved extremely plastic. The longer the ramets of this species were cultivated in a particular type of soil, the greater their divergences – differences appearing in such features as hairness and time of flowering, as well as in habit and size of parts.

Agriculturists have similar procedures for animals: Jersey cows produce large quantities of high fat milk – but they only do this if fed well; Ayrshires do not produce such a high quality milk, but maintain their yield under much harsher conditions than would be tolerated by Jerseys.

There are two points to make about the gene-environment interaction:

1. The genetical composition of an individual is distinct from and almost always unaffected by its environment. This realization dates from the work of embryologists (especially Weismann) at the turn of the century, showing that the cells which are going to produce the reproductive elements are distinct at an early stage of development from those which form the bulk of the body. This means that they are not influenced by the catastrophic or glorious events which mould the body. The sperm of a deep sea fisherman or an income tax clerk are protected from and are independent of the body and way of life of their bearers.

Yet less than a century before Weismann, Lamarck was basing the whole of biological diversity on modifications undergone by the reproductive cells during life: the puny weed who underwent a body-building course would be expected to sire an infant Atlas; the ancestral giraffe who triumphed in life by stretching upward and lengthening his neck would produce a longer-necked baby than he was; and so on. Indeed, one has only to go back a few years earlier than Lamarck to find the doctrine of spontaneous generation fully accepted, and this means that inherited factors would have no place at all. Gilbert White's frogs arising *de novo* in each generation would have had to have all their characteristics produced by their environment.

There have been numerous attempts to make the inheritance of characters acquired during life into orthodox doctrine. Koestler has described in *The Case of the Midwife Toad* (1971) the tragic history of the Austrian biologist

Kammerer who thought he had evidence of the environmental modification of inherited traits, and committed suicide when he discovered that some of his results had been faked by a faithful technician trying to help his master.

More recently, the Russian Lysenko claimed that many characteristics of plants could be affected permanently by environmental means (seed treatment or grafting). This was too tempting for his Marxist masters who wanted to believe that good proletarians could be produced by appropriate administrative decrees, and they boosted Lysenko and persecuted his more orthodox colleagues. Unfortunately for Lysenko no independent worker has been able to repeat his results, and he has become a figure of political rather than scientific history.

Even in Britain, Sir Cyril Hinshelwood, a distinguished chemist and President of the Royal Society, claimed in the 1950s that bacterial action was the result of the microbe's surroundings and not its constitution. This proved to be due to ignorance about the biology of bacteria; but it illustrates how easy it is to become confused in what appears to be a simple problem. The closest to a 'proof' for the inheritance of characters acquired during life has come from C. H. Waddington (1953*a*), who pointed out that characteristics arising during life from environmental pressures in fact show an inherited possibility of the organism responding to those pressures, and that 'responders' will tend to leave more offspring than 'non-responders'. This may lead to a character which at first appeared only in stressful situations, becoming manifest whatever the environment. Waddington called this 'genetical assimilation'. It could be particularly important in changing instinctive behavioural patterns.

A further complication is that there is increasing evidence of *some* reactions to external influences coming to be inherited. The most likely mechanism for this is a class of virus which can be incorporated into the constitution of the host, and become inherited instead of infectious. The virus is then called an *episome*. Nobody knows how important this process is, because of the difficulty of experimentation. Nevertheless, the general rules of inheritance (pages 36–9) certainly hold in the majority of cases.

2. It is largely meaningless to make a sharp distinction between 'inherited' characters and others determined 'environmentally'. The fact is that *any* characteristic of an organism must be based on the raw material provided by the inherited constitution. However, there are many possibilities for the gene products which are the raw material of an individual. They are like butcher's meat: it can be stewed, roast, fried or grilled; under-done or over-done; seasoned or plain; minced, sliced or served whole. The permutations are enormous. But no amount of culinary art will transform beef into lamb or bad meat into good. All our characteristics are the results of an interaction between genes and the environment, and usually between different genes as well. There is a gene in rabbits which produces yellow fat in its possessors, but only white fat develops unless xanthophyll is provided in the diet. Con-

sequently, a rabbit with white fat may be the result of a genetical or an environmental influence. Flamingos are only pink if they feed on and extract pigment from small Crustacea. Himalayan rabbits and Siamese cats develop pigment in the colder parts of their anatomy – their toes, nose and ears. If these regions are kept warm they will remain unpigmented; conversely if an area of skin in another part of the body is shaved it will cool and grow black hair.

The classical stupidity of making genes and the environment alternatives is illustrated by the argument about the inheritance of intelligence in man. The evidence is overwhelming that certain environmental factors (such as family size, parental attitude, exposure to an extended vocabulary) can influence ultimate 'intelligence', but also that the performance of children is highly correlated with that of their close relatives. Genes and environment interact together. The real difficulty is that examination success depends on a range of traits, such as application, neatness, and physical health, as well as on the maligned quantity we call I.Q. Until we know more about the chemistry of intelligence, we are unlikely to progress far in the semantics of the argument. The practical answer for the time being ought to be to maximise the intellectual response of every child. This is more likely to come from diversifying educational opportunities rather than homogenizing them. The reasons for not attempting this are political and doctrinaire, rather than biological and humanitarian.

If a group of organisms is reared in a uniform environment any differences between them can be attributed to inherited factors. However, it is extremely difficult in practice to ensure that a particular environment is uniform and always has been for the organisms living in it. The upper leaves of a plant may get more sunlight but less nutrients than the lower ones. Even a culture of insects reared on an artificial diet in a bottle may have experienced different conditions during development – differences due to the different amount of food available to early and late hatching eggs, of humidity as more eggs hatch, of progressive crowding and possible cannibalism of larvae, of the site where the eggs were laid, even the age of the mother and the amount of food. These may lead to differences among the young in size and any characters associated with it.

This brings us to the second problem associated with estimating the amount of variation in a group of individuals. Few of the characters which can be easily counted or seen are determined by only one gene. Most are affected by many genes and by a variety of environmental effects; conversely a single gene may influence the expression of a number of different characters. For example, nearly all the genes affecting the colour of mice seem to have some effect on body size; of 17 genes affecting eye-colour where mutations were induced by X-rays in *Drosophila melanogaster*, 14 showed effects on such apparently unrelated traits as the shape of the spermatheca (= sperm store) in the female. Taken together, these facts mean that differences between two groups in a character which can be measured (such as bone length or flower spike

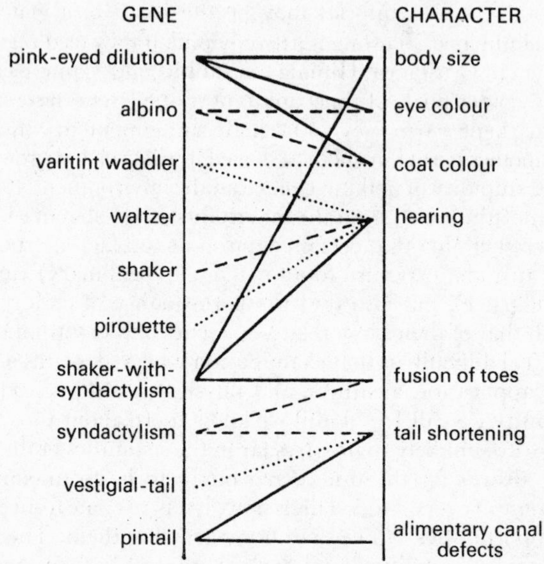

FIG. 1. The relationship between some genes and characters in the mouse: the connecting lines indicate the influence of a gene on a character. The product of one gene may influence many characters; a character may be influenced by the product of many genes (from Berry, 1965*a*).

height) may be inherited but it is very difficult to ascertain how many gene differences are involved.

This brings us back to electrophoresis. Technically, all that electrophoresis involves is running an electrical current through a protein-containing extract from an organism. The protein is put onto a gel (usually made of starch or a polyacrylamide ester), on which its position can be seen from its colour (in the case of a substance like haemoglobin) or by the use of an appropriate stain.

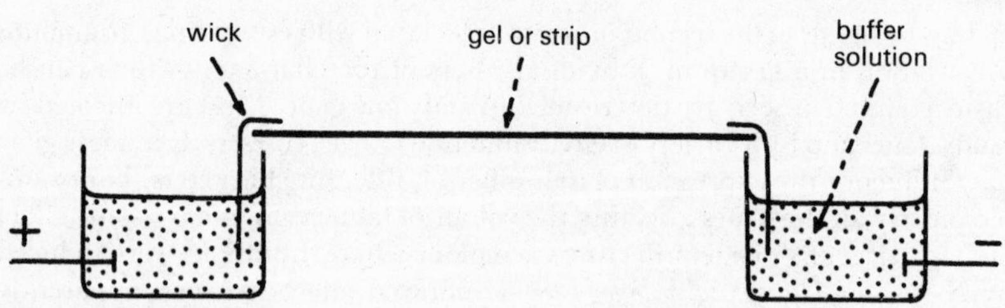

FIG. 2. Electrophoresis: current flows through the buffer solution, and along the gel or strip on which a protein extract is placed. The distance the protein molecule(s) moves in a given time is proportional to the net charge on the molecule.

The net electrical charge on the protein molecule will determine its direction and speed of movement when a current is applied. If two proteins are the same except for a particular amino acid which affects the net charge, they will move different distances in the gel in a given time. In this way they can be simply recognized as distinct.

Variation detected by electrophoresis has two virtues:

1. The proteins are primary gene products (or close to them), and they have a much lower probability of being dependent on more than one gene than a shape or size trait, and
2. If the proteins for electrophoresis are chosen at random, an estimate of genetical variation based on them should be unbiassed. In practice the proteins are chosen primarily if convenient detection methods for them are available, and in the past more 'functional' than 'structural' proteins have been looked at. Structural proteins may be less variable than functional ones, which could mean that current estimates of variation are too high (page 298f).

Notwithstanding any reservation that we have to make about the absolute level of genetical variability in most species, it is clearly very great – too great according to the earlier theoreticians for adapted individuals ever to occur, and for species to survive. A Nobel Prize winner, H. J. Muller put forward the idea in 1950 that there was a *genetical load* for any species dragging it towards extinction (pages 288–90). We shall have to consider in detail the forces operating to maintain the observed variation in any species. Suffice for the moment to note that there have been two major revisions of fact in population biology in the past two decades, and that the implications of neither has yet been intellectually assimilated. The first of these revisions is the fact we have discussed – the enormous amount of variation; the second is that strengths of natural selection are several orders of magnitude greater than used to be believed. Taken together these facts must colour our attitudes to a great spread of ecological phenomena.

POPULATIONS AND TYPES

The better we know any species, the more individuality we recognize. This is particularly obvious for humans. Our family and closest associates we tell apart easily; floating acquaintances have to be remembered in some identifying context (their dress, car, voice, smell, place of work); whilst whole classes of people are completely indistinguishable from each other (every Englishman knows that Chinamen all look alike, and cannot understand why Chinamen think all Englishmen look alike). Once we step outside the limits of our own species, we have to be specialists to be able to tell individuals apart. To the uninitiated all members of a herd of cows or a flock of sheep are identical, but the stockmen who look after them know most of them individually by small variations of pigmentation, shape, or behaviour. A rose grower will easily identify many breeds of roses, but a non-gardener will only be able to distinguish between different coloured flowers, or perhaps between plants with different growth habits.

Now if we transfer our thoughts from man and his commensals to the thousands of species which live in our surroundings, it will not be surprising if most of them are no more than anonymous groups. For example, the common Field (or Wood) Mouse (*Apodemus sylvaticus*) exists in a number of different forms in the British Isles. In past times 3 different species and 14 subspecies have been described: Devon mice have been said to be different from Derbyshire ones; Irish from English; Shetland from Hebridean; Lewis from Uist; and so on. Most of the subdivisions into these small taxonomic groupings have been made by over-eager systematists, and we shall have to consider how real these microgeographic groupings are. Nevertheless, there is no doubt that differences do exist between mice collected from different areas: the back colour may be redder or browner; the belly white or merely greyish; the brown spot between the fore-limbs large or virtually absent; the skull flat or high. But for most people the problem is not to distinguish between local races, but to be able to tell a Field Mouse proper from a Yellow-necked one (*A. flavicollis*) or even from the common House Mouse (*Mus musculus*). A few years ago there was a scare that rats had got onto the outlying Hebridean island of St Kilda, where they would have been likely to kill most of the ground-nesting birds. They were identified by an army sergeant recently posted from the Far East where rats are small, whilst on St Kilda the Field Mice are twice as big as mainland mice. Fortunately he was wrong; his 'rats' were the distinctive St Kildan Field Mice, *Apodemus* (*hirtensis*) *sylvaticus hirtensis*. But the confusion here was not between individuals, but between species. Notwithstanding differences probably exist between individuals of virtually every species. Even small flies may be different from each other: between 2.8% and 8.2% of *Drosophila melanogaster* individuals in France and 6.3–9.5% of *D. subobscura* in Greece have been shown to carry visibly detectable abnormalities, many of these identical with mutant genes studied in the laboratory (*q.v.* Dobzhansky, 1970).

The practical problem is that only if we study a group intensively will we be able to distinguish individuals apart. Differences between individuals are usually much less than ones between species. This is one of the main reasons why the study of genetics in natural populations has tended to be the preserve of the specialist. Nevertheless stamp collectors progress from merely collecting stamps to identifying watermarks, perforation aberrations, printing errors, and so on, and in the same way many naturalists graduate from identifying species to recognizing local and even individual differences in the forms they are particularly interested in. Read Lockley (1942) on his shearwaters or Jane Lawick-Goodall (1971) on her chimpanzees if you find this hard to believe.

Regrettably variant or local form description too often becomes an end in itself. Variants in nature are one manifestation of the inherited variation that occurs in all species. In the past, these variants have tended to be described as local forms or subspecies, or even aberrant individuals; hopefully we shall want to ask why these variants occur, and not stop at the collecting stage.

Firstly, however, let us look at some examples of local forms and their characteristics. Among zoologists, probably the most assiduous describers of geographical variants have been lepidopterists (see the two *New Naturalist* volumes by E. B. Ford, *Butterflies* and *Moths*). For example, professional collectors have been visiting the Shetland Islands (the northernmost part of the United Kingdom, lying relatively isolated in the North Atlantic) since the end of the nineteenth century, because a third of the moth species have developed local melanic forms which used to be marketable in much the same way as tiger-skin rugs. All these melanics have their own name in the scientific literature. The advantages of butterflies and moths for the study of variation are that variants of wing pattern or colour can be relatively easily recognized and displayed, and that many species can be reared to find out the method of inheritance of a particular variant.

It would be impossible to summarize adequately the extent of local variation in different groups; every widespread species has local forms. We can only describe genetical mechanisms and leave specific interpretations to specialists in particular groups. Fortunately, the nature of the inherited material and its transmission is such that we can make generalizations even though genetical studies on natural populations have developed in a piecemeal fashion. Bateson in his classical *Materials for the Study of Variation* (1894) and Huxley in *Evolution: the Modern Synthesis* (1942) describe a wealth of situations which could be investigated. I have made no attempt to catalogue all such situations in this book.

1. Monotypic and polytypic species

No one doubts the reality of the species as a basic biological unit and it was all very well for Linnaeus to give a detailed description and name to a species when he had only half-a-dozen specimens. The species as originally understood by Linnaeus was an abstraction, a pattern in the mind of God. (But even Linnaeus became less certain towards the end of his life, and dared not decide 'whether all species are the children of time, or whether the Creator from the very beginning of the world had restricted the course of development to a definite number of species'.) Any deviant specimens he labelled 'varieties'. This term he used for geographical races, for domestic animals and plants, and for horticultural 'sports'. For example, Linnaeus recognized six varieties of man, four of them geographical races (*americanus*, *europaeus*, *asiaticus*, *afer*) and the other two monstrosities or mythical. All this fitted the prevailing deistic ideas of creation theology. Linnaeus was as much in the scholastic tradition of the mediaeval writers of bestiaries, as Vesalius was a disciple of Galen.

However ideas of the species have developed, particularly by recognizing the possibility of variation. We have replaced the old 'type' with the idea of a *polytypic species*, which Ernst Mayr has described as one of the most important advances of modern biology (Mayr, 1963, 1969). This is because intra-specific diversity can be included, and brings the static formal description of a species closer to the truer, variable situation (see also Mayr, 1959). A polytypic species is simply defined as 'a species that contains two or more subspecies'. In other words a species may contain individuals which are not all alike. We are back with the problem that increasing study leads to increasing diversity.

Just as Renaissance craftsmen brought a return from the cloying thralls of sophisticated pedantry to observation and reality, so the modern recognition of variability has imposed a change in traditional taxonomy. Take the North American Song Sparrow: four similar species were discovered in eastern North American by the early explorers and pioneer ornithologists – the Fox Sparrow (*Passerella iliaca* 1786), the Swamp Sparrow (*P. georgiana* 1790), the Song Sparrow (*P. melodia* 1810), and Lincoln's Sparrow (*P. lincolni* 1834). During the exploration of the west in the mid-nineteenth century, several additional forms were discovered and named: *insignis* on Kodiak Island, *mifina* in Alaska, *gouldi* in California, *fallax* in Arizona, and others. These forms were described as 'species' because to their describers they seemed as different from each other as the four original species. However, as the ornithological exploration of North America continued, additional populations were found intermediate between the four western 'species' and the Song Sparrow (*melodia*) of eastern North America. As a result, all five 'species' were reduced to the rank of subspecies and combined into a single polytypic species with more than 30 subspecies. Sharpe (1909) listed 19,000 bird species throughout the world. The arrangement of these species (plus the many hundreds discovered since 1910) into polytypic species has reduced the total number to about 8,600 (Mayr & Amadon, 1951).

The European Hedgehog (*Erinaceus europaeus*) is another example of the same simplifying process. Its distribution and habitat requirements are known fairly accurately, and hedgehogs seem to be distributed fairly continuously throughout Europe and western Asia (with a separated group in eastern Asia). However the number of specimens available for detailed study has been few, and differences have been described both in the continental range and in various isolates (in Ireland, Britain, Scandinavia, and the Mediterranean islands). The difficulties of describing variation by 'sample taxonomy' is shown by the twelve names which have been proposed as species or subspecies within the continental range. Really the only distinction that can be asserted with any confidence is between a western European form, *europaeus*, with dark breast and short snout, and an eastern form, *concolor* (= *roumanicus*), with white breast and long nose. The boundary between the two has been shown to be objective, at least in Czechoslovakia, and additional differences in the skull and chromosome complement suggest that these forms may be separate species. The situation is complicated by the fact that animals

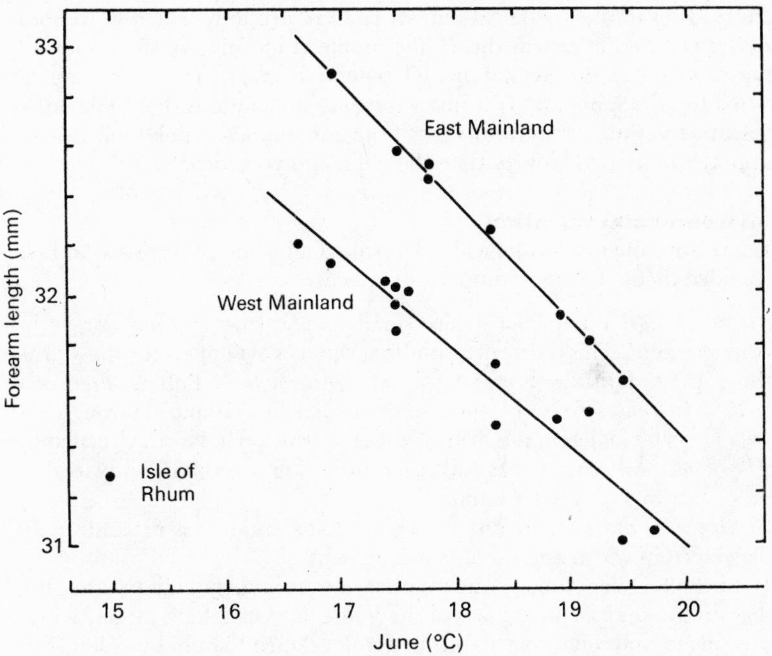

FIG. 3. At least three groups of Pipistrelle Bats occur in the British Isles, each with different associations between forearm mean length and summer temperature. Each dot represents the average measurement of about eighty bats from a single locality (after Stebbings, 1973).

from northern Russia seem to belong to the western 'race' (Corbet, 1970).

The opposite situation occurs in Pipistrelle Bats (*Pipistrellus pipistrellus*) in Britain. No one doubted that all Pipistrelles were alike until Stebbings (1973) started studying individual roosts in detail. He caught and measured the forearm length in nearly 2000 female bats from 21 colonies. Bats from the island of Rhum in the Inner Hebrides were smaller and darker than mainland animals and did not fit into the general pattern of variation. On mainland Britain there were differences between colonies from the west and east of the Pennines. On both sides the length of the forearm (and thus the overall size of the bats) increased in a northerly direction, and was significantly correlated with temperature, relative humidity, and wind-speed (which are all interrelated factors of the climate), and fitted Bergman's rule (see page 30). But the eastern forms were consistently bigger than the western ones for a given environmental variable. There are thus at least three races of Pipistrelles in the United Kingdom, probably representing different inter-breeding groups. If they remain separated from each other they may in time develop into distinct species.

This leads us to the biological unit we shall return to repeatedly. In both ecological fact and genetical theory the practical meaning of species for an animal or plant is the availability of potential mates. These are largely restricted to its neighbours. It is inter-breeding populations that make up a species, and these may or may not differ from one another. A polytypic species incorporates more real biology than the old monotypic idea.

2. Environmental variation

There are a number of 'ecological rules' called after the zoologists who first propounded them. The more important ones are:

i. *Bergman's rule:* related forms are smaller in warmer regions, larger in colder regions. This is true of altitudinal changes of temperature as well as those due to latitude. For example, the wing span of Puffins (*Fratercula arctica*) increases from 5½ inches in the Balearic Islands, through an average of 6½ inches in the British Isles, to 7½ inches in North Greenland.
ii. *Allen's rule:* tails, ears, beaks and other projecting parts of the body tend to be shorter in a cool environment.
iii. *Glober's rule:* dark colour due to melanin pigmentation is more strongly developed in warm and humid environments.
iv. In birds of colder climates the number of eggs in a clutch is greater, the digestive and absorptive parts of the alimentary canal are larger, wings are longer, and migratory habits better developed than in birds living in warmer areas.

The difficulty is to know how much of these differences are genetical and how much a direct result of the environment. For example, a laboratory mouse reared in a cold room has a shorter tail than if reared in a hot one. In this it parallels the trend in wild mice, in which Scottish mice have a tail 20% shorter than southern English mice. This difference persists if the mice are

TABLE 3. Sea Plantain (*Plantago maritima*) seeds from different habitats around the Forth estuary grown under identical conditions (Gregor, 1938, 1946)

Site from which seeds were collected	*Average length of flowering stems (cms)*	*% of sample in arbitrarily defined groups of growth form (1 = erect; 5 = recumbent)* 1	2	3	4	5
Waterlogged mud zone (salt concentration 2.5%)	23.0 ± 0.58	74.5	21.6	3.9	0	0
Intermediate habitats with intermediate salt concentrations	38.6 ± 0.57	10.8	20.6	66.7	2.0	0
Fertile coastal meadow above high tide mark (salt concentration 0.25%)	48.9 ± 0.54	0	2.0	61.6	35.4	1.0

bred in a laboratory, *i.e.* it is inherited. We have already noted that environmental modification of characters is more marked in plants than animals. Nevertheless undoubted inherited and continuous changes do occur in widely distributed species. They are called clines (pages 160–75). For example the dry matter in Scots Pine (*Pinus silvestris*) needles increases in a northerly direction throughout the range of the species, thus enabling the more northern trees to survive the harsh winters more easily (Langlet, 1959). The significance of such variation is easier to see if the material comes from different habitats in the same area: Table 3 shows the results of growing Sea Plantain (*Plantago maritima*) collected from different sites around the Firth of Forth, and grown under similar conditions in an experimental garden (Gregor, 1946).

3. Age and seasonal heterogeneity

A small sample from any locality may not be typical of the variation in the place. This is important if the characters under study change with age or season. For example, the average size of individuals in a mammal population will be larger at the beginning than the end of the breeding season, since a sample collected at the end will contain a high proportion of young animals.

A classical example of local variation is in the flower parts of the Lesser Celandine (*Ranunculus ficaria*). Ludwig (1901) found that plants from different localities had different numbers of carpels and stamens, and he tried to define local races in terms of these differences. Unfortunately, differences of the same magnitude can be found between early and late flowering plants from the same locality (Table 4). Since similar seasonal changes occur at different sites,

TABLE 4. Variation in the Lesser Celandine (*Ranunculus ficaria*) (after Briggs and Walters, 1969)

	N	*Number of Stamens*	*Number of Carpels*
Locality 1 (Gais)	80	23.83 ± 2.89	18.11 ± 4.29
Locality 2 (Trogen)	385	20.37 ± 3.82	13.26 ± 3.06
Early flowers } from one locality	268	26.73 ± 3.76	17.45 ± 3.89
Late flowers } from one locality	373	17.86 ± 3.30	12.15 ± 3.39
Plants under trees: 3 March	32	22.87	13.41
16 April	75	19.49	11.95
Plants on open cliffs: 31 March	100	38.24	32.32
4 May	43	30.67	25.72

there may not be distinct local races of Celandines. This means that care is needed in making the observations upon which any claim of population differences is to be based. It is not enough merely to collect data from a group of animals or plants, and not know how the characters being studied are controlled.

INHERITANCE IN FAMILIES: GENE ACTION

So far we have looked at the prevalence of variation in natural populations, and some of its consequences for traditional classification. We must not blind ourselves here: natural history for too many is merely being able to give names to species, and knowing where to find them. Let us now eschew this simplism and go on to the mechanisms of inheritance.

The bodies of both animals and plants are made up of *cells*. Plant cells have a wall made of cellulose while animal cells do not, but essentially all cells are similar: with few exceptions (such as mammalian red blood cells), they have a *nucleus* surrounded by cytoplasm. The cytoplasm is very variable – it may be little more than a store (as for secretions in gland cells), or it may contain all the machinery for digestion, recognition and reaction (as in the simplest animals and plants which consist only of one cell). Because we are concerned with inheritance, we are concerned mainly with the nucleus, and this seems to be similar in all organisms. Essentially it is an envelope containing a number of thread-like *chromosomes*. The wall of the envelope (or *nuclear membrane*) is a two-layered structure permitting the passage of small molecules between nucleus and cytoplasm.

The number and shape of the chromosomes are fairly characteristic of a species. At one time it was thought that the chromosomes were diagnostic of a species, but when good techniques became available, it became clear that they may vary as much as any other trait. We shall describe some of these consequences in Chapter 2. Normal humans have 46 chromosomes, 23 coming from the father and 23 from the mother. Forty-four of the 46 can be arranged in identical pairs. They constitute the *autosomes*. The other two chromosomes are a pair in females, but in males one of the pair is replaced by an unpaired element. These are the *sex chromosomes*. Females have two X chromosomes; males one X, one Y.

With few exceptions, the chromosomes of all 'higher organisms' can be arranged in pairs in this way. One chromosome set is contained in the male gamete, and the other in the female. The chromosomal constitution of the body is said to be *diploid* (= two sets of chromosomes), while the gametes have the *haploid* number. Fertilization of an egg unites two haploid complements, and all the body cells are formed by *mitotic* cell division which produces an identical set of chromosomes in each cell. The cell divisions immediately preceding gamete formation are *meiotic*, and result in the gametes having only one member of each chromosome pair (see Appendix).

The chromosomes themselves are made of a backbone of simple protein (mainly histone), around which is arranged a double helix of desoxyribose nucleic acid (DNA). DNA is a long chain polymer of four different nucleotides, two containing purine bases (adenine and guanine), two with pyrimidines (thymine and cytosine). The double helix is arranged rather like a spiralled rope ladder, with the bases linked together by hydrogen bonds to

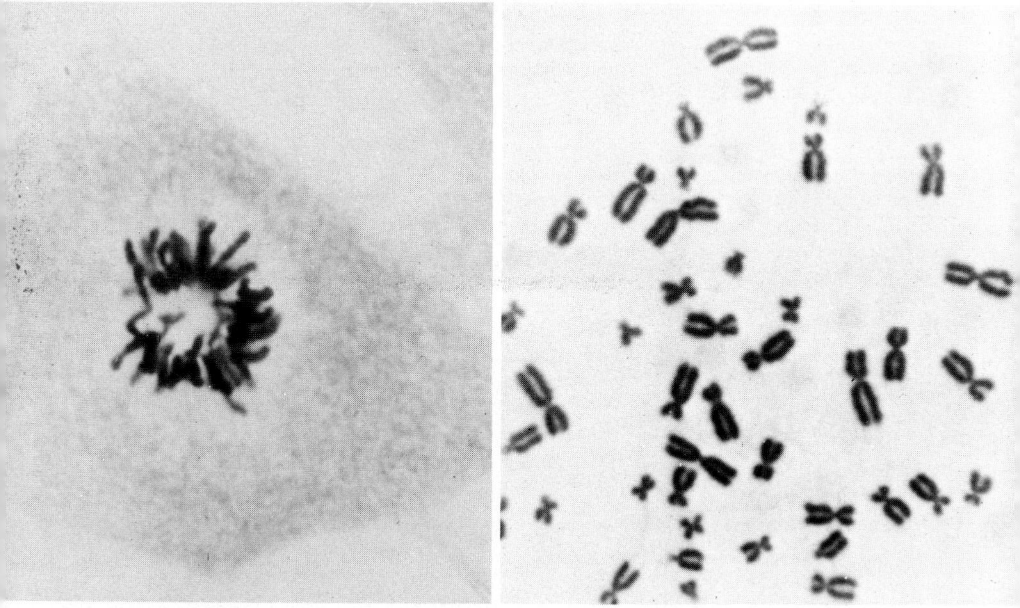

PLATE I. *Human (normal male) chromosomes (pages 32, 308)*

Above left, Cell division (mitotic metaphase) as seen under the microscope without special treatment. *Right*, metaphase in a cell treated with hypotonic saline which separates the chromosomes from each other.

Below, chromosomes from the cell shown above right cut out, paired and numbered to show the normal karyotype.

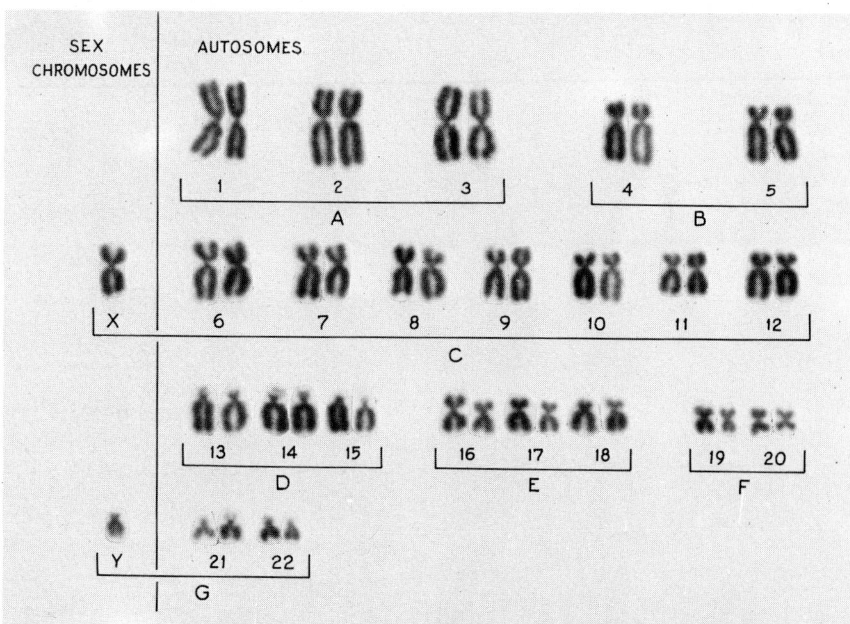

PLATE 2.
Biochemical variation

Separation of enzymes by electrophoresis to show inherited variation between different individuals (page 24).

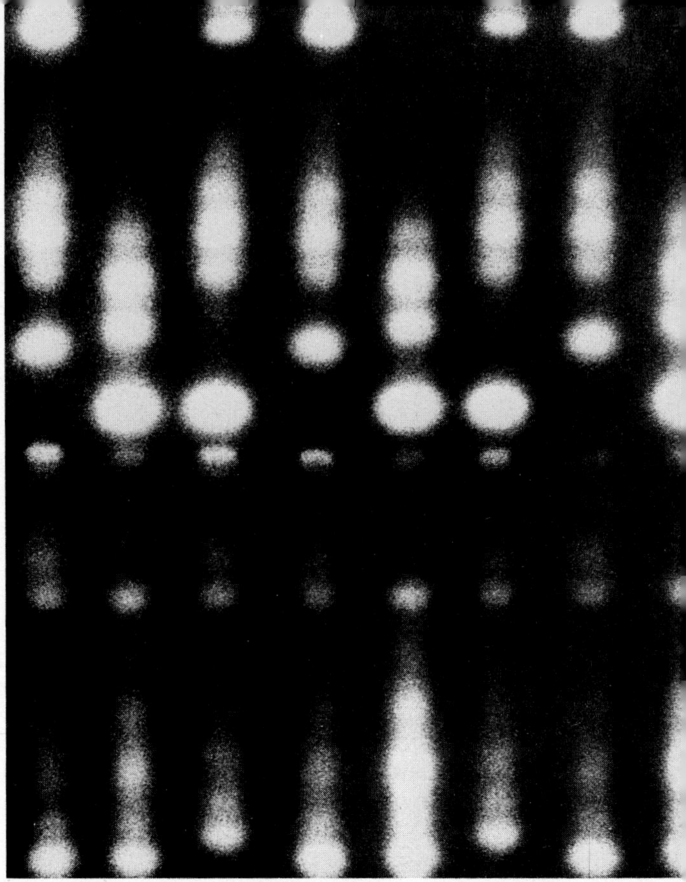

Eight House Mouse *(Mus musculus)* individuals, showing esterase variation (in liver extracts). Seven loci are represented by bands at different positions of each column; four of these loci are variable between the specimens.

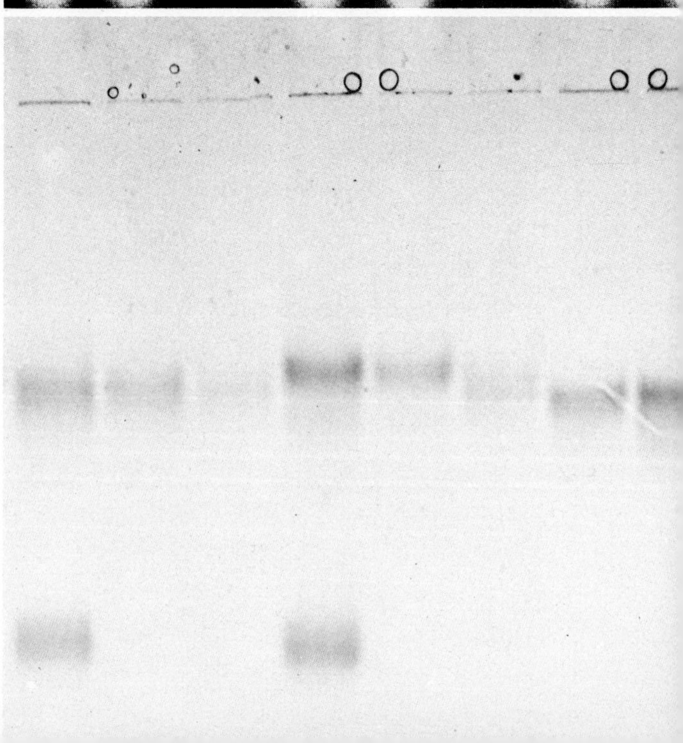

Eight Peach Aphid *(Myzus persicae)* individuals, also showing variation in esterases. The two specimens with two bands are resistant to organophosphorus insecticides; the others are susceptible.

FIG. 4. The structure of DNA, showing the complementary paired strands (top), linked together by hydrogen bonds; and the chemical structure of the molecule (bottom). Adenine (A) and thymine (T) always pair together, as do guanine (G) and cytosine (C). The backbone of the DNA molecule consists of phosphate (P) and deoxyribose sugar (S) groups.

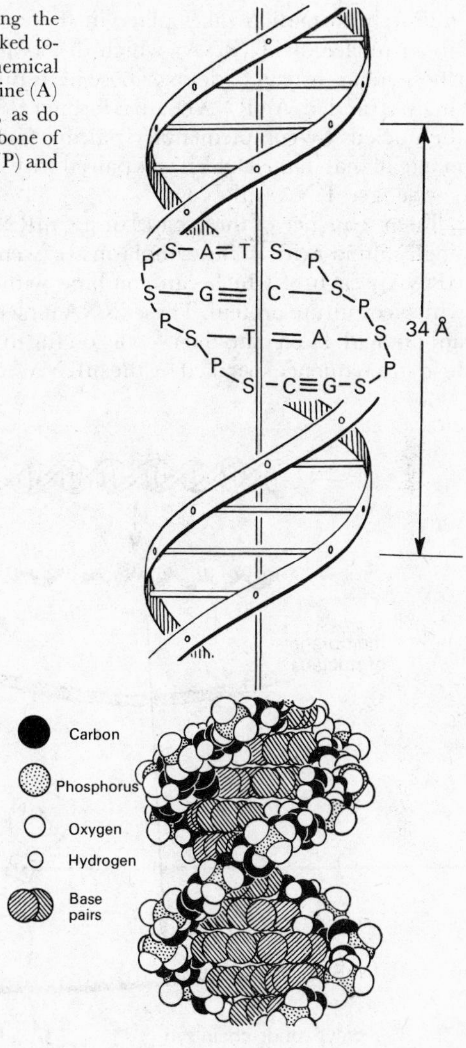

form the 'rungs' of the ladder. The spatial geometry of the paired polymers is such that adenine can only pair with thymine, and guanine with cytosine. This means that the two chains are entirely complementary. It was this complementarity which was the essential part of the elucidation of DNA structure by Watson & Crick (1953: see Watson's story of it in his book *Double Helix*, 1968). DNA is the actual hereditary material, and directs the formation of the specific proteins which are characteristic of an individual.

Protein formation takes place in the cytoplasm through the medium of ribose nucleic acid (RNA) which differs from DNA mainly in containing ribose sugar instead of desoxyribose in its nucleotides, and which is normally single-stranded. An RNA chain is formed alongside one of the DNA strands in the nucleus by complementary pairing (*i.e.* its order of bases is the 'mirror image' of that in the DNA it was paired with), and passes into the cytoplasm as 'messenger' RNA (mRNA).

Each sequence of three bases in the mRNA represents or is the 'code' of a single amino-acid. In the cytoplasm are twenty different sorts of transfer RNA (tRNA), each of which can combine with one of the twenty amino-acids which constitute protein. These tRNA molecules, each linked with its specific amino-acid, latch onto their 'code' on the mRNA so that the amino-acids are held in a sequence specified by the mRNA (and the DNA). Finally the amino-

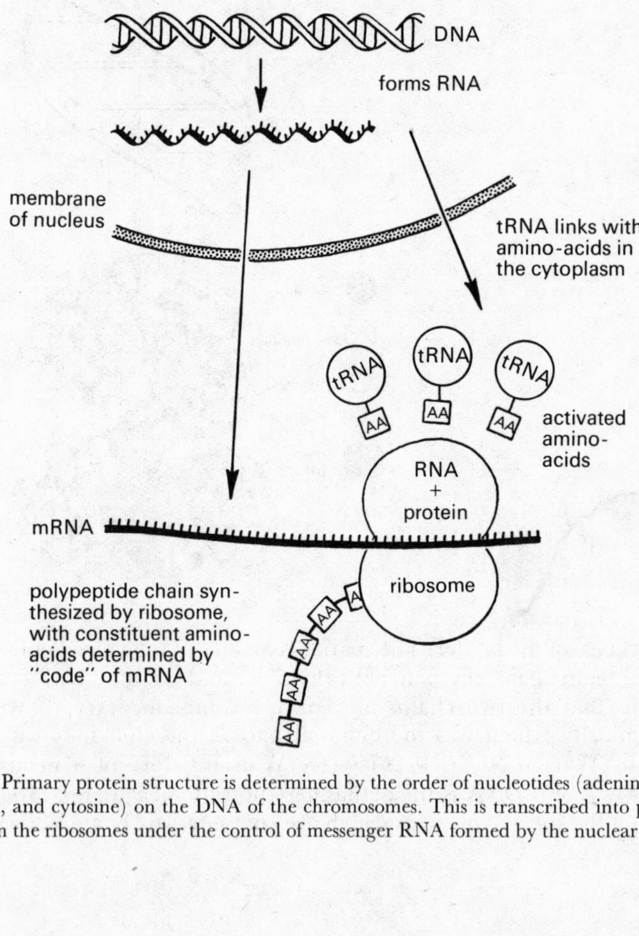

FIG. 5. Primary protein structure is determined by the order of nucleotides (adenine, thymine, guanine, and cytosine) on the DNA of the chromosomes. This is transcribed into polypeptide chains in the ribosomes under the control of messenger RNA formed by the nuclear DNA.

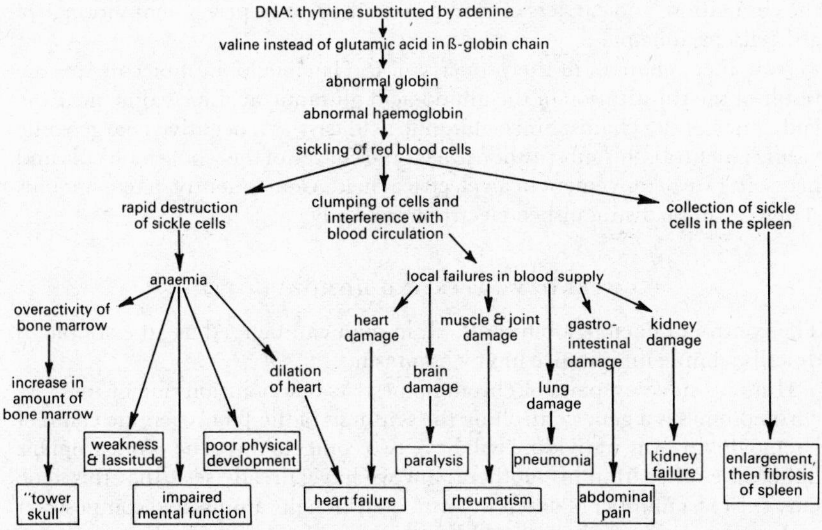

FIG. 6. The 'pedigree of causes' from a single DNA change in the gene coding for one of the polypeptides making up the haemoglobin of human blood. A variety of symptoms may arise.

acids join together to form a polypeptide chain, and are freed from the tRNA 'adaptors'. This final stage takes place in association with yet another sort of RNA, this time in association with protein forming ribosomes.

This description of the relation of genes and proteins has one aim: to point out that change in a single base in a DNA chain will produce a protein with one amino-acid substituted for another, and this may affect its properties.

For example, probably the best known protein in the animal kingdom is haemoglobin. A human haemoglobin molecule has four polypeptide chains (two α chains with 141 amino-acids each, and two β ones with 146 each) and an iron-containing porphyrin ring. Over a hundred different haemoglobins are known in man, virtually all of them differing from normal haemoglobin in one or two amino-acids in the α or β chain. Knowing the 'genetic code', it is possible to deduce in every case the base change which has occurred to produce the changed haemoglobin. One haemoglobin is of particular interest, because the length of DNA determining it (or, to use normal language, the gene controlling it) occurs at a high frequency in parts of Central Africa, and in North American blacks who originated from that area. This gene produces a haemoglobin (haemoglobin -S) which is 25 times less soluble than normal haemoglobin (haemoglobin -A) when it is not carrying oxygen. This leads to some of the red blood cells becoming sickle-shaped in the veins, as the haemoglobin 'tries to precipitate'. These cells are likely to be removed from the circulation, and carriers of the *S* gene suffer from a permanent shortage of red cells, or anaemia.

Now these changes in the property of the haemoglobin molecule are the result of the substitution of the amino-acid glutamic acid for valine near the end of one of the chains. Since glutamic acid carries a negative charge while valine is neutral, this substitution affects the charge of the whole molecule and hence its rate of movement in an electrical field. Consequently, haemoglobins *A* and *S* can be distinguished electrophoretically.

HETEROZYGOTES AND HOMOZYGOTES

The control of haemoglobin synthesis in man can be used as an example to describe simple inheritance in any organism.

Humans have 23 pairs of chromosomes. At one point on one of these 23 chromosomes is a gene controlling the synthesis of the β polypeptide chain of haemoglobin. Any individual will have two copies of this gene – one from his father, the other from his mother. Now we have already seen that this gene may exist in a number of different forms on which the amino acid composition of the polypeptide chain depends. For simplicity let us consider the possibility of only two forms of the gene – one coding for A haemoglobin, one for S. The forms of a gene in this sense are called *allelomorphs*, usually abbreviated to *alleles*. Moreover since 'gene' can have several meanings, it is best to describe the place on the chromosomes which codes for the β-haemoglobin chain as a *locus* rather than a gene. In the days at the turn of the century when Bateson invented the term gene it was a unitary inherited factor. We now know so much about its fine structure that the word 'gene' has come to be used as a portmanteau lacking the precise meaning of allele or locus.

If there is more than one allele possible at any locus, an individual may have the same allele on both chromosomes (in which case he is said to be a *homozygote*, or to be *homozygous* at the locus in question), or different alleles (when he is a *heterozygote*, or *heterozygous* at the locus). There will be two different sorts of homozygote: an individual may have either the haemoglobin -A or the haemoglobin -S allele on both chromosomes. We can write his genetical constitution or *genotype* at the locus as *AA* or *SS*. There can only be one sort of heterozygote *AS*.

Haemoglobin synthesis is a useful example, but we do not have to know the molecular mechanism of a gene to describe its possibilities in this way. If there is inherited variation at *any* locus, we can ignore all the other thousands of loci and concentrate only on the effects of the locus which interest us. It is not irrelevant to note that the Czech monk, Gregor Mendel, who discovered the whole business of single factor inheritance, did so by considering one factor at a time and not confusing himself with the whole organism as was the wont of his contemporaries.

GENE TRANSMISSION

When the gametes are formed – whether the male ones are pollen or sperm – the chromosome number is reduced, and each gamete contains only one of each pair of chromosomes. (This does not mean that the *haploid* chromosomes sets so formed are the ones received from the grand-parents. There is considerable reorganization (pages 45, 308–11).) If an individual was homozygous at a particular locus, *all* the gametes must carry the same allele; if heterozygous *exactly* half the gametes will carry one allele and half the other allele. In the general case of a locus *A* with two alleles *A* and *a*, individuals will be *AA*, *Aa*, or *aa*. Matings can be one of six types:

$$AA \times AA$$
$$AA \times Aa$$
$$AA \times aa$$
$$Aa \times Aa$$
$$Aa \times aa$$
$$aa \times aa$$

These can be divided into four groups:

1. Where both parents are the same homozygote ($AA \times AA$ or $aa \times aa$), there is no variation possible and all the offspring will be genetically identical with their parents.
2. Where the parents are the different homozygotes ($AA \times aa$), all the gametes from one parent will carry the *A* allele and all those from the other will carry *a*, and all the offspring must be heterozygous (*Aa*). This is the classical case of two fairly distinct parents producing children intermediate between them – as in skin pigmentation in a marriage between a black and a white.
3. Where one parent is homozygous and the other is heterozygous ($AA \times Aa$, or $aa \times Aa$), the heterozygous parent will produce equal numbers of *A* and *a* carrying gametes. These will have an equal chance of combining with a gamete from the homozygous parent (which will be either *A* or *a*), so equal numbers of homozygous and heterozygous offspring will occur – 'just like' the mother or the father respectively.
4. Finally when both parents are heterozygous, the *A* and *a* gametes from each will have equal chances of combining with *A* or *a* from the other. This means there will be equal numbers of four genotypes produced: *AA*, *Aa*, *aA*, *aa*. *Aa* and *aA* are genetically identical, so that the progeny will be in the ratio of 1*AA* : 2*Aa* : 1*aa*. A characteristic which is 'lost' in the offspring of a couple because the children are heterozygous and like neither parent, may recur in the second generation as homozygotes appear again. This is the explanation of a trait 'skipping a generation' so that a child resembles his grandparents more than his parents in a particular aspect.

There is a limerick of doubtful morality and incorrect genetics which illustrates the simple inheritance of single gene segregations:

> There was a young woman called Starkey
> Who had an affair with a darkey
> The result of their sins
> Was quadruplets not twins
> One black, one white, and two khaki

There are two genetical errors in this immoral tale. Firstly, skin colour is not controlled by a single locus like a protein or blood group. It is affected by a number (probably about six: Stern, 1970) of different loci which act together to produce the trait which we call 'skin colour'. The simple segregation of black, white and khaki does not occur.

Secondly, let us assume for simplicity that skin colour is controlled by a single locus. Both parents married to someone of their own colour would be pure-breeding, *i.e.* a white would only produce whites, and a black only blacks. This means we can assume both Ms Starkey and her paramour were homozygotes. We can designate their genotypes *WW* and *BB* respectively. All their children must be the same (*BW*), and all will be khaki however many there are.

If we compound the sin of the parents and let two of the khaki children commit incest, we then have two heterozygotes mating ($BW \times BW$), and *their* children will be in the ratio one black (*BB*), one white (*WW*) and two khaki (*BW*). The limerick writer left out a generation.

There is a simple rule that can make sense of any family segregation: work out the genotypes of the gametes produced by one parent and the proportion of each in the total gametic output, and write these down at the head of a table. Make the same calculation for the other parent, and write these down at right angles to the first. Then the cells in the table can be filled in with the genotypes of the progeny, and the frequency of the progeny types will be the product of the frequencies of the gametes determining that type.

For example with our incestuous khaki couple, each parent will produce $\frac{1}{2}B$ gametes and $\frac{1}{2}W$ gametes. Then:

		Father's gametes	
		B $\frac{1}{2}$	W $\frac{1}{2}$
Mother's gametes	$B\frac{1}{2}$	$\frac{1}{4}BB$	$\frac{1}{4}BW$
	$W\frac{1}{2}$	$\frac{1}{4}BW$	$\frac{1}{4}WW$

This may seem to make heavy weather of an extremely simple situation, but it becomes the key to understanding more complicated problems when more than one locus or more than two alleles are involved.

Understanding of even the most complicated developments in genetical processes follows directly from the simple segregations described above; if the workings of the 2 × 2 table above is still mysterious, try reading the previous couple of pages again. The whole of genetics is nothing more than 2 × 2 tables written many times and superimposed on one another.

Since it is so important to understand simple inheritance in families, let us look at another example, a real one this time. Antirrhinum flowers may be white, pink or red; white and red-flowered plants breed true if crossed with flowers of the same colour. They are therefore both homozygotes. We can call them *WW* and *RR* respectively. White crossed with red produces all pink plants. These will be *WR*. Finally, pink plants crossed among themselves produce white, pink, and red plants in the ratio of 1 : 2 : 1. The pink × pink situation is:

		pollen	
		W $\frac{1}{2}$	*R* $\frac{1}{2}$
ova	*W* $\frac{1}{2}$	$\frac{1}{4}$*WW* (white)	$\frac{1}{4}$*WR* (pink)
	R $\frac{1}{2}$	$\frac{1}{4}$*WR* (pink)	$\frac{1}{4}$*RR* (red)

COMPLICATIONS

There are three immediate complications of the simple genetical system that we must consider.

1. Dominance

Frequently a heterozygote is not intermediate between the two homozygotes, but is indistinguishable from one of them. This happens for human eye-colour: a blue-eyed person is always homozygous for an allele which produces no pigmentation in the iris. (The blue colour and finer shades of hazel, grey, turquoise and so on are caused by refraction in the superficial layers of the eye, and not by pigmentation); a brown-eyed person may be homozygous or heterozygous for the pigmenting allele. (Let me hasten to add that there are a large number of exceptions to this simple pattern. No one should start questioning his or anyone else's paternity solely on the basis of eye-colour.) A black or brown sheep is usually a homozygote, but a white one may be either heterozygous or homozygous. This is one reason why black lambs are often seen in a flock of entirely white sheep (some black lambs grow up into white adults).

If the heterozygote and one of the homozygotes at a locus are indistinguishable, the character produced by those genotypes is said to be inherited as a *dominant*, while the character which is only manifest when the alleles are homozygous is said to be *recessive*. These descriptions have nothing whatsoever to do with vigour or strength: blue-eyed people are not noticeably weaker than brown-eyed ones, while the primitive tough hill sheep of the north and west of Scotland are all brown (although they are increasingly being replaced by or crossed with southern white breeds) (Ryder, Land & Ditchburn, 1974).

We shall see later that dominance and recessivity are properties of a character and not usually of an allele (pages 188–93). It is possible to change the dominance of a character by appropriate breeding. Although we frequently speak of a dominant or recessive allele this is really short-hand, and we should bear in mind that there is a distinction between gene and character.

Since two people may have the same appearance but different genotypes, we need to be able to distinguish between appearance and genetical constitution. We do this by referring to appearance as *phenotype*. We can always know phenotype by observation; we may only be able to tell genotype by breeding. There was a notorious Holstein bull who sired 6,000 young by artificial insemination. He was heterozygous for an allele which caused a lethal abnormality of bone growth when homozygous. Although he was completely healthy and normal himself, more and more of his offspring were defective as his sperm were used to inseminate his daughters who had inherited the allele from him.

Traditionally our modern understanding of inheritance is reckoned to stem from Gregor Mendel, abbot of the Augustinian monastery in Brno (now in Czechoslovakia). Mendel discovered dominance. Among the crosses of pea plants he carried out in his monastery garden was one between unpigmented plants (white peas and flowers) and pigmented ones (green peas and flowers with violet standards (centres) and purple wings). All the progeny were pigmented, *i.e.* pigmentation was dominant to lack of pigmentation. In the next generation, three-quarters were pigmented, one-quarter unpigmented. The 1 : 2 : 1 ratio of genotypes in the offspring of heterozygotes had become a 1+2 : 1 or 3 : 1 ratio of phenotypes. The non-pigmented ones were pure-breeding if self-fertilized; some of the pigmented ones bred true, others segregated.

The actual numbers that Mendel obtained in the second generation of his experiment were 705 pigmented plants and 224 unpigmented ones. This is 3.15 pigmented to 1 pigmented but it is obviously 'close enough' to 3 : 1 for it to be accepted as such. Statisticians have devised tests to measure the probability of observed results (in this case 705:224) deviating from expected ones (3 : 1). The usual convention is to accept that observations with a probability of 5% or less (*i.e.* 1 in 20) of being different from the expected results are likely to mean that the expectation is wrong and that some other hypothesis must be devised for the observations.

Ironically Mendel's results were too good. They do not deviate from expectation as much as would be expected from the laws of probability (Fisher, 1936; Bennett, 1965). It could be that the noble monk faked his data. A more charitable (and likely) explanation is that Mendel had carried out preliminary experiments and knew what his results were going to be. His published figures were therefore only a demonstration of his theory, and he made sure he had numerical data sufficiently close to his theoretical treatment to make the latter acceptable. The double irony is that Mendel's work (published in the *Proceedings of the Brno Natural History Society* in 1865) was ignored for over 30 years. It was rediscovered by three botanists independently (Correns, De Vries, and von Tschermak) in 1900 when their own work was approaching similar ideas.

2. Multifactorial inheritance

Most characters are controlled by many genes working together. The effect of this is that most inherited factors do not produce two or three clearcut classes like the flower colours we have been considering but a range of expressions of a character. The size of a breed of dogs, the rate of growth of a strain of pigs, the yield of a tomato plant, the time of flowering of an apple tree are all inherited characters, but ones where a number of genes interact to produce the phenotypic character. Most characters of interest and importance to a naturalist are multifactorial in this way, although sufficiently few genes are involved too rarely for the contribution of each gene to be worked out.

For example, some strains of bees are resistant to foulbrood disease which kills bee larvae between hatching and their emergence as adults. Protection against the disease seems to result from 'hygienic' behaviour in worker bees, who uncap cells containing dead larvae and remove them from the hive. In a

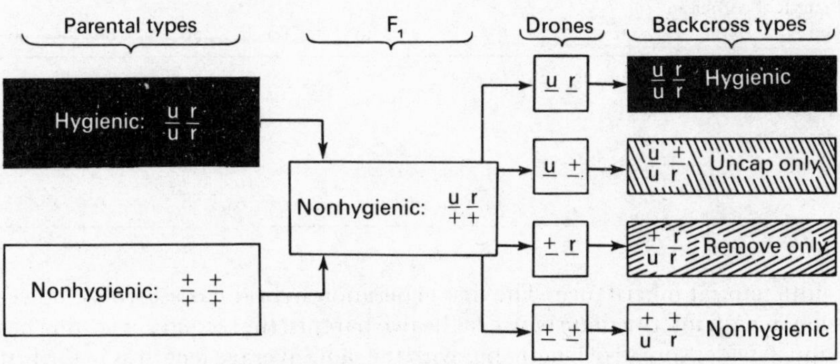

FIG. 7. Interaction of two gene loci determining resistance to foulbrood disease in honey-bees: only the doubly recessive workers will both uncap cells containing dead larva, and also remove the larva. *u*, *r* are the recessive alleles at loci controlling the 'uncapping' and 'removal' behaviours respectively; the normal alleles at both loci are indicated by + (after Rothenbuhler, 1967).

cross between a resistant strain and a susceptible (and 'unhygienic') one, Rothenbuhler (1967) reported that all the first generation hybrids were susceptible to the disease, *i.e.* resistance is a recessive character. However when drones* produced by hybrid queens were crossed back to a resistant strain, 29 colonies were produced, but only 6 showed 'hygienic' behaviour instead of the 50% expected if the behaviour pattern was determined by a single gene. In the other 23 colonies, the workers in 9 would uncap cells containing dead larvae but not remove them, while dead larvae would be removed in a further 6 colonies if the cells were uncapped by a human. Rothenbuhler suggested that hygienic behaviour was the result of the interaction of behaviour controlled by two independent genes – one determining 'cell-uncapping', one 'larva-removal'.

A more complex but more typical situation emerged from work by an American biologist, F. B. Sumner, who carried out an extensive series of breeding experiments on races and species of the North American Field Mouse, *Peromyscus*, in order to prove that differences in size, colour, body proportions and so on, were the result of 'direct responses to local and unstable environmental conditions' and were not inherited in the same way as Mendel claimed for his peas. He failed; geographical and ecological variation in *Peromyscus* is inherited in exactly the same way as eye colour in man or enzyme variation in the mouse (reviewed Sumner, 1932). The only difference was that the inheritance patterns had to be analysed in other ways.

Table 5 shows data on the length of flowers in two species of Tobacco plant (*Nicotiana*) taken from the classical work of Jones & East who first analysed

TABLE 5. Frequency distribution of corolla length in the cross *Nicotiana forgetiana* × *N. alata* var. *grandiflora* (data of East 1913)

Length of corolla in mm	20	25	30	35	40	45	50	55	60	65	70	75	80	85	90
N. forgetiana	9	133	28												
N. alata var. *grandiflora*										1	19	50	56	32	9
F1				3	30	58	20								
F2		5	27	79	136	125	132	102	105	64	30	15	6	2	

multifactorial inheritance. The first generation hybrid has a corolla (flower tube) length intermediate between the two parents; the second generation has a much wider spread of length, but with the same average length as in the first generation. At first sight this may seem to indicate that the parental characters have blended, but further generations of breeding show this not to be so,

* Drones hatch from unfertilized eggs, and therefore have only genes carried by their mother.

but that the character was the result of many genes interacting together.

The easiest way to approach the question of gene interaction is in terms of loci which depend on each other for their effect. For example, an animal (rat, rabbit, mouse, or man) homozygous for an albino allele is unpigmented whatever other colour genes are present. If we cross an albino rabbit with a

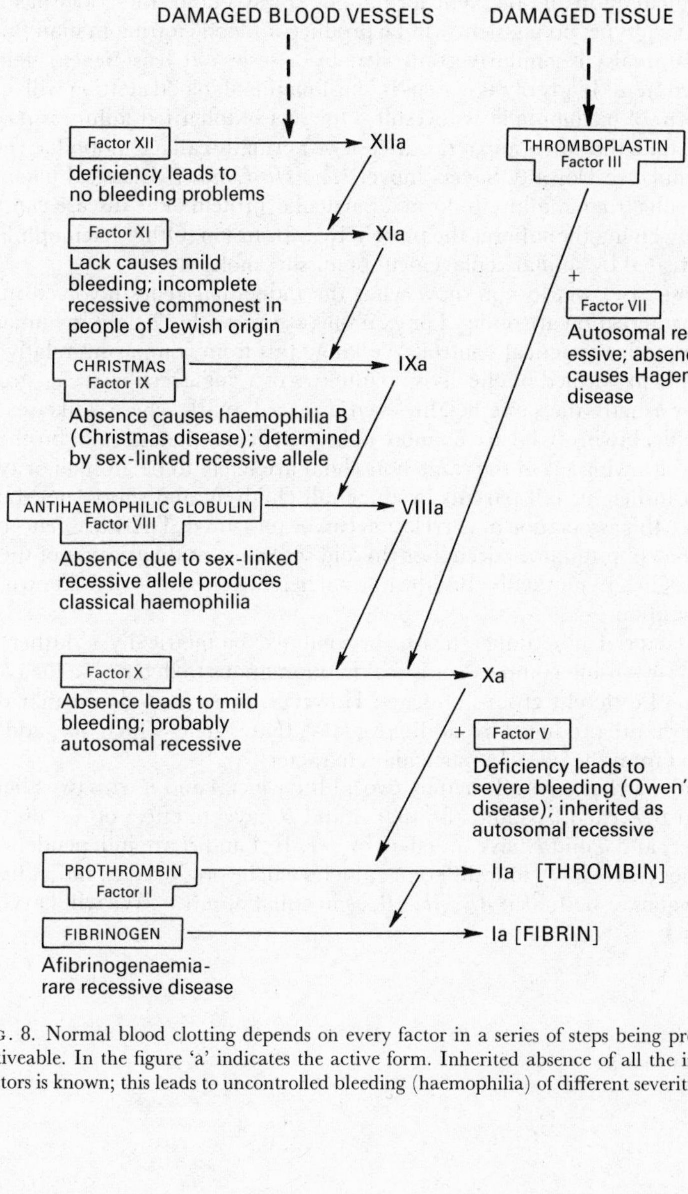

FIG. 8. Normal blood clotting depends on every factor in a series of steps being present and activeable. In the figure 'a' indicates the active form. Inherited absence of all the individual factors is known; this leads to uncontrolled bleeding (haemophilia) of different severities.

normal brown ('agouti') one, the young may all be brown or may include browns, blacks, chocolates and whites. If an animal cannot make any pigment, all the genes affecting the distribution or intensity of pigment are unable to express themselves.

A similar example from plants is the purple anthocyanin pigment characteristic of many flowers and seeds. This depends on three loci, each producing an intermediate in the pigment formation. If any one of these loci has an inactive genotype, no pigment will be produced. Blood clotting in man (and other mammals) is similarly controlled by a series (at least seven) gene-controlled steps. If any of these steps is non-functional, blood clotting will fail, and a form of haemophilia will result. One sort of inherited failure is more common than the others and is the disease we normally call haemophilia. (For more detail, see Dorothy Sayers' novel *Have His Carcase*.) Classical haemophilia results from a failure to form a particular protein. The disease can be treated by giving the sufferer the protein by transfusion. Other haemophilias can be treated by similar replacement of missing molecules.

In most cases we do not know what the individual genes involved in a multiple interaction are doing. For example, we know that height in humans is largely under genetical control. We know this from comparing relatives: identical twins formed by the division into two of a single fertilized egg, grow to almost exactly the same height, even if reared apart, whereas like-sexed non-identical twins (who are no more alike genetically than normal brothers or sisters) growing up in the same household are likely to be an inch or two different in height; tall parents produce tall children, and short ones short children – this association or correlation can be measured. There are gene loci which control pituitary, adrenal or thyroid secretion, and variation of them may affect height markedly. But the normal height genes are still unknown in their operation.

Multifactorial inheritance has to be analysed biometrically (Mather & Jinks, 1971) – using complex statistical manipulations to determine the contribution of different groups of genes. However, there is no doubt that the main effects are produced by 'additive genes', that is genes which may add to or detract from the size of a particular character.

Consider a character affected by two additive loci A and B with two alleles (A,a and B,b) each. Assume also that A and B have an effect of $+1$ on the character, and a and b have an effect of -1. If A and B are independent of each other (see below), four different gametes will be produced by individuals heterozygous at both loci: AB, Ab, aB, ab in equal numbers. We will have 16 genotypes:

		Effects of gamete	Gametes from 1st parent *AB* +2	*Ab* 0	*aB* 0	*ab* −2
Gametes from 2nd parent	*AB*	+2	AABB +4	AABb +2	AaBb +2	AaBb 0
	Ab	0	AABb +2	AAbb 0	AaBb 0	Aabb −2
	aB	0	AaBB +2	AaBb 0	aaBB 0	aaBb −2
	ab	−2	AaBb 0	Aabb −2	aaBb −2	aabb −4

This will give rise to 5 'phenotypes': +4, +2, 0, −2, −4 in the relative proportions 1 : 4 : 6 : 4 : 1; the average type will be the most common, and the extremes (which are multiple homozygotes) the rarest.

The more loci and the more alleles that are involved, the more phenotypic classes will result. Thus 3 loci each with 2 alleles will produce 7 phenotypic classes in the proportions 1 : 6 : 15 : 20 : 15 : 16 : 1. Each class will be less distinct from the next, and a histogram of the distribution of the character will approach a smooth curve.

3. Linkage

Genes are carried on chromosomes. So far we have assumed gene-loci segregate independently of each other. But this is not completely true – genes on the same chromosome will be inherited together, because they will go together into the same gamete. In fact even this is not completely accurate because of recombination during the first part of the first meiotic division when the gametes are formed. (A full understanding of meiosis is unnecessary for the purpose of this book. A summary of the main stages in cell division is given in the Appendix, pages 308–10.) Homologous chromosomes (*i.e.* members of the same pair) twine round each other and interchange segments. An individual may receive a pair of chromosomes from his parents carrying the alleles *A B C* on one and *a b c* on the other. As the chromosomes cross-over, the genes *recombine*. If crossing over takes place between *A* and *B*, the chromosomes which go into the gametes will be *Abc* and *aBC*; if the crossing-over is between *B* and *C*, the daughter chromosomes will be *ABc* and *abC*. More rarely, crossing-over may take place in both segments at the same time, and two double recombinant chromosomes will be formed (*AbC* and *aBc*). This will mean eight possible gametes: commonest will be the two parental or non-cross-over types; then will be the four single cross-overs; and finally the two double cross-overs. The amount of crossing-over between two loci is approximately proportional to their physical distance apart on the chromosome. Loci on a chromosome are said to be *linked*; loci which are near to one another will be closely linked and only rarely separated.

Linkage is not the same thing as *association*. Two inherited characters may always (or never) occur together, but this will not be because they are on the same chromosome. For example, there are a number of different patterns of baldness in man, and certain patterns run in families (weddings are good opportunities for observing this). The genes determining the pattern are transmitted by both men and women, but only men usually go bald and show the pattern. The effects of the genes are associated with maleness, but not linked to it.

CHAPTER 2

POOLS OF GENES

IN families – human, domestic and farm, we can observe the transmission of inherited traits. However, in groups we do not usually know the parents of individuals. We only have a number of individuals with characteristics we can determine. The group may be mice, grass-hoppers, gorillas, clovers, Red Indians or black-birds. Some years ago my wife and I worked on ancient Egyptian skulls in the British Museum. A number of cemeteries had been excavated and the skulls were all labelled with their origins. Knowing the method of inheritance of certain skull characters (such as extra bones in the sutures, extra foramina, missing wisdom teeth) it was possible to calculate the genetical differences between different cemeteries from the frequencies of the characters in each sample, and measure how much the Egyptian population had changed with time, even though we knew nothing about any family groups in the remains (Berry, Berry & Ucko, 1967).

FAMILIES AND FREQUENCIES

It is with frequencies and not ratios that we have to use when we are dealing with populations. Int the first instance it is easier to see the implications of this if if we can recognize heterozygotes (*i.e.* if we are dealing with a character in which the heterozygote is intermediate between the two homozygotes. This is sometimes said to be a state of *semi-dominance*, but more correctly it shows no dominance).

Let us take an example from human blood frequencies. There are 13 or more human blood systems known, and probably over a hundred alleles altogether at the 13 loci. It is well known that different races have different frequencies of the different groups. For example, a group of Icelanders were typed for the MN system, where individuals may react to anti-M antibodies (which means they are carrying only the *M* antigen, and are thus homozygous *MM*), to anti-N (*NN* people), or to both (*MN* people). Out of a total of 747 people, 233 were *MM*, 385 *MN*, and 129 *NN*. In this sample we can count the number of *N* alleles, since *NN* people will carry two each, and the heterozygotes (*MN*) one. There were $(2 \times 129) + 385$ *N* alleles out of 1494 alleles in the sample (since every person of the 747 has two alleles).

The *frequency* of *N* is

$$[(2 \times 129) + 385]/1494$$
$$= 0.430$$

Similarly the *frequency* of *M* is

$$[(2 \times 233) + 385]/1494$$
$$= 0.570$$

Since the frequency of all alleles at the locus must add up to 1.00 (or 100%), we could more easily calculate the frequency of *M* as 1.000 − 0.430. Conventionally the frequency of alleles in a sample is represented as p, q, etc. In our example, $p = 0.430$, $q = 0.570$; $p + q = 1.00$.

What happens to these frequencies when mating takes place? For simplicity we start with two assumptions:

1. That there is random mating, *i.e.* that any individual has an equal chance of mating with any other individual in the population or, put slightly differently, that the chance of fusion between any two gametes is proportional only to their relative frequencies.
2. That the population is 'large' in a statistical sense. In practice this means it contains more than about a hundred individuals.

We shall see later the effects of not making those assumptions.

Consider the case of a locus with two alleles *A*, *a*, such that the frequency of *A* is p, and of *a* is q or $(1-p)$. There are three stages in the argument:

1. The 3 parental genotypes (*AA*, *Aa*, *aa*) will produce gametes carrying either the *A* or *a* allele, and the frequency of each gamete will be p and q respectively.
2. The gametes will combine to form zygotes, whose frequencies we can calculate.
3. Knowing the genotype frequencies in the children, we can calculate the gene (or allele) frequency in them.

We begin with a 2 × 2 table constructed in the same way as discussed on page 38:

		Father: A p	Father: a q
Mother	A p	AA p^2	Aa pq
	a q	Aa qp	aa q^2

There will be p^2AA, $2pq$Aa, and q^2aa. As this includes all zygotes in our

sample $\quad p^2 + 2pq + q^2 = 1$

and since $\quad p^2 + 2pq + q^2 = (p + q)^2$

and $p + q = 1$,

this is reasonable.

PLATE 3. *Peppered Moths (Biston betularia) on different backgrounds (page 120)*

Above left, an oak tree covered with lichens in an unpolluted area in mid-Wales, with (top) typical, (centre) *f. insularia*, and (bottom) *f. carbonaria* moths. *Right,* a tree in central Birmingham with no vegetative lichens, on which are (top) *f. carbonaria* and (bottom) *f. typica*.

Geographical variation

Melanic moths from Shetland (page 27).

Row 1. Ingrailed Clay *(Diarsia mendica)*: *left*, typical form from Caithness; *right*, *f. thule* from Shetland.

Row 2. Autumnal Rustic *(Amathes glareosa)*: *left*, typical and *right*, melanic *(f. edda)* forms from Shetland.

Row 3. Marbled Coronet *(Hadena conspersa)*: *left*, typical form from Oxfordshire; *right*, *f. hethlandica* from Shetland.

Row 4. Northern Spinach *(Lygris populata)*: *left*, typical and *right*, melanic *(f. fuscata)* forms from Shetland.

Row 5. Ghost Moth *(Hepialus humuli)*: *left*, typical form from Caithness; *right*, *f. thulensis* from Shetland.

PLATE 4. Looking north across Loch Tay to two forestry plantations, that on the right (Boreland) being younger than that on the left (Balnearn). For some years Balnearn was occupied by a population of Bank Voles *(Clethrionomys glareolus)* markedly different from the more typical animals found in Balnearn (page 59).

Aerial view of Milford Haven from the north-west, showing the sheltered water of the Haven where Dog-whelks *(Nucella lapillus)* experience little stabilizing selection of shell shape traits, in contrast with the strong pressures exerted on populations living on exposed rocks outside the Haven (page 138). In the foreground is Skokholm, site of an intensive study of genetical changes in House Mice *(Mus musculus)* (Chapter 7).

One point worth making at this stage, is that we can calculate gene frequencies for a recessive character where we are unable to 'count' alleles as in our Icelanders. If we can identify only two phenotypes (dominant homozygotes plus heterozygotes, and recessive homozygotes), the frequency of q is simply the square root of the frequency of recessive homozygotes in the sample. Thus in our Icelandic example, the frequency of *MM* was $233/747 = 0.312$. The frequency of M would be $\sqrt{0.312} = 0.559$. Our earlier estimate of *M* frequency is more accurate, because it was based on no assumptions about population size or random mating. However, we are forced to use this method of gene frequency estimation if we cannot identify heterozygotes with confidence.

To return to our general example: we know the genotype frequencies in the children's generation. We can calculate p and q by 'counting'. Thus the *A* allele will be carried by p^2 homozygotes and $2pq$ heterozygotes. But only half the alleles carried by the heterozygotes will be *A*, so frequency of *A* in children's generation

$$\begin{aligned} &= p^2 + \tfrac{1}{2}(2pq) \\ &= p(p+q) \\ &= p \end{aligned}$$

since $p + q = 1$.

This leaves us with an apparently trivial conclusion that allele frequencies remain constant from generation to generation. Furthermore, genotype frequencies are determined in a random mating population entirely by allele frequencies, so these also will remain constant.

This result is in fact extremely important. To restate it: *in a large random-mating population, gene and genotype frequencies remain constant in the absence of migration, mutation and selection.* This statement is usually dignified by being called the *Hardy-Weinberg* principle. One of its advantages is that changes in frequencies can be attributed to one of the five disturbing agencies included in the definition above (population-size, mating-choice, migration between groups, mutation, and natural selection).

The Hardy-Weinberg principle is simply a particular expression of the binomial theorem, but it is easier for most people to follow its derivation in genetical than in algebraic terms. However, in its original form it is said to derive from the mathematical understanding of G. H. Hardy, one-time Professor of Mathematics at Cambridge University. The story is that R. C. Punnett (afterwards Professor of Genetics at Cambridge) gave a talk to the Royal Society of Medicine on 'Mendelian Heredity in Man'. In his own words: 'I was asked why it was that, if brown eyes were dominant to blue, the population was not becoming increasingly brown-eyed: yet there was no reason for supposing such to be the case. I could only answer that the heterozygous browns also contributed their quota of blues and that somehow this must lead to equilibrium. On my return to Cambridge I at once sought out G. H. Hardy with whom I was then very friendly, for we had acted as joint Secretaries to the Committee for the retention of Greek in the previous examination and we used to play cricket together. Knowing that Hardy had not the slightest interest in genetics I put my question to him as a mathematical problem. He replied that it was quite simple and soon handed me the now well-known formula. Naturally pleased at getting so neat and prompt an answer I promised him that it should be known as "Hardy's Law" – a promise fulfilled in my next edition of *Mendelism*. Whether the battle of Waterloo was won on the playing fields of Eton is still, I gather, a matter for conjecture: certain it is, however, that "Hardy's Law" owed its genesis to a mutual interest in cricket' (Punnett, 1950).

That was in 1908. In the same year, a German physician by the name of Weinberg published the same result, and we have had the Hardy-Weinberg principle ever since.

FREQUENCY OF HETEROZYGOTES FOR RARE ALLELES

The connection between genotype and allele frequencies given by the Hardy-Weinberg principle (or the binomial theorem if preferred), gives a number of insights into population structure. Perhaps the most important is that the great majority of rare alleles will be carried in the heterozygous condition. This will be clear from an example: albinism is a recessive condition, so that all albinos are homozygotes. The frequency of albinism in man is about 1 in 20,000, *i.e.*

$$\begin{aligned} p^2 &= 0.00005 \\ p &= \sqrt{0.00005} \\ &= 0.0071 \end{aligned}$$

The expected frequency of heterozygotes $(2pq = 2 \times 0.0071 \times (1-0.0071)$

$$= 0.0141 \text{ or } 1 \text{ in } 71$$

In other words, heterozygotes are 280 times commoner than homozygotes.

The chance of two heterozygotes marrying by chance will be very small: it will be $(0.0141)^2$ or 0.0002, *i.e.* one chance in 5,000. (One in 4 of their children will be albinos, which brings us back to 1 in 20,000 children born – the population frequency of the condition.) There is one important qualification to this risk of two heterozygotes marrying: if two people share a common ancestor in their recent history, they will have a much greater chance than average of being heterozygotes for the same allele. Although the frequency of cousin marriages in Britain is less than 1%, 8% of albinos are the children of cousins.

At once we are introduced to two important features of natural populations: firstly, uncommon alleles will exist almost entirely in the heterozygous state, and this means their effect in the population as heterozygotes will be proportionately much greater than as homozygotes. Secondly, alleles are unlikely to be distributed at random through a species because individuals are more likely to mate with a neighbour than a far distant member of the species, and the neighbour is proportionately more likely to share a common ancestor.

GENE FREQUENCIES IN CATS

Searle (1949, 1964) examined 700 cats destroyed at three of the main animal clinics in London, and classified them for the main coat colour phenotypes. Subsequently he compared gene frequencies in London cats with those in cats from other parts of the world (Table 6). Knowing the inheritance of the various patterns, he was able to work out the frequencies of the main alleles (Searle, 1964). This is reasonable, because a city's cat population is a less artificial assemblage than, say, that of dogs. As Darwin commented in the *Origin of Species* cats pair at random 'owing to their nocturnal rambling habits', and although man reduces the male breeding population by about

TABLE 6. Gene frequencies in European and European-derived populations of cats (after Todd, Fagen and Fagen, 1975)

	Orange (*o*)	*Non-agouti* (*a*)	*Blotched* (t^b)	*Dilute* (*d*)	*Long hair* (*l*)	*White* (*W*)
Cyprus	0.21	0.72	0.25	0.35	0.50	0.001
Chios	0.20	0.70	0.23	0.28	0.13	0.00
Athens	0.13	0.72	0.26	0.34	0.13	0.009
Argolis	0.18	0.74	0.18	0.20	0.00	0.00
Vienna	0.10	0.57	0.29	0.22	—	0.001
Venice	0.06	0.58	0.48	0.35	—	0.007
Rome	0.07	0.65	0.47	0.32	—	0.02
Chamonix	0.10	0.75	0.69	0.40	0.32	0.014
Marseille	0.06	0.70	0.69	0.30	0.28	0.02
Paris	0.06	0.71	0.78	0.33	0.24	0.011
Mayenne	0.15	0.64	0.61	0.29	0.17	0.006
The Hague	0.21	0.65	0.66	0.26	0.16	0.02
Southern England	0.19	0.80	0.84	0.26	0.31	0.01
London	0.11	0.76	0.81	0.14	0.33	0.004
York	0.20	0.81	0.78	0.27	—	0.01
Reykjavik	0.13	0.60	0.53	0.44	0.14	0.01
Iceland – rural	0.20	0.64	0.30	0.34	0.10	0.06
Boston	0.19	0.64	0.44	0.43	0.30	0.022
New York	0.14	0.75	0.47	0.44	0.13	0.013
Philadelphia	0.26	0.70	0.45	0.48	0.18	0.012
Columbus	0.29	0.64	0.31	0.50	—	0.01
Chicago	0.22	0.71	0.34	0.45	0.37	0.02
Bloemfontein	0.08	0.78	0.48	0.49	0.13	0.02
Adelaide	0.19	0.78	0.71	0.33	0.38	0.023
Brisbane	0.14	0.81	0.68	0.45	—	0.022
Melbourne	0.14	0.85	0.61	0.47	—	0.039
Hobart	0.22	0.76	0.81	0.25	0.46	0.04
Dunedin	0.20	0.81	0.87	0.30	0.47	0.036

half through castration (and the female by a smaller proportion), gene frequencies estimated from recessive homozygotes are close to the Hardy-Weinberg expectation based on random mating. Black kittens are more common than expected (which probably indicates more about human superstition than cat character). However dark cats are commoner in lower socioeconomic areas of Glasgow than in higher ones, suggesting that other factors than human activity are important (Clark, 1975).

Some comments on the distribution of cat alleles are relevant here, although the main factors determining the distribution of alleles are discussed later in the book. Wild cats have a 'striped tabby' phenotype. These are commonest in the Singapore region. There is another allele at the tabby locus called 'blotched tabby' which reduces the number of stripes, and breaks them up. Linnaeus gave the name *Felis catus* to cats with the blotched tabby pattern, suggesting that the allele was very common in Sweden in his time. However, the allele is rare outside Europe, although they occur occasionally: Colonel Meinertzhagen shot one 'in very wild country at Doab in the Hindu Kush'. Two of the three cats from Australia in the Natural History Museum are blotched tabby. This is likely to be due to the spread of an allele which originated in Europe. White cats are rare everywhere, presumably because they are deaf and thus at a disadvantage. The highest frequencies of the yellow allele are in eastern Asia, the converse of the blotched allele situation.

UPSETS OF THE HARDY-WEINBERG PRINCIPLE

The Hardy-Weinberg principle is a statistical abstraction. Its advantage is that we can describe the assumptions necessary for it to hold. Since one or more of the five main disrupting agencies are acting in most natural situations, we must describe them in some detail so that we have the knowledge available for discovering the genetical forces acting on any population. The five upsetting agencies are: sampling error in small populations, mutation, natural selection, migration between groups, and mating choice.

1. Small populations

The gametes that transmit genes from one generation to another carry a sample of the genes in the parent generation. If the sample is not large, gene frequencies are liable to change between the generations. This can be understood by thinking about the human sex ratio. There are very nearly the same numbers of boys and of girls born in the world. Yet we all know of single sex families – two, three, four, or even more children of the same sex. However, for every four boy family there is somewhere a four girl one, so that the sex ratio evens out. The larger the group we are considering, the more likely it is that we shall find equal numbers of boys and girls. The important question from our present point of view is what happens in small populations, because however many individuals there may be in a species, in a local mating group, there may only be comparatively few breeding pairs. For example, the common Wren (*Troglodytes troglodytes*) is widespread and fairly common throughout northern Europe and western Asia, but it has a number of small isolated populations. Williamson (1958) estimated there were 45–50 pairs on Fair Isle and 120 on St Kilda.

Gene frequencies in small populations like these may behave in the same way as the sex ratio in humans. We have already seen that when Mendel crossed the first generation progeny of pure-breeding pigmented and non-pigmented pea plants, he obtained 705 pigmented and 224 non-pigmented ones. This was not quite a 3 : 1 ratio, although gametes carrying the pigmented and non-pigmented alleles would have been produced in *exactly* equal numbers. If he had bred very small numbers, he might have found even greater 'sampling error'.

The theory behind this sampling error is simple. In any generation the frequency of an allele will be distributed about the mean value (p), with a variance $p(1-p)/2N$, where N is the number of organisms in the population (and $2N$ the number of gametes from which they were formed). The standard deviation of p will be the square root of the variance.

Consider populations with 2, 20 and 200 individuals respectively, each with a frequency of p of 50%. The standard errors of p will be respectively 25%, 7·9% and 2·5% – as the population size increases, the value of p will show less and less fluctuation. In the smallest population, p may change from

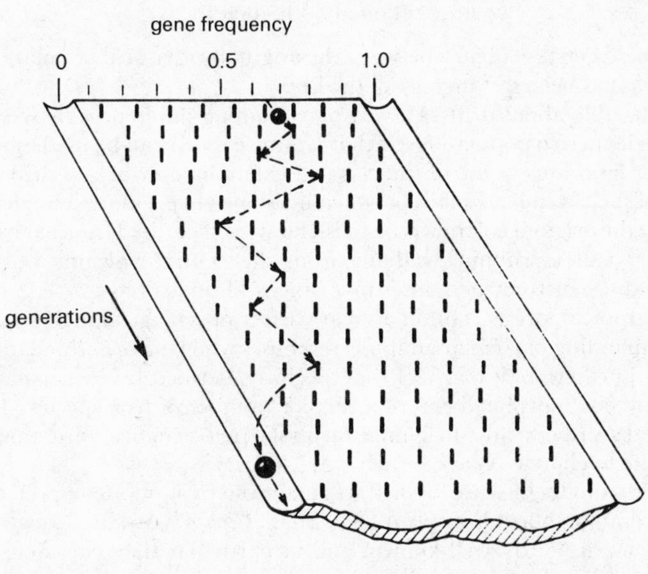

FIG. 9. 'Bagatelle' changes in allele frequency in a small population: in the absence of mutation and migration one allele will inevitably be lost (from Ehrlich and Holm, 1963).

50% to as low as 25% or as high as 75% by chance; whereas in the 200 population, p will only vary between 47.5% and 52.5%.

This fluctuation from generation to generation is called *genetic drift* (or simply *drift*), or sometimes the Sewall Wright effect after the American geneticist who derived its theory in great detail. It has three consequences:

i. If a population remains small, frequencies will change erratically in a 'bagatelle' manner, and the rarer allele at any locus will inevitably be lost in due course. The probability of an allele being lost in any generation is $1/2N$. This means that genetical variability will decrease.
ii. As alleles are lost there will be fewer heterozygotes; and as inbreeding increases, the proportion of homozygotes will rise.
iii. A series of small populations formed from a single large one will inevitably diverge, even if they were initially identical.

These effects have been shown experimentally. Kerr & Wright (1954) set up 96 lines (or populations) of *Drosophila melanogaster* each with 50% of an allele called flexed (symbol f) which produces a crumpling of the wings, and 50% of the normal allele at the locus. They took four males and four females at random in each generation to be the parents for each line in the next. After 16 generations

29 lines had only the f allele
41 lines had only the normal allele
26 lines retained both alleles

There had been divergence between the originally identical populations, and variation had been lost in 73% of the lines.

The trouble about drift as a way of explaining differences between allele frequencies in two populations, is that it is too easy. It can be used merely as a cloak for ignorance. One of the classical examples ascribed to drift was the origin of species in land snails on volcanic islands like Tahiti. The suggestion was that the original colonizers of an island were separated from each other in the steep valleys running radially from the central volcanic peaks, and became distinct from each other by chance. This idea has proved much too simple: tropical forests contain a great variety of ecological niches for snails, each demanding particular adaptive responses which have nothing to do with chance. Furthermore, many of the 'species' described by earlier naturalists are really polymorphic forms of a smaller number of true species. Tropical land snails present an intriguingly complicated genetical situation which owes little to chance (Clarke & Murray, 1969).

A more convincing case for drift was made out by Lamotte (1951), working on the common north European Field Snail, *Cepaea nemoralis*. The genetics of this species are fairly well-known, and in particular the genes determining colour and banding of the shell. Lamotte counted the proportions of unbanded snails in 826 colonies in France. He divided the colonies into three size

FIG. 10. Proportions of banded snails (*Cepaea nemoralis*) in a series of colonies along the Ariège River in the Pyrenees (after Grant, 1963).

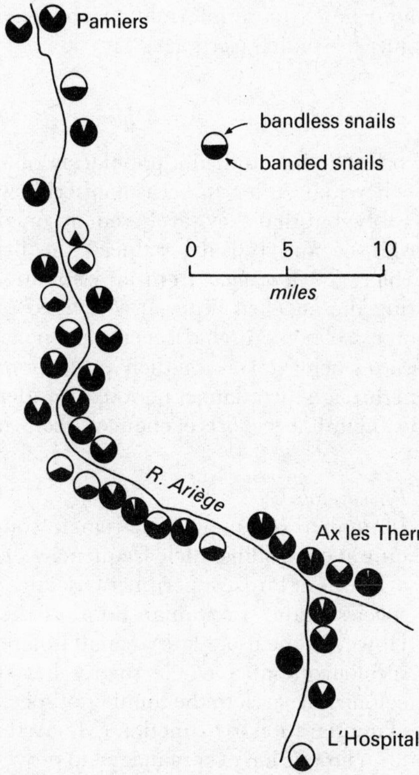

classes (large: 3,000 to 10,000 individuals; small: 500 to 1000; and intermediate) and compared the colony to colony diversity among the large populations with that of the small populations. He found the frequency of the allele determining bandlessness lay between 1% and 30% in most large colonies, and neighbouring large colonies tended to have similar frequencies. All large colonies contained some unbanded individuals. In contrast the frequency of the bandless allele in small colonies varied considerably more, and some had no bandless individuals. Marked frequency differences sometimes occurred between neighbouring colonies. Along the Ariège River in the Pyrenees, colonies with no bandless individuals were flanked by ones with 20 and 21% unbanded, and a colony with 74% unbanded was surrounded by ones with 7, 4, and 32% unbanded.

Notwithstanding all these data on population sizes which provide the pabulum for armchair scientists, both colour and banding in *Cepaea nemoralis* are now known to be affected by bird predation and by differential survival through the climate, and Lamotte's claim for drift cannot be accepted without much more information about ecological differences between the colonies he studied (see pages 147–56).

Different types of drift

Anyone who has studied a population of animals or plants in nature knows that they may suffer an occasional crisis which reduces their numbers considerably, but that they rarely remain uncommon for long. The cold winters of 1961–62 and 1962–63 reduced the British population of the Dartford Warbler (*Sylvia undata*) from an estimated 450 pairs to 10 recorded pairs. During the next ten years, it recovered to something like its former level. Numbers of some orchid species are now very low indeed, but the active measures being taken for their protection and spread will hopefully bring them through their danger period. Consideration of situations like this help us to distinguish three sorts of chance effect which tend to be lumped together as drift:

i. *Persistent drift*
   If numbers of a population remain consistently low, drift may be important in controlling allele frequencies. This may happen in small colonies of insects or birds, or peripheral isolates of a species (such as the rare clover species of the 'Lusitanian flora' of the Lizard Peninsula in Cornwall). However, the usual fate of small isolates may be repeated extinction and recolonization: a whole theory has sprung up relating the size of a colonizable area to the number of species it can support, and the balance of immigration to extinction (MacArthur & Wilson, 1967). More important, no-one has ever managed to prove that gene frequency changes with chance in any small population which has been intensively studied. The classical investigation of this nature was on morphs* of the Scarlet Tiger Moth (*Panaxia dominula*) at Cothill, near Oxford (Fisher & Ford, 1947). Frequencies of the morphs fluctuated considerably from year to year, but they were not correlated with population numbers and anyway were too great to be explained by allele sampling error (page 225).

ii. *Intermittent drift*
   An occasional drastic reduction in numbers is a more likely event to affect a population than a continued low level. When such a catastrophe occurs, there may be so few survivors that sampling error becomes important. Charles Elton (1930) once believed that the periodic population 'crashes' in species like voles, lemmings, snowshoe hares, lynx and the like, played an important role in their evolution.

*Morph is a convenient word to describe any of the different types coexisting in a polymorphic population.

A particularly good example of intermittent drift occurred in *Cepaea nemoralis* in East Anglia after big floods in 1947. The New Cut is a canal straightening the course of the Bedford River near Ely. Alongside the canal is a bank about 20 feet high running parallel to the canal for about two miles. The steep slope of the bank is grass covered. Below the bank is a road, and below the road a verge bordering fields. The verge varies in character depending on its neighbouring field. The bank forms an apparently uniform habitat for snails for a two mile stretch.

The floods of 1947 were the first serious ones for over a century, and the bank was almost completely submerged for three weeks. Snails could only have survived in isolated pockets. Five years later Goodhart (1962) collected samples of *C. nemoralis*. He found the whole bank and road verge area supported snails, presumably the descendants of the flood survivors. He counted several thousand individuals, collected at furlong intervals.

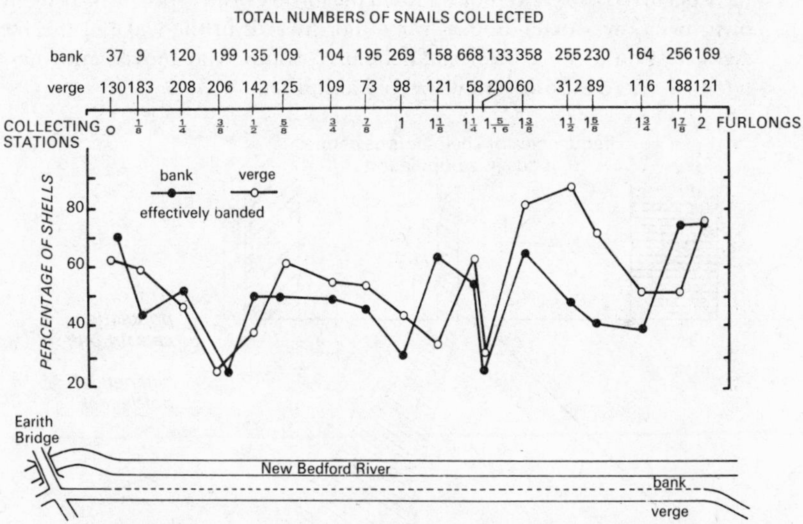

FIG. 11. Distribution of frequencies of banded snails (*Cepaea nemoralis*) at furlong intervals along the 100-foot Bank near Ely (from Berry, 1967, after Goodhart, 1962).

The frequency of banded snails varied between 20 and 80% over short distances – distinctly unconstant. Moreover the allele frequencies on the verge closely followed those on the bank, even though the verge was ecologically different and more variable than the bank itself, suggesting that it had been recolonized by snails moving across the road from the bank. In a second sample, collected eight years after his first, and a third eight years later again, Goodhart (1973) found that the extremes of fluctuation were decreasing slightly. This suggested that the first sampling represented a genetically unstable situation and hence an equilibrium distribution of allele frequencies would only be approached as individuals moved about the bank. Possibly the overall allele frequencies are the same as prior to 1947, but this cannot be known.

iii. *Founder effect*

The individuals who colonize an empty habitat must usually be few in number. Sometimes a species may rapidly extend its range, as happened with the Collared Dove (*Streptopelia decaocto*) across Europe in the 1940s and 1950s*, but more often only a handful of animals or seeds will manage to breach a barrier into new territory.

Usually this colonizing event will be speculative – no-one doubts that the first Galapagos Finch reached the islands from South America, but when it was and how many birds were involved can only be guessed – nevertheless there can be no doubt that similar episodes must have occurred time and time again in the history of any species. When our own species was extending its range northwards in the wake of the Ice Ages, it would have been small family groups who moved into new territory and established new loci of habitation.

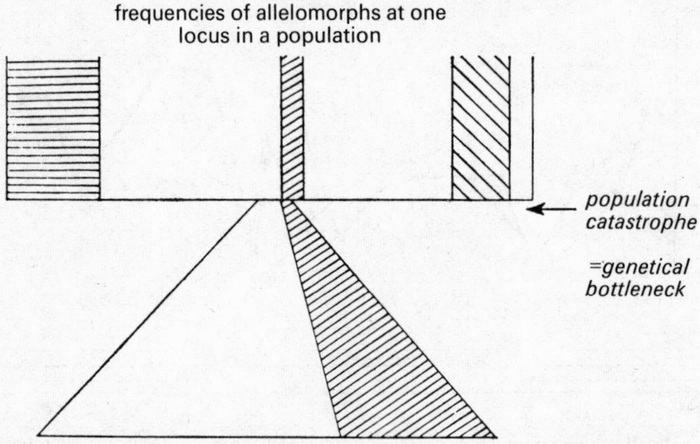

FIG. 12. The founder effect – a drastic method by which allele frequencies may change and variation be reduced.

Now a colonizing event has similarities to intermittent drift, but shows important differences. It is like because it is effectively an acute bottleneck in numbers which will lead to a drastic change in both allele frequencies and alleles present in the descendant population; it is unlike

*Mayr has suggested that this spread after a period of at least four centuries of relative stability might have been due to a genetical change which suddenly rendered the species less restricted in its requirements – such as its temperature tolerance: *q.v.* Murton, 1971.

because the whole future reaction and adjustment of the colonizing group will be dependent on the alleles represented in the original members. Even if immigration brings in fresh variation at a later date, by that time the founders will have increased in numbers and range and fresh individuals will be unlikely to contribute significantly (Berry, 1975*a*).

A small but elegant example of the operation of this effect was described by Corbet (1963) in the Bank Vole (*Clethrionomys glareolus*). He found that voles in a recently planted Forestry Commission plantation (Boreland) in Perthshire, had a tooth-character (presence of a fourth loop on the inner side of the third upper molar, sometimes called 'complex' in distinction to the 'simplex' pattern) carried by 92% of individuals, although it was possessed by only 29% of the animals in a neighbouring plantation and throughout the rest of the Highlands. The Boreland wood was planted on a hillside unsuitable for Bank Voles, with its nearest point about 150 yards away from a nature plantation (Balnearn) where voles have presumably been long established. The new plantation must have been colonized by a few wandering animals – perhaps only a single pregnant female – unrepresentative of the area as far as their molar tooth pattern was concerned. However this genetical situation was not stable. Fifteen years later bracken had grown up between the two plantations allowing voles to move about more, and the atypical Boreland population had reverted to the usual Highland situation (Corbet, 1975). It is unfor-

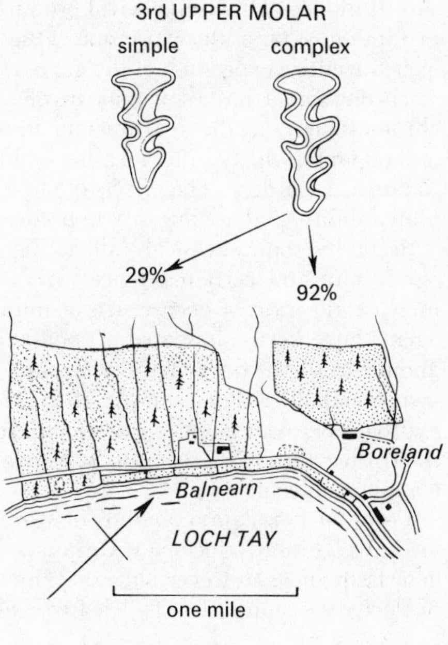

FIG. 13. Map of an area on the north side of Loch Tay to show two plantations inhabited by Bank Voles (*Clethrionomys glareolus*) differing in the frequency of an inherited tooth character (from Berry, 1967, based on Corbet, 1963).

tunately not possible to know whether the change was through selection or immigration.

An additional interest in this example is that different areas of the United Kingdom have different frequencies of complex molars: in the south of England there are about 16% with complex teeth, in north England and mainland Scotland 33 to 41%, but those on the islands of Skomer, Raasay and Jersey have *c.* 80%.

The *founder effect* is the most powerful way known of changing allele frequencies. Although it is a chance phenomenon, it differs from the other forms of drift, and has certainly contributed much more to present allele frequencies than classical drift. We shall return to it in Chapter 6.

2. Mutation

So far we have said nothing about how different alleles arise. Yet the large number of variable loci means that alleles must frequently change from one form to another. The change is called a *mutation*, and a newly arisen allele (or the individual carrying it) may be referred to as a *mutant*. Mutation is often described as random and spontaneous. Both these statements are out-dated confessions of ignorance, since we now know a great deal about the mechanism of mutation.

In the first place, a mutation may be genic or chromosomal. A change in a gene involves a change in DNA base sequence which will affect the polypeptide chain produced (page 35). Some chemical agents will produce certain substitutions in DNA, others will produce other substitutions. The result will be random as far as the expression of that locus is concerned, but not random over the entire expression of the genome (*i.e.* all the genes of the organism). A chromosomal mutation may involve the addition of a chromosome or chromosomes, or the rearrangement of chromosomes through breaking and rejoining with another member of the set (or in its original position but in the inverted order). The effects of chromosomal mutations vary from none to inviability, and are less easy to predict than a gene mutation.

Secondly, some chemicals (alkylating agents, purine and pyrimidine analogues, etc) and certain physical agents (notably ionizing radiation) may increase the rate of occurrence of mutation. When all known mutagenic agents have been eliminated a 'spontaneous' mutation rate at any locus of about 1 in 10,000 per generation persists. Some of this is caused by the body's own metabolites. For example, if the enzyme catalase which breaks down hydrogen peroxide is added to a bacterial culture, the mutation rate is reduced; the truth is that DNA seems to be as reactive and error-prone as any similar complex molecule.

From the population point of view, mutation is continually occurring at a low rate. This introduces new variation into the population, and changes the gene frequency. However, the rate of mutation is so low that mutation can be neglected as a controller of allele frequencies. Only in a few special situations does mutation become important. These are the attempted induction of new mutations for economic reasons (this applies particularly to plants), and in man where new cases of disease arise through recurrent mutation. Fresh gene mutations occur in about one in 200 births; chromosomal mutations in about 1%.

3. Natural selection

Natural selection is the result of reproductive success (or failure); it need have nothing to do with death. Since genes are functional entities – producing or controlling proteins and enzymes – different alleles at a locus are likely to affect the survival, health or behaviour of their carriers (although see Chapter 9). If an allele increases the chance of its bearer leaving children it will be selected for and increase in frequency. Conversely an allele which has the effect of reducing the number of children produced will not be represented in the next generation by as many copies as in the parental generation, and thus will have been selected against.

We can define selection formally as the *differential change in relative frequency of genotypes due to differences in the ability of their phenotypes to obtain representation in the next generation*. This variation in competence may result from different abilities in direct competition with other genotypes, differential survival under the assault of parasites or predators, changes in the physical environment, variable reproductive competence, variable ability to penetrate new habitats, and so on.

It is only phenotypes which are subject to selection; strictly speaking it is impossible for selection to act on an allele, because it is phenotypes not genotypes which have children. The reproductive success of a phenotype is its *fitness*. Conventionally the fittest phenotype at a locus is given a fitness of one, and other phenotypes a fitness less than one – the reduction being measured by a factor indicating the selective intensity acting on the phenotype.

A classical example of selection was found by Brown (1965) on a farm. A barn was overrun by House Mice, 47% of which were light-coloured due to a recessively inherited allele. A hole was made to allow cats to enter the barn and hunt, and very rapidly the population was reduced to a third of its former level – with all the light animals gone. Some months later the cats were shut out again. The mouse population began to increase, and light animals reappeared, eventually making up 5% of the population. The allele which produced light colour when homozygous had remained in the population in the heterozygous state whilst predation was occurring; any light-coloured young born during that period would have been quickly spotted and eaten.

There is a mutation in man (Recklinghausen's neurofibromatosis) which causes growths on the brain and spinal cord and some skin mottling. Carriers of the allele are completely fertile. Its effect is the same in both sexes. However, the *effective* fertility of affected men is reduced to 41.3% while that of afflicted women is 74.8%. In this case the reduction in fertility (the fitness loss) is due to an interplay of psychological and sociological factors, and not to simple inability to breed (Crow, Schull & Neel, 1956).

Natural selection is by far the most important agency regulating allele frequencies, and we shall have to come back to it at length, especially in Chapter 4.

4. Migration between groups

Two populations are unlikely to be genetically identical, but the more interchange of individuals between them, the more will they come to be alike. Put another way, a population may gain an allele either by mutation in one of its own members or by immigration from a population which contains it. Differential migration is often called *gene flow*.

The effect of gene flow can be recognized most easily by its prevention of local adaptation. One well-known example concerns the Galapagos Finches. On the Galapagos archipelago 13 species of locally adapted finches have evolved (6 ground, 6 tree, and one warbler finch), but they have only been able to do this because selection for adaptation to a particular niche could be effective in isolated populations on different islands, *i.e.* where gene flow was small. Once an efficient adaptation had developed, its bearers could spread to other islands and successfully compete with the sitting tenants for a specialized niche – whilst remaining adapted because they had diverged so far from the ancestral stock that they no longer interbred with it. On the single Cocos Island miles away to the north of the Galapagos there is only one all-purpose finch. Gene flow has prevented adaptation and divergence.

Camin & Ehrlich (1968) investigated the reason for the low proportion of striped Water Snakes (*Natrix sipedon*) on the islands of Lake Erie. They found that gulls ate the more conspicuous snakes, and on all the islands the proportion of banded snakes was lower in older than in younger individuals. However, there was a high frequency of banding on the mainland, and banded snakes frequently swam out from the mainland (where there were more snakes than on the islands), and 'topped up' the banded frequencies there. This effect was more marked on islands close to the shore than those further out.

The effect of selection overcoming gene flow is rather beautifully illustrated by Snaydon's work on the 'Park Grass' plots at the Rothamsted Experimental Station. These are plots in a long thin field which have received different fertilizer treatments since 1856. The grass Sweet Vernal (*Anthoxanthum odoratum*) occurs in most of the plots, and usually comprises 0–10% of the hay yield. Snaydon (1970) showed the grass on different plots had adapted to the particular environmental conditions of its plot, in that growth was most rapid when seed was grown in soil of the same mineral composition as that of the plot whence its parent came. The significant point for the present was the situation at the boundary between two adjacent plots. Gene flow (measured by eleven morphological characters having different values in different plots)

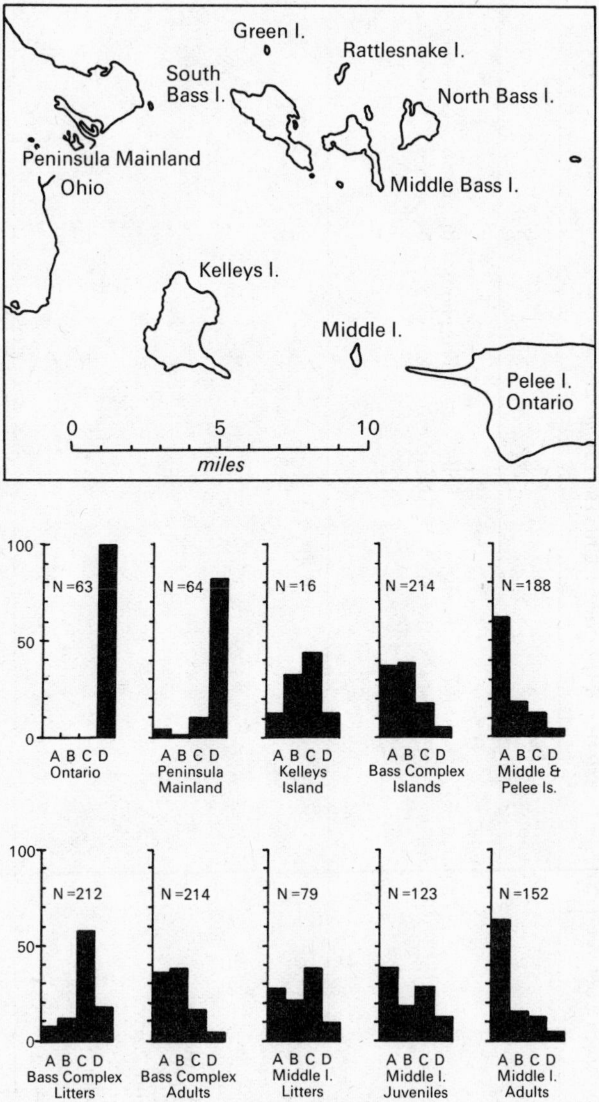

FIG. 14. Frequency of banding in Water Snakes on islands in Lake Erie – A is unbanded, D completely banded. The mainland population is highly banded, but bird predation eliminates the more obvious banded individuals on the islands. Migration from the mainland reservoir of banded snakes affects the frequencies on the islands, particularly those near the mainland; only the less banded snakes tend to survive to adulthood (lower histograms) (after Camin and Ehrlich, 1958).

TABLE 7. Compatibility between different varieties of the Sweet Cherry (*Prunus avium*)

		Genotype	S_1S_2	S_1S_2	S_1S_3	S_1S_3	S_3S_5	S_1S_4
		Male parent variety	Bedford Prolific	Early Rivers	Belle Agathe	Frogmore Early	Bigarreau Napoleon	Governor Wood
Genotype	Female parent variety							
S_1S_2	Bedford Prolific		−	−	+	+	+	+
S_1S_2	Early Rivers		−	−	+	+	+	+
S_1S_3	Belle Agathe		+	+	−	−	+	+
S_1S_3	Frogmore Early		+	+	−	−	+	+
S_3S_5	Bigarreau Napoleon		+	+	+	+	−	+
S_1S_4	Governor Wood		+	+	+	+	+	−

was detectable up to two yards down-wind of the boundary, but was only appreciable up to 18 inches upwind. Notwithstanding the characters changed to their characteristic average values within *two inches* of the boundary – and this in a wind pollinated plant!

5. Mating choice

One of the assumptions made in deriving the simple Hardy-Weinberg situation was the random union of gametes – that the formation of zygotes is uninfluenced by any characteristic of the parents. The most obvious exception to this is sex. In species with approximate equality of sexes, any individual is able to mate with only half the members of the species. Among flowering plants, incompatibility is a common mechanism to prevent inbreeding; certain classes of pollen grain do not germinate effectively, or grow quickly down the styles of certain other genotypes, although the pollen and styles are perfectly compatible in other combinations. In the simplest situation, a species may have an incompatibility locus with up to 100 alleles. A pollen grain carrying a particular allele will not grow on a style containing the same allele. This precludes self fertilization, and reduces mating between close relatives. It also promotes genetical diversity. This can be illustrated by work on the Sweet Cherry (*Prunus avium*) (Crane & Brown, 1937). In this species, only 0.06% of nearly 50,000 self pollinations carried out at the John Innes Research Institute gave rise to mature fruit. Different varieties could be arranged in groups according to the chance of successful fertilization and fruit production (Table 7). Research on one of the Evening Primroses (*Oenothera organensis*) has shown that the incompatibility gene is really at least two genes close together on a chromosome – one determining the reaction of the female parent and the other the male (pollen) specificity (Lewis, 1954). Similar mechanisms are found in many species from many flowering plant groups.

Assortative mating in animals has the same effect in encouraging union between dissimilar (and hence probably unrelated) or similar individuals, and thus maintaining population variation. Scarlet Tiger moths 'prefer' to mate with a morph (phenotype) other than their own. This means that a rare morph has a higher chance of finding a mate and transmitting its genes than a common morph. In humans the opposite happens, people of the same height, or with musical talents tend to marry each other.

A solution to the problem that we do not seem to be declining in intelligence despite more intelligent people having small families and less intelligent ones large ones, has been suggested by Penrose (1955) on the basis of an observation that people tend to marry someone of similar intelligence to themselves.

He imagined a population in which intelligence is determined solely by a pair of additive alleles A and a. The meaning of 'additive' here is that heterozygotes (Aa) will have an intelligence halfway between the two homozygotes (AA and aa). If we make an extreme assumption that one-tenth of the

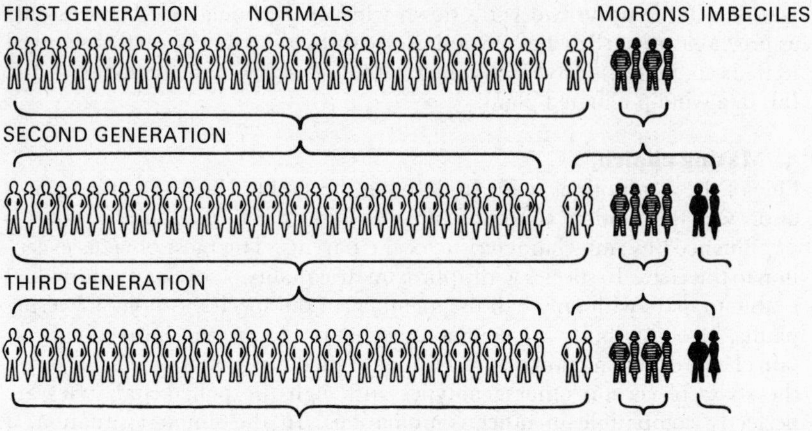

FIG. 15. The intelligence of a population may remain unchanged if there is strong assortative mating between people of like intelligence, even if less intelligent parents have twice as many children as intelligent ones. Normals (white) have genotype *AA*, less intelligent people (shaded) *Aa*, and imbeciles (who leave no children) (black) *aa* (from Penrose, 1959).

British population have an inferior intellect, we can divide everyone into separate 'intelligence' groups. In this population, assume that the less intelligent people are heterozygous and also have a high fertility, so that their families are more than twice the size of those of 'superior' people.

Suppose that the two groups (normal and inferiors) do not mix at all, so that like always marries like. Normal homozygotes will produce only normal children, the 'inferior' will have children like themselves (*Aa*), but also normals (*AA*), and a new genotype *aa* which may be presumed to be worse affected than their parents. We can call these imbeciles or idiots. (Before the

TABLE 8. Imaginary population with completely assortative mating: intelligence level determined by a perfectly additive gene pair (after Penrose, 1963)

Types of mating	% frequency of mating pair	Relative birth rate per family	Offspring: AA (I.Q. 103), Superior	Offspring: Aa (I.Q. 73), Inferior	Offspring: aa (I.Q. 43), Sublethal
AA × AA	90	1.89	170	—	—
Aa × Aa	10	4.00	10	20	10
All types	100	2.10	180	20	10
% parental pairs in next generation			90	10	–

TABLE 9. Correlations between married couples for various traits. (If mating was completely at random, the correlation would be zero)

Trait	*% correlation*
Memory	57
Intelligence	47
Ear length	40
Waist circumference	38
Neurotic tendency	30
Height	28
Eye colour	26
Neck size	20

Mental Health Act of 1959, imbeciles were classified as having an I.Q. of less than 50, and idiots one of less than 20. All types are now grouped together as varying grades of mental subnormality.) Imbeciles are quite infertile (Table 8). This situation can easily produce a stable situation with the excess fertility of heterozygotes supplementing the low fertility of more intelligent people (*AA* homozygotes). Defectives do not increase in numbers because they produce so many offspring of low mentality who do not breed.

This model is not to be taken too literally. The genetics of intelligence are far more complicated than it assumes. Nevertheless, there is a strong tendency of like humans to marry like (Table 9), and the model suggests that the human race may not be deteriorating genetically as rapidly as some of the more doctrinaire eugenists have believed.

EQUILIBRIA

We have seen that there are a number of factors tending to upset the simple stability of allele frequencies expected on the Hardy-Weinberg principle. Notwithstanding, frequencies tend to remain stable – not because upsetting agencies are not operating, but because they tend to cancel each other out. The result is an equilibrium between two opposing forces. There are two important equilibria:

1. Mutation v. selection

If a mutation is deleterious to its carrier, it will be transmitted to the next generation at a lower rate than the normal allele, and will in due course be eliminated from the population. This is easiest to see for alleles producing human disease. Take a condition like achondroplasia which is a dominantly inherited condition in which the growth regions of the long bones are abnormal. Achondroplasics have short arms and legs, and a high head because the base of the skull is reduced in size. A similar mutation produces the bulldog condition in dogs, and in sheep was once the basis for a special breed (the

Ancon) which was prized because it could not easily jump walls. Human achondroplasics have normal fertility and intelligence, but have on average only one-fifth of the number of children when compared with ordinary people. The effect of this is that 20% of achondroplasics are born to an affected parent, the other 80% are the result of fresh mutation. We can think of a pool of achondroplasic alleles, which is losing its contents as achondroplasics fail to reproduce, but maintaining its volume as fresh alleles are added by mutation (rather like an 'O' level maths problem about the rate of filling of baths when the plug has been pulled out).

We can generalize this equilibrium with a piece of extremely simple algebra:

Let N be the total number of individuals in a population
x be the frequency of an inherited anomaly
f be the fitness of the abnormal phenotype
($= 1-s$, where s is the selection coefficient)
m be the net frequency of mutation to the abnormal allele

If the condition is dominantly inherited, there will be
xN affected parents with
xN normal and xN abnormal alleles
$N-xN$ normal parents with
$2(N-xN)$ normal alleles

The sum of normal alleles in the population will be $xN + 2N - 2xN$

$$= 2N - xN$$
$$= N(2-x)$$

Now the number of mutations giving rise to abnormal births

$$= mN(2-x)$$

since a mutation must take place in a normal allele to produce the abnormality.

x is usually small, so this is approximately $2mN$

In each generation the number of alleles eliminated = (number of abnormalities present) × (proportion eliminated by reduced fitness)

$$= xN(1-f)$$

At equilibrium

number of alleles produced = number eliminated

$$2mN = xN(1-f)$$
$$m = \tfrac{1}{2}(1-f)x$$

For a recessive condition, two alleles are lost for each case eliminated so

$$m = (1-f)x$$
$$= sq^2$$

(where q is the allele frequency of the mutant)

Many recessive conditions in man are so deleterious that the affected homozygotes never reproduce, *i.e.* $s = 1$ and $m = q^2$. For example, phenylketonuria is an inborn error of metabolism in which homozygotes are unable to use the amino-acid phenylalanine and consequently this accumulates in their blood. Probably because they are poisoned by this accumulation, the patients were always grossly mentally defective, and never married until modern treatment (a diet containing no phenylalanine) was applied. The frequency of the condition in Caucasian populations varies from 2 to 20 per 100,000. If the incidence is taken as 1 in 40,000, the mutation rate will also be 1 in 40,000.

2. **Heterozygous advantage**

Before dealing with our second equilibrium situation, let us recapitulate on the effects of selection on phenotypes determined by different modes of inheritance. Selection may act against one homozygote only (in a recessive condition), against one homozygote and the heterozygote equally (for a fully dominant condition), or against the heterozygote and homozygote differently (for an additive or incompletely dominant condition). Now if we tabulate these possibilities we see there is a fourth situation when selection acts against both homozygotes:

		A_1A_1	A_1A_2	A_2A_2
Fitness:	Recessive	1	1	$1-S$
	Dominant	1	$1-S$	$1-S$
	Dominant with heterozygous manifestation	1	$1-S_1$	$1-S_2$
	Heterozygous advantage	$1-S_1$	1	$1-S_2$

Heterozygous advantage (also known as heterosis or over-dominance) may seem rather an abstract possibility, but it frequently crops up in nature. The easiest situation to envisage is where both homozygotes are lethal. In this case all matings must be between heterozygotes, and the heterozygote will have a fitness of one, both homozygotes a fitness of nought.

An approach to this extreme may occur under primitive conditions in malarious parts of tropical Africa. Here heterozygotes for sickle cell and normal haemoglobin are protected against cerebral malaria, the most frequently fatal malaria of childhood. The probable mechanism of this is that the red cells have a life span too short to allow the malarial parasite to complete its reproductive cycle.

Allison (1954) pointed out that the distribution of haemoglobin S (and some other abnormal haemoglobins) was virtually the same as that of malaria, and that in malarial areas the frequency of parasites in the blood was higher in sicklers than non-sicklers. He innoculated malaria into two groups: 14 out of 15 non-sicklers developed clinical malaria, but only two out of 15 sicklers. It is an advantage to be a heterozygote, but a disadvantage to be either homozygote, since sickle cell homozygotes die of anaemia and normal homozygotes have a high chance of contracting malaria (this will not kill all of them, but will severely reduce their potential for reproduction).

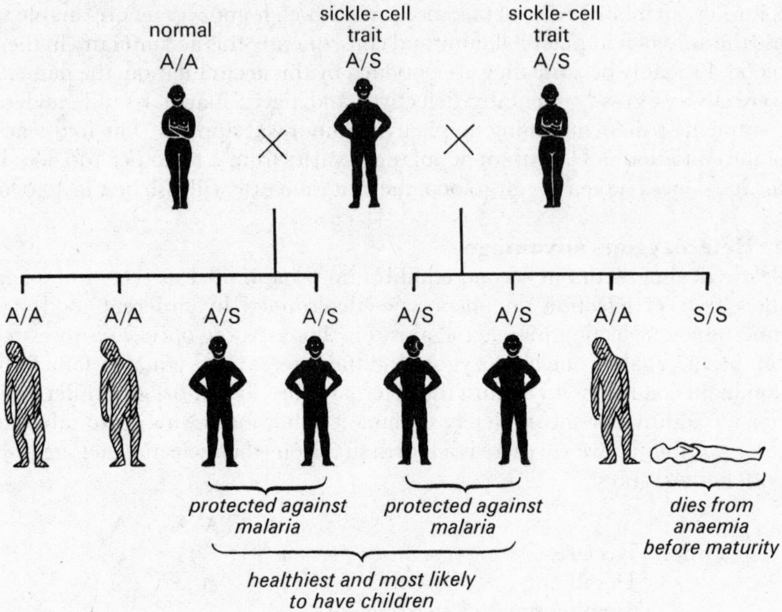

FIG. 16. Heterozygous advantage of sickle cell anaemia: only heterozygotes are healthy when living in a highly malarial environment: the distributions of the sickle cell allele (together with thalassaemia, another haemoglobin 'defect') and cerebral (falciparum) malaria are almost the same.

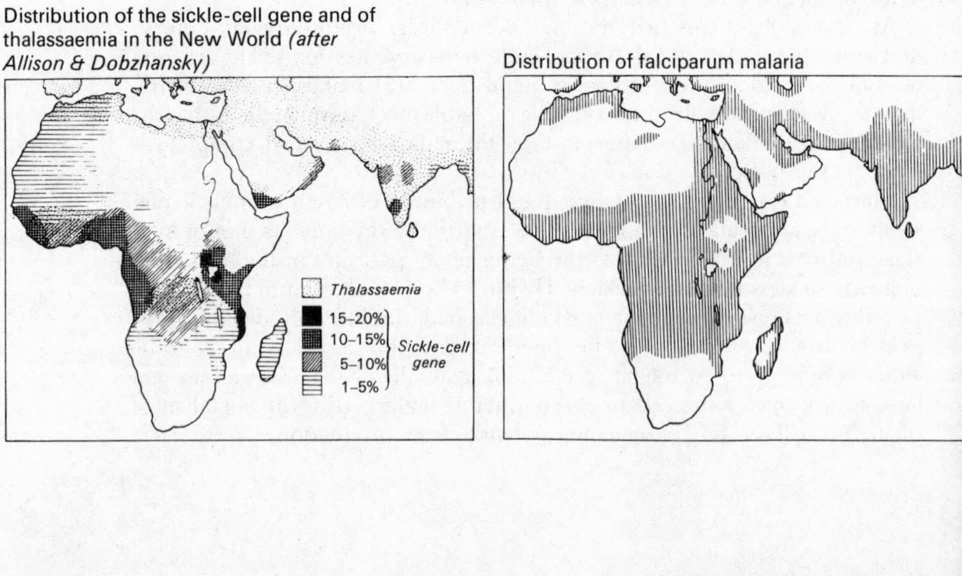

Heterozygous excess also apparently occurs at an amylase (an enzyme catalysing starch breakdown) locus with two alleles (B,C) in milk cows (Friesians, Ayrshires, Guernseys and Jerseys). In a group of 13 Friesian herds assuming the genotypes were classified correctly, the observed numbers of the genotypes at the locus, and their numbers if the Hardy-Weinberg expectations were met, were (Spooner, *et al.*, 1973):

	BB	BC	CC
Observed	86	402	74
Expected	146	281	135

Young calves had the expected proportions of homozygotes, but by the time they had entered the milk herd, sufficient had died or been culled for there to be a 43% deficiency of homozygotes (or excess of heterozygotes).

We can generalize the conditions for heterozygous advantage:

	A_1A_1	A_1A_2	A_2A_2
Initial frequency	p^2	$2pq$	q^2
Fitness	$1-s$	1	$1-t$
Frequency after selection	$p^2(1-s)$	$2pq$	$q^2(1-t)$

In the first generation, population composition

$$= p^2 + 2pq + q^2$$
$$= (p+q)^2 = 1$$

In the second generation, population composition

$$= p^2(1-s) + 2pq + q^2(1-t)$$
$$= p^2 + 2pq + q^2 - sp^2 - tq^2$$
$$= 1 - sp^2 - tq^2$$

Now initial value of q

$$= q_0$$
$$= (pq + q^2)/1$$

Value of q after selection

$$= q_1$$
$$= \frac{pq + q^2 - tq^2}{1 - sp^2 - tq^2}$$
$$= \frac{q(p+q) - tq^2}{1 - sp^2 - tq^2}$$
$$= \frac{q - tq^2}{1 - sp^2 - tq^2}$$

The change in allele frequency due to selection will be

$$q_1 - q_0 = \frac{q - tq^2}{1 - sp^2 - tq^2} - q$$
$$= \frac{q - tq^2 - q - sp^2q - tq^2(1-p)}{1 - sp^2 - tq^2}$$
$$= \frac{pq(sp-tq)}{1 - sp^2 - tq^2}$$

At equilibrium, there will be no frequency change, and

$$q_1 - q_0 = 0$$

This is fulfilled when

$$sp = tq$$
$$s(1-q) - tq = 0$$
$$s - sq - tq = 0$$
$$s = q\,(t + s)$$
$$q = \frac{s}{t + s}$$

If q is between its equilibrium value and one, p will be below its equilibrium value, and sp will be less than tq. Consequently q will decrease until it reaches its equilibrium value. When q has any value (except 0 and 1), selection will force it back to equilibrium value. The equilibrium value of q is independent of mutation.

As originally shown by R. A. Fisher (1922) in a remarkable theoretical paper which dealt with many of the problems of population genetics before anyone realized they were problems, heterozygous advantage leads to a balanced polymorphism: a situation where two (or more) alleles are maintained in a population by a balance of selective forces. It brings us back to Ford's definition of polymorphism (page 18). Where we have coexisting forms, either there is a balanced polymorphism or a transient one where a new morph is in the process of replacing another. In both situations, selection is regulating the allele frequencies. When electrophoresis showed objectively the amount of inherited variation in a species (Table 2), it implied an enormous amount of selection. This led to controversy because theoreticians insisted that so much variation could only arise if many of the alleles concerned had no effect on their carriers, *i.e.* were neutral. This is really the reopening of an old argument, because as long ago as 1930 R. A. Fisher showed that an allele would spread even if its advantage was as small as 1% (*i.e.* if $s = 0.01$), and he pointed out that even if an allele was neutral in one particular environment, it may not be in another one, and few organisms live in an unchanging environment for many generations (*q.v.* Chapter 9).

Heterozygous advantage is not the only way to maintain a polymorphism. Williamson (1958) listed seventeen other methods without claiming to be exhaustive. The most important ones involve situations where selection acts against one extreme (or homozygote) at one part of the life-cycle or part of the range, and against the other at another time or place. This may be difficult to distinguish from true heterozygous advantage:

	A_1A_1	A_1A_2	A_2A_2
Stage or place I	$1-s_1$	1	1
Stage or place II	1	1	$1-s_2$
Overall	$1-s_1$	1	$1-s_2$

Fluctuating or *endocyclic* selection of this nature occurs in many organisms, where variation is lost during life, but reconstituted to reappear in the developmental or young phases (see pages 124, 132 138 and 247).

Polymorphism may also arise through frequency or density dependent mechanisms. For example, birds get 'searching images' when looking for their prey and it can be an advantage to be a rare morph and get overlooked (Thompson & Vertinsky, 1975). Harvey, Jordan & Allen (1974) put out pastry baits for Blackbirds (*Turdus merula*) on a lawn, and dyed them brown or green. They offered the birds nine times as many brown baits as green, or vice versa. Consistently the birds ate 15 or more times as many of the commoner baits whether it was brown or green in the particular experiment. This was a highly artificial situation, but it suggests important possibilities for a polymorphic prey species.

One final point about heterozygous advantage: examples involving a single locus like the sickle-cell haemoglobin situation are not common. What are common are examples where extreme phenotypes are eliminated. This is similar to the simple heterozygous advantage situation, and may be identical if the trait subjected to selection is controlled by additive genes (page 44). We shall return to the action of selection in Chapter 4.

CHROMOSOMES

There was a time not very long ago when chromosome form and number (the *karyotype*) was believed to be as diagnostic of a species as morphology. This was partially a technical problem since chromosome preparations were difficult to make, and only in especially favourable material was it possible to make large numbers of preparations (such as the polytene chromosomes of Diptera, where the original chromosomes have repeatedly divided but have remained attached to each other longitudinally). Indeed, it was only in 1956 that man was given his correct chromosome number of 46. Prior to this he was generally believed to have 48. The study of mammalian chromosomes generally was revolutionized by the simple expedient of putting the preparation into hypotonic saline for a few minutes. This breaks the nuclear membrane and allows the chromosomes to separate from each other. We know now that chromosomes may show as much variation as any other system. Just as with genes, there may be differences between individuals of a population within a species; geographical variation within a species (*i.e.* chromosomal races); or karyotypically abnormal individuals. Unfortunately – and ironically since chromosomes carry the genes – the significance of this variation is known in all too few cases.

The karyotype of a species may be looked upon as the way in which the total DNA of the nucleus is broken up into separate chromosomes. 'Normal' nuclei will contain two chromosome sets (the *diploid* number), and all have the same amount of DNA. *Haploid* nuclei (the sperm and egg nuclei of animals, or the spore nuclei of higher plants) have half the amount. The human haploid karyotype contains about 113 cm of DNA, corresponding to about 7,000 million nucleotide pairs. Mice and dogs have approximately the same amount, small flies like *Drosophila* and the midge *Chironomus* about one-twentieth, locusts two to three times as much. The common Vetch (*Vicia sativa*) has a similar length of DNA to men and mice; and the Broad Bean (*Vicia faba*) has five and a half times more.

Haploid chromosome numbers range from one in the Horse Round-worm (*Parascaris equorum*) to 630 in the fern *Ophioglossum reticulatum*, but here there are repeated chromosome sets (probably 42 of them). The highest known chromosome number in animals is 233 in a Butterfly (*Lysandra atlantica*) from the Atlas mountains. However, the great majority of animal species have

TABLE 10. Chromosome numbers (somatic) in familiar organisms

FUNGI	
Bread mould (*Neurospora crassa*)	14
Yeast (*Saccharomyces cerevisiae*)	10 or 12
HIGHER PLANTS	
Pines, Larches, Birches	24
Juniper (*Juniperus* spp.)	22
Barley (*Hordeum vulgare*)	14
Sugar Cane (*Saccharum officinarum*)	80 (octoploid)
Bread Wheat (*Triticum aestivum*)	42 (hexaploid)
Onion (*Allium cepa*)	16
Cucumber (*Cucumis sativus*)	14
Garden Pea (*Pisum sativum*)	14
Potato (*Solanum tuberosum*)	48 (tetraploid)
Tea (*Camellia sinensis*)	30
Oaks (*Quercus* spp.)	24
ANIMALS	
Fruit fly (*Drosophila melanogaster*)	8
Housefly (*Musca domestica*)	12
Honey bee (*Apis mellifica*)	32
Garden snail (*Cepaea nemoralis*)	44
Chicken (*Gallus domesticus*)	78
House Mouse (*Mus musculus*)	40
Brown Rat (*Rattus norvegicus*)	42
Rabbit (*Oryctolagus cuniculus*)	44
Gorilla (*Gorilla gorilla*)	48
Chimpanzee (*Pan troglodytes*)	48
Man (*Homo sapiens*)	46

haploid numbers between six and thirty. About a third of higher plant species are *polyploid* (*i.e.* have more than two chromosome sets), but again very high numbers are uncommon (Table 10).

Polyploidy

The vast majority of animal species and about two-thirds of higher plant species are diploid. Polyploids may have three (triploid), four (tetraploid), five (pentaploid) or more sets of chromosomes. They often have a low fertility because when individual chromosomes have more than one partner they do not segregate evenly during gamete formation.

The classical example of the origin of a polyploid was the formation at Kew in 1900 of *Primula kewensis* by crossing two commonly cultivated *Primula* species, *P. floribunda* and *P. verticillata*. The hybrid was intermediate in appearance between the parents and had the same chromosome number (18). It was usually completely sterile. However, on three occasions hybrid plants were observed to set good seed; in each case they were shown to be tetraploid (with 36 chromosomes). The fertile *P. kewensis* had arisen by doubling of the chromosomes from each of the original parents so that all the chromosomes had a homologue to pair with. It was effectively a new species, distinct from both parents and unable to produce fertile seed with either.

Red Clover (*Trifolium repens*) is extremely common and variable. It has 32 chromosomes. A low growing form found in shallow soil on the cliffs of south-west England and north-west France has been described as a separate species, (*T. occidentale*) and only has 16 chromosomes (Coombe, 1961).

The probable reason polyploidy is comparatively rare in animals is that it is difficult to establish itself with the sex determining mechanisms and obligate out-crossing which are usual in most animals (Chapter 3). Only in animal groups which reproduce asexually is polyploidy common (certain insects, crustaceans, earthworms, fish and amphibia). When polyploid species occur they may have different tolerances. For example, there is a psychid Moth *Solenobia* in southern Europe in which there is a sexual diploid, a parthenogenetic (self-fertilizing) diploid, and a parthenogenetic tetraploid. The sexual form is found mainly near the limits of the last glaciation and along the nunataks of the northern slope of the Alps; the tetraploids are found in the upper valleys of the northern Alps, and along the slopes of the southern Alps; the Swiss Jura is inhabited by the parthenogenetic diploids.

But polyploidy is really a plant phenomenon, and it is so common among plants that we must consider it in some detail. Take the docks for example. The basic chromosome number is 10, but diploid numbers vary from 20 in the Red-veined Dock (*Rumex sanguineus*), through 40 in the tetraploid Broad-leaved Dock (*R. obtusifolius*), 60 in the Curled Dock (*R. crispus*) up to the decaploid 200 in the Great Water Dock (*R. hydrolapathum*).

Traditionally chromosome workers have distinguished between autopolyploidy and allopolyploidy. Autopolyploidy involves doubling the diploid number in the same species. Temperature shocks, for example, may disturb normal cell division and lead to one cell with four sets of chromosomes, rather than two cells each with two sets. The same result can be produced by treating a plant with the drug colchicine (an alkaloid extracted from *Arum colchicum*) which stops spindles being produced in cell division, and thus chromosome sets separating. Allopolyploidy is more important in nature and may arise following hybridization. If two diploid species have chromosome sets *AA* and *BB*, a hybrid between them will be *AB*. This is likely to be almost completely infertile since the chromosomes cannot pair properly, and hence do not separate into two chromosomally balanced haploid cells at meiosis. However it may occasionally produce a small number of gametes with the full *AB* complement. Fused with another gamete of the same type they will give a tetraploid *AABB*. *Primula kewensis* is of this type. The difficulty of distinguishing allo- from autopolyploidy is that no two chromosome constitutions in a population or species are identical and there is a continuum connecting truely infertile species hybrids to ones with slightly differentiated chromosome sets which are almost fully fertile.

The best known example of a probable natural allopolyploid is Cord-grass (*Spartina townsendii*). This is abundant round British coasts and adjacent shores of Europe, and has been widely planted to help stabilize mud-flats. It apparently originated in Southampton Water around 1870 where an introduced American species, *S. alterniflora* (first recorded in 1830 on the River Itchen, which flows into Southampton Water) hybridized with the native British species, *S. maritima* (Hubbard, 1965; Marchant, 1968). Populations of the hybrid may have 62 chromosomes and be sterile (the putative parents have diploid numbers of 62 and 60 respectively), or may have 124 and be fertile. The fertile plants can readily be distinguished from their sterile 'sisters' by their long anthers. The success of *S. townsendii* in exploiting a natural habitat where previously there was little competition, emphasizes the importance of hybridization and polyploidy in conditions when human activities may bring together separate chromosome sets.

Commercial cereals are often polyploids. Wheat (*Triticum aestivum*) is a hexaploid of 42 chromosomes formed from three ancestral grass species. One of these was an Einkorn wheat of the Middle East where it was used as a cereal in Neolithic times. The Einkorn wheats are 14 chromosome diploids whose chromosome complement may be symbolized *AA*. The Emmer, *durum* and macaroni wheats (*T. dioccum*, *T. diococcoides*, *T. durum*, etc.) are all 28 chromosome tetraploids with the formula *AABB*. It was previously thought that the *B* chromosome set came from *Triticum speltoides*, but this now seems unlikely. The bread wheats (*T. aestivum*) have the hexaploid formula *AABBDD*, with the *D* component coming from a hybridization with *T. tauschii*. Fertility in *T. aestivum* depends on a locus in the *B* set: an allele at this locus controls regular chromosome pairing at meiosis; when this allele is not present, many three and four chromosome associations are formed due to pairing between chromosomes derived from different ancestral species.

The New World cultivated cottons are also polyploids. They have arisen by hybridisation between one of the Old World cultivated species and a form related to the South American wild species *Gossypium raimondii*.

Chromosome number: aneuploidy and centric fusion

Chromosome numbers may vary in ones and twos as well as in whole sets. Chromosome breakage can lead to a particular chromosome being so damaged that it is not transmitted in the next cell division. Cells lacking a chromosome often fail to survive. Another way of losing a chromosome at cell division is by *non-disjunction*. If a pair of chromosomes fail to separate into the two daughter cells at a cell division, one of the daughters will have an extra chromosome, and the other will lack one. The chromosomally deficient cell is likely to die; the cell with an extra chromosome is more likely to survive but be abnormal. If the cell division which involved non-disjunction was a meiosis leading to a gamete formation, an individual may be formed in which all the cells have an extra chromosome. This situation is known as *aneuploidy* (literally, not having the true number of chromosomes). Since there are normally two chromosomes of any one type in a diploid, an organism with an extra chromosome is *trisomic* for the chromosome in question; an organism lacking one is *monosomic*.

Non-disjunction gives rise to most of the chromosome diseases recognized so far in man. Five of the 23 human chromosomes are most frequently involved in trisomy: the commonest condition in Down's syndrome or mongolism, in which over 95% of cases are caused by trisomy of one of the smallest chromosomes. (The other 5% are produced by a translocation of the major part of the same chromosome – see below.) The incidence of Down's syndrome is about one in 600 overall, but increases from about 1 in 2000 in young mothers to about 1 in 40 in mothers aged 45. A number of agents have been implicated in the causation of non-disjunction (*e.g.* large doses of X-rays, viral hepatitis) but these are hints rather than proved mechanisms. Most of these chromosomally abnormal foetuses spontaneously miscarry early in pregnancy; those which survive to be born are merely the tip of an iceberg.

A large survey of any species will almost certainly show some aneuploids. For example, a study of Water Voles (*Arvicola terrestris*) in Sweden revealed one aberrant individual – a young female with an unusually short tail and abnormal teeth. She had 37 chromosomes instead of the normal 36, the smallest chromosome being represented three times instead of two. All other voles examined were chromosomally normal (Fredga, 1968).

Aneuploids are also reasonably common in some plant taxa. A survey of Hawk's Beard (*Crepis tectorum*) revealed 18 among 4000 plants, which is the same sort of order as aneuploids occur in human populations. In the Thorn-

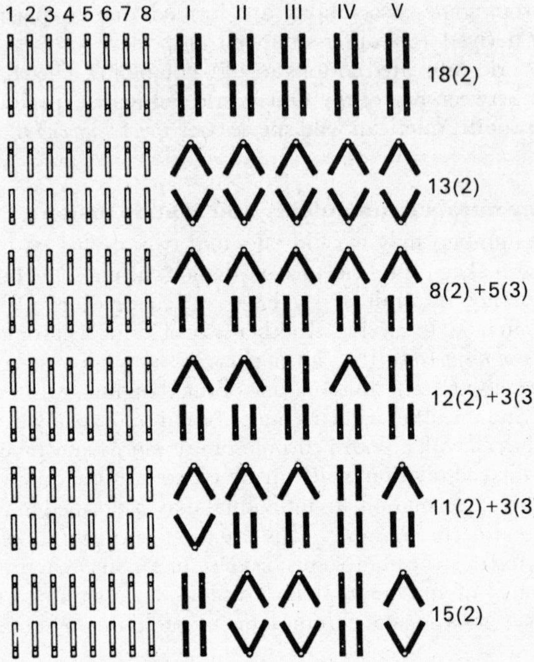

FIG. 17. The chromosomes of the Dog Whelk (*Nucella lapillus*). Individuals may have 36 chromosomes (top two rows), 26 (2nd two rows), or any intermediate number. The differences are due to the loss (or gain) of centromeres, producing long chromosomes with a central centromere (metacentric) from two short chromosomes with a centromere near one end (acrocentric) (after Staiger, 1954).

Apple (*Datura stramonium*) there are 12 chromosomes in the haploid set, and trisomics are known for all of them, each affecting the appearance of the seed capsule in a different way.

A much more regular way of variation in chromosome numbers is so-called Robertsonian variation or centric fusion. Somewhere along its length all chromosomes have a *centromere* which becomes attached to the spindle during cell division. If the centromere is near the middle, the chromosome is called *metacentric*, if it is near one end it is *acrocentric*. Two acrocentric chromosomes can fuse end to end with no effective loss of chromosomal material; all they lose is one centromere, and become a single metacentric chromosome. Conversely a metacentric chromosome may gain a centromere and become two acrocentrics. This means that a species may have different chromosome numbers, but the same amount of chromosomal material. The number of chromosome arms (the fundamental number or N.F.) is apparently a more important characteristic than the number of chromosomes. Hence, it is often more profitable to compare species in terms of the numbers of chromosome arms rather than chromosomes themselves. For example, chimpanzees have 48 chromosomes, and two of these are acrocentrics which are almost identical with one of the metacentrics in man if they are joined end to end. There are also small differences between the species in the ratios of the long and short arms in four other chromosomes suggesting that pericentric inversions (see below) have occurred since the species separated from their common ancestor, but they are small differences.

However, to return to intra-specific variation in chromosome number: one good example occurs in the common inter-tidal snail, *Nucella lapillus*, the Dog Whelk. In most parts of its range, this species has 26 chromosomes. In the early 1950s it was discovered (Staiger, 1957) that whelks living on exposed shores near Roscoff in Brittany had 36 chromosomes, although sheltered

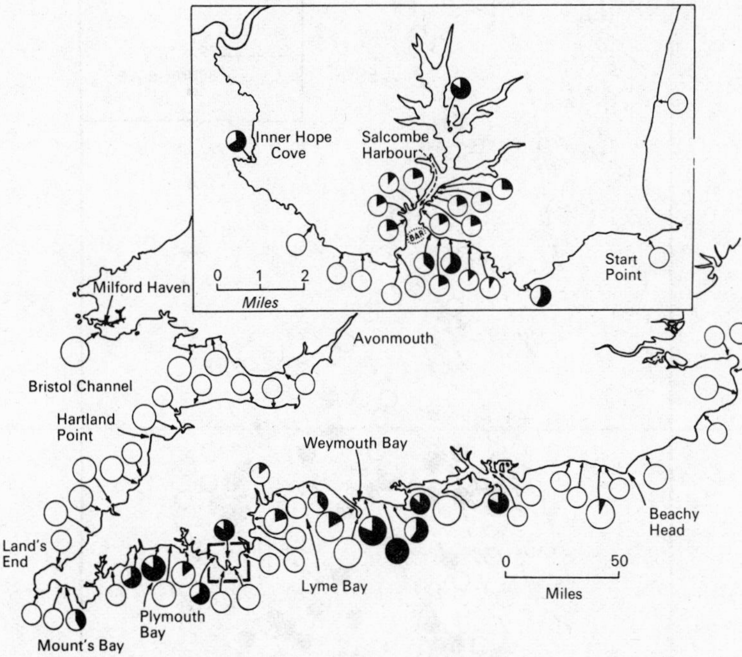

FIG. 18. Chromosomal polymorphism in Dog Whelks in southern Britain: the black sector in each sample represents the frequency of acrocentric chromosomes as the proportion of the maximum possible frequency (due to centric fission in chromosomal elements I to V). The Salcombe Harbour region is drawn in greater detail, showing the high incidence of acrocentric chromosomes in the sheltered waters of the creek; there are strong tidal flows into and out of the harbour (after Bantock and Cockayne, 1975).

shores in the same area had the more usual 26 chromosomes. The 26 chromosome karyotype has 5 pairs of metacentric chromosomes which are represented by 10 pairs of acrocentrics in the 36 chromosome form. The two forms seem to be fully infertile and all chromosome numbers between 26 and 36 occur.

The same polymorphism occurs on the south coast of England (Bantock & Cockayne, 1975). In the same way as in France, fewer chromosomes are found in populations living on sheltered shores, although the highest proportion of acrocentric chromosomes are found in localities which are pro-

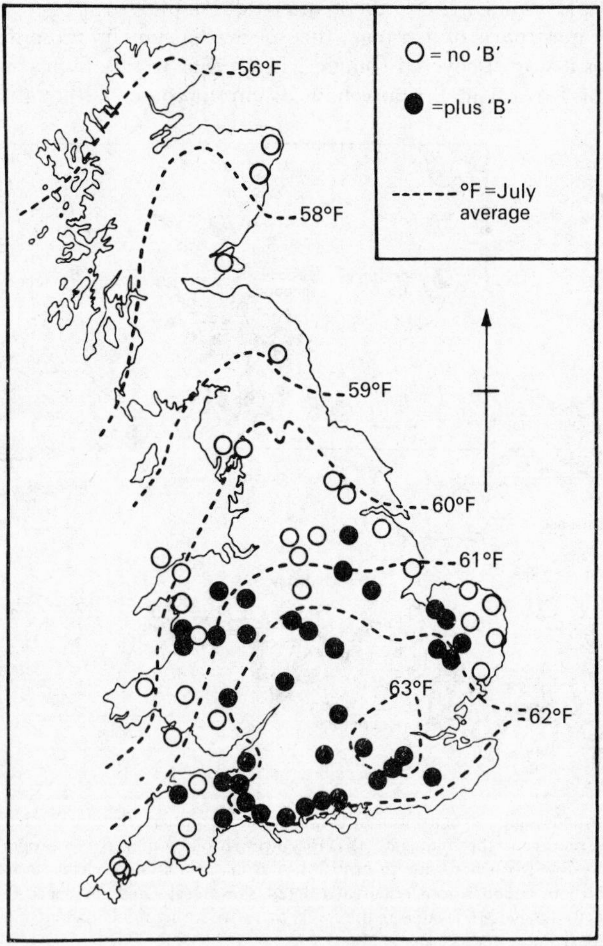

(for caption see opposite)

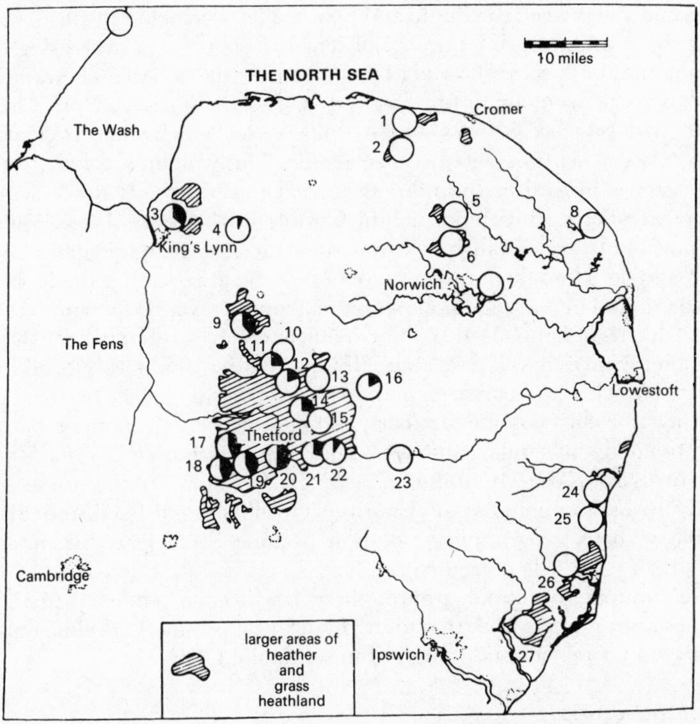

FIG. 19. The distribution of supernumerary (or B) chromosomes in Mottled Grasshoppers (*Myrmeleotettix maculatus*) in Britain (*opposite*), and (*above*) the frequencies of B-containing individuals in East Anglia (from Hewitt and Brown, 1970).

tected against both wave action and tidal currents. For example, the quiet waters of Weymouth Bay harbour the highest chromosome numbers in Britain, whilst lower numbers are found in the shelter of Salcombe Harbour, Devon, which has little exposure to the open sea but a strong current through it.

Why this variation? No one knows. Genic variation is 'locked up' in chromosomes, and the fewer chromosomes there are, the less will this be reassorted during reproduction to produce fresh combinations. In other words, organisms with a small chromosome number are likely to be less variable than organisms with a large number. The paradox with *Nucella* is that it is the exposed shore populations with the lower chromosome number which tend to be more variable in shell characters.

There are other examples of similar centric fusions or fissions. A species of the common littoral Isopod, *Jaera syei*, has six chromosomes races of 28, 26, 24,

22, 20 and 18 between the Baltic and Spain. The coccinellid Beetle *Chilocorus stigma* shows an increasing number of centric fusions (*i.e.* a decreasing chromosome number) across Canada from Nova Scotia to Saskatchewan, most populations being polymorphic for one or more fusions. A House Mouse isolate (*Mus poschiavinus*) from a high alpine valley which was recognized by taxonomists a century ago because of its slightly lighter colour, has 26 chromosomes instead of the normal 40 in House Mice. It has a reduced fertility when crossed with normal mice. Mole Rats (*Spalax ehrenbergi*) in the Lebanon and Israel exist in four chromosome races, with diploid numbers, 52, 54, 58 and 60. Here the number increases southwards, so that the 60 karyotypes are found in marginal populations extending towards the southern and eastern deserts of Israel. Wahrman, Goitein and Nevo (1969) believe that the chromosome races in this case are really species: they find few hybrids in the wild; given a choice of partners, mating more often takes place between pairs with the same chromosome numbers; and fighting is more common between chromosomally dissimilar animals. They suggest that *Spalax* has spread southwards from Asia Minor into increasingly arid areas. In this process, new selectively superior homozygous karyotypes would become fixed, and this was possible without geographical isolation because Mole Rats live in small populations with little movement.

In a number of animal groups, there has been a tendency to reduce chromosome numbers from that found in the most primitive species. This has happened in copepods and isopods. But we do not know why.

B chromosomes

Apparently superfluous (or B-) chromosomes occur in some species. These are small elements consisting of little more than a centromere and seem to be genetically inert. Individuals can exist perfectly well without them, although in some cases they seem to have an adaptive function.

For example, the frequency of individuals of the Mottled Grasshopper (*Myrmeleotettix maculatus*) with B-chromosomes is generally correlated with latitude: most northerly populations do not have any B-chromosomes while southern ones do. However, there are some southern populations which do not possess a B-chromosome: the ecologically marginal populations on Plynlimon Mountain in central Wales have none, but the frequency increases as one proceeds eastward or westward into areas of lower rainfall. In East Anglia the Mottled Grasshoppers occur on heath-land where sandy soil is exposed, and on large sand-dune systems. B-chromosomes occur in two of these areas: at up to 50% on the breckland around Thetford, and at lower frequencies around King's Lynn. Local frequencies have been shown to be unchanged for up to 3 years (Hewitt and Brown, 1970). No B-chromosomes occur on the dunes bordering the North Sea; probably on-shore winds make these a more marginal climate than the inland sites with a more continental climate. However, in the Thetford area there are regions of a very rapid change in

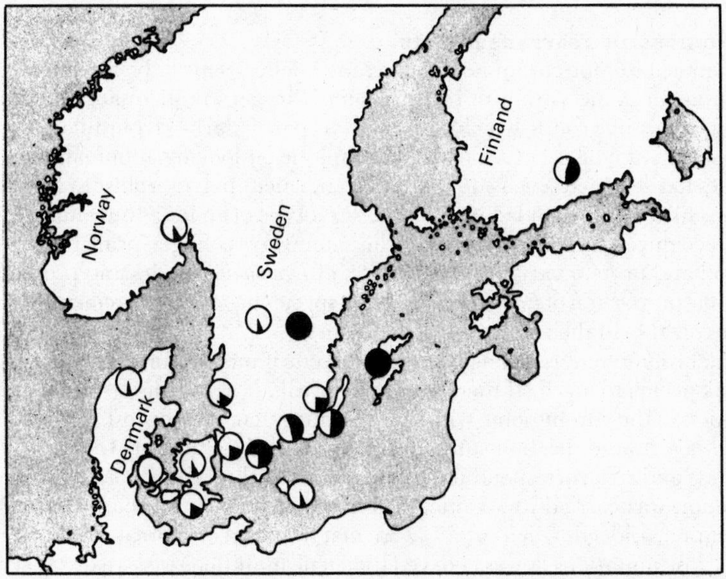

FIG. 20. Accessory chromosomes in Knapweed (*Centaurea scabiosa*); populations with an average of two extra elements per plant are shown by a solid black circle (after Frost, 1958).

frequencies from 40% to 0% in a few miles, and obviously strong genetical factors are acting in such a situation. We shall return to this later.

The general correlation between a favourable environment and a high frequency of B-chromosomes has been found in other organisms. For example, the common Knapweed (*Centaurea scabiosa*) has a high frequency in southern Sweden and Finland, much less in more western wetter populations. Similar associations have been found in the grasses Meadow Fescue (*Festuca pratensis*) and *Phleum phleoides*. No B-chromosomes are found in commercial strains of Rye (*Secale cereale*), but they are common in primitive varieties from Asia.

Multiple B-chromosomes tend to have a deleterious effect. In some maize strains large numbers (10 or more per cell) reduce plant growth significantly. Pollen fertility is reduced in some species. Such supernumerary chromosomes may be lost from tissues like root tips and leaves, although retained (and transmitted) in the germinal tissues.

The presence of B-chromosomes raises the frequency of crossing over in the normal (A) chromosomes at meiosis and with it the variability of the progeny. This has been found in barley, maize and the Mottled Grasshopper. If this is a general rule they would presumably have a similar function to Robertsonian variation of chromosome number.

Chromosome rearrangements

Polymorphism for chromosomal rearrangements seems to be almost as common as for genic variants. The problem is to get an adequate estimate of frequencies in groups which do not have particularly favourable chromosomes for study. Recent advances of technique for looking at human chromosomes has revealed seven common rearrangement polymorphisms, and there seems little doubt that before long the karyotype of an individual human will be recognized as characteristic in the same way as finger-print patterns or blood and protein variants. About 80% of *Drosophila* species and probably a similar proportion of *Chironomus* (Midge) species have rearrangements readily recognizable in their polytene chromosomes.

A chromosomal rearrangement may occur if more than one chromosome break occurs in a cell. If there is a single break, it may either restitute or the portion of the chromosome without the centromere may be lost. This latter is a *deletion*. Large deletions are effectively the loss of almost whole chromosomes, and are correspondingly deleterious to the cell (or organism if the deletion was carried in a gamete). Small deletions may behave like a gene mutation. Indeed, when X-rays were first discovered to break chromosomes *and* cause mutations, it was believed that gene mutations were really 'scars' on

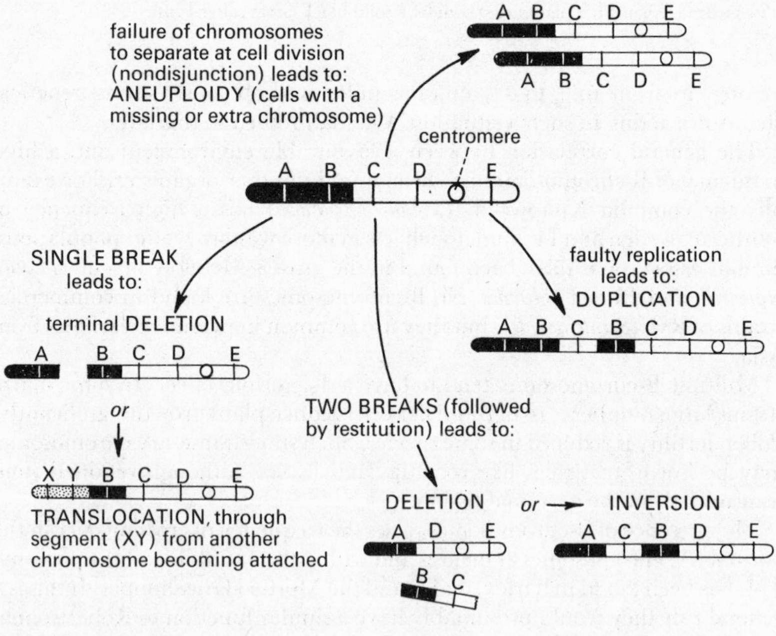

FIG. 21. Different chromosomal anomalies, and how they arise.

the chromosomes where breaks had healed. The opposite of a deletion is a *duplication* where a segment of chromosome is repeated. This probably arises from faulty chromosome pairing rather than chromosome breakage.

If more than one break occurs there are two additional possibilities:

1. A piece of one chromosome may become stuck on to another chromosome. This is *translocation*. It may be reciprocal, *i.e.* a segment of chromosome A may become attached to chromosome B, and a segment of B to A. Translocation heterozygotes usually have reduced fertility, since a translocated chromosome will be homologous with two different chromosomes. This will lead to quadrivalent associations in meiosis, and the only gametes without deletions will carry *either* the translocated *or* the original chromosomes. Nevertheless this effect on fertility may be more than counterbalanced by a beneficial effect of the heterozygote. For example, Romney rams carrying translocations have been shown to have a higher fertility than normal, although the reason for this is not known (Bruere, 1974). If a translocation brings together on a chromosome loci which complement each other, the rearranged chromosome may be selected in the same way as an allele producing an improved phenotype. This is one of the ways in which supergenes arise, such as the complex locus controlling the pin-thrum polymorphism in primroses (pages 186–7).

2. Two breaks in the same chromosome may lead to the segment being reinserted the 'wrong way' round. This is an *inversion*. Thus an inverted segment in a chromosome ABCDE would lead to ABDCE. If the inverted segment includes the centromere it is said to be *pericentric*; if the inversion is outside the centromere, the inversion is *paracentric*. Pericentric ones are easier to detect because they often lead to a change in the relative length of the chromosome arms.

A number of pericentric inversions have been found in humans – some associated with normal individuals, others with abnormal. However, the general importance of inversions is that they lead to the suppression of all crossing-over in the inverted segment, and thus the preservation of particular gene combinations.

Inversion polymorphism

When inversions were discovered to be widespread in natural populations of flies, there seemed to be no rhyme or reason for their distribution. It was assumed that they were neutral in their effect, and that they had attained their commonness by drift (pages 54–6). This belief has proved so wrong, that it has influenced future biologists to be sceptical whenever drift is claimed to explain an observed genetic frequency or distribution. As far as inversions in *Drosophila* are concerned, it was ignorance about their biology which led to drift being assumed.

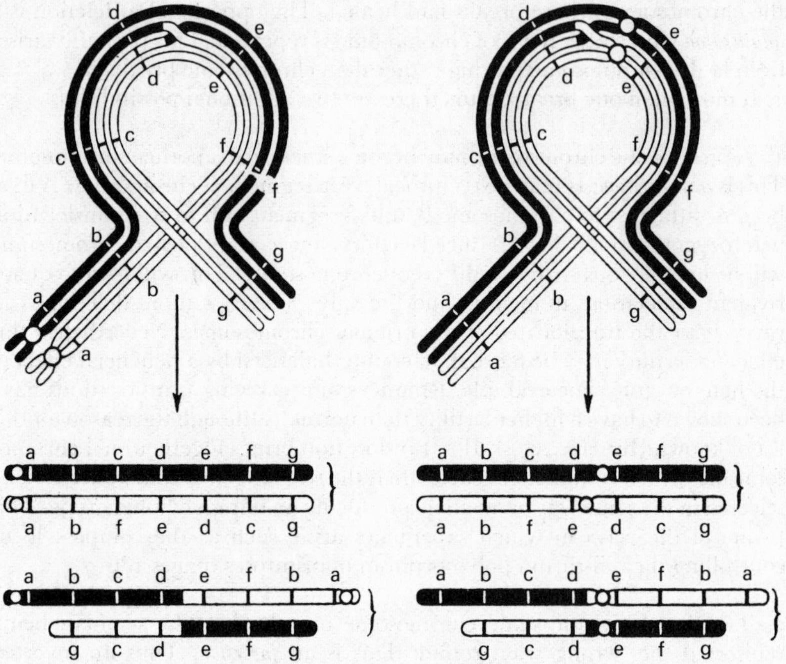

FIG. 22. Crossing-over within chromosomes, one of which has an inverted segment, differs according to whether the centromere is outside (*left*) or within (*right*) the inversion. In both cases it leads to *either* a preservation of the alleles in the inversion *or* (bottom) unbalanced chromosomes (from Berry, 1965*a*).

The most studied inversions are those in the third chromosome of the North American species *Drosophila pseudoobscura*. The species beloved in laboratory and classroom (*D. melanogaster*) has no widely occurring inversions. Sixteen inversions are known in *D. pseudoobscura*, recognizable by the transverse bands which appear in stained preparations of polytene chromosomes (preparations are most commonly made from the salivary glands of well-grown larvae). They are referred to by names – usually abbreviated to two letters – Standard (ST), Arrowhead (AR), Chiricahua (CH), etc.

The original observation was that certain inversions undergo regular changes in frequency with height above sea level in California and Tennessee. At the same time it was discovered that they also showed regular seasonal changes in frequency which were in general repeated every year. The results suggested that different genetical combinations were superior to others under different environmental conditions: as the environment changed, so there was selection for a different equilibrium.

It followed that it should be possible to repeat the different equilibria

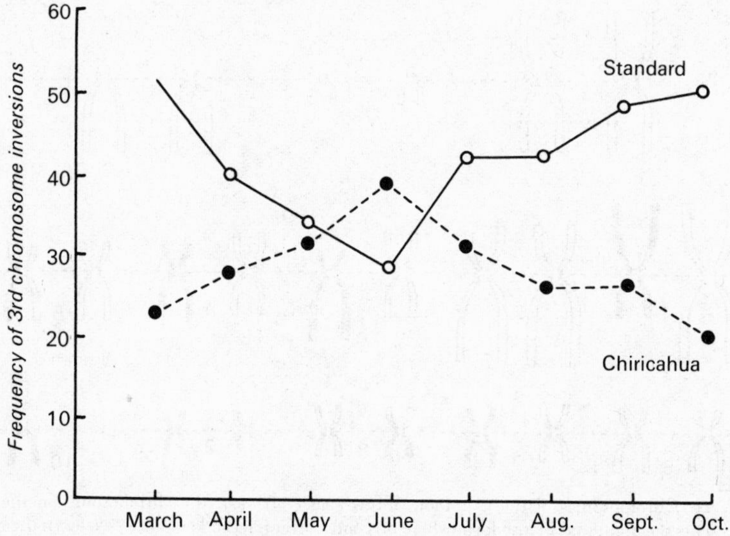

FIG. 23. Changes in the frequencies of two chromosomal inversions in a population of *Drosophila pseudoobscura* in the San Jacinto Mountains during the summer (average data 1939–1946) (after Grant, 1963, from data of Dobzhansky).

artificially by rearing flies under different culture conditions. This has been done (Dobzhansky, 1961). Stocks containing inversions can be maintained with the inversions at almost any frequency if the culture can be manipulated adequately. The equilibrium frequencies depend on the inversions present, the temperature, degree of larval crowding, the availability and type of food.

Midges have many inversions. The most chromosomally polymorphic species known is a midge, *Simulium vittatum*, in which 134 different inversions have been recorded. *Chironomus tenans* is polymorphic in Britain, and many populations have more heterozygotes than expected from the Hardy-Weinberg principle (Acton, 1957). Blaylock (1965) investigated this species in an area in Tennessee where the forest was being cleared around a stream in which the midges bred. There were three genotypes (or more strictly, karyotypes): *I/I*, *I/S*, and *S/S*. *I/I* larvae were at an advantage in the potholes left by the heavy clearing equipment, but both *I/I* and *I/S* phenotypes were selected against in the stream. The pupae hatched into flies which mated at random. If the habitat remained constant (which it did not, as clearance proceeded), this would lead to a balanced polymorphism maintained by different selection pressures in different niches.

Inversion polymorphisms have been detected in a wide range of other groups – in mosquitoes, simuliids, and other flies; in grass-hoppers; and among plants in Herb Paris (*Paris quadrifolia*), paeonies, and lilies.

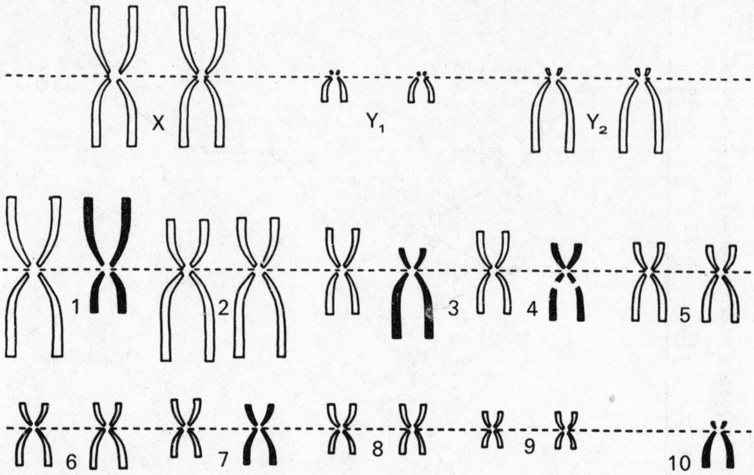

FIG. 24. Chromosomes of the Common Shrew (*Sorex araneus*): the chromosome on the left represents the mainland British form where no centric fission has taken place (see text); the right hand element in each pair is that in the Jersey form of the species. Chromosomes in black indicate those of the Jersey form that differ significantly from the mainland form in length or arm ratio, or both (from Ford and Hamerton, 1970).

It is relatively easy to classify chromosome changes in theory, but in practice different sorts may occur in the same species. The Common Shrew (*Sorex araneus*) illustrates this. The basic diploid chromosome complement of this species in Britain contains 18 metacentric autosomes plus 3 sex chromosomes in the male and two in the female. Ford and Hamerton (1970) found a population near Chiltern in Berkshire where metacentric chromosomes 6, 7 and 8 and rarely 4 were sometimes replaced by two acrocentric chromosomes each, heterozygotes being more common than expected. In south-east England from Kent to Hampshire, elements 7 and 8 may be represented by twin acrocentrics; in Devon all the chromosomes are metacentric, but twin acrocentric 8 occurs again in the west of Devon and in Cornwall. Shrews in Jersey and north France have a karyotype distinguished from the main British one by at least three pericentric inversions and a translocation. This karyotype occupies lowland western Europe; the other one (the 'British' form) has a more northern, eastern and alpine distribution. Zones of overlap without evidence of hybridization have been found in Switzerland.

Just as Pipistrelle Bats proved to be divided into several races when studied by morphological means, so Common Shrews seem to include two virtually indistinguishable species recognizable only by their chromosomes.

CHAPTER 3
SEX AND BREEDING

ALTHOUGH this chapter is about sex, it covers entirely different ground from the *Kama Sutra* side of things. We have to go behind the glorious solemnities and sordities of sex as understood by human beings, or the practical necessities of fertilization as recognized by gardeners to ask a seemingly naïve question: Why does sex exist? Like so many ultra-simple questions, it is not too easy to answer.

Obviously sex is about reproduction. Leaving aside a few technological aberrations there can be no reproduction without sex in the organisms most familiar to us (cows, guinea pigs, cabbages and ourselves). Nevertheless, other organisms almost as familiar (such as apples or potatoes) usually reproduce entirely vegetatively. The simplest unicellular organisms (*Amoeba* and the like) always increase their numbers by simply dividing by two. This is a simpler and less wasteful method than sexual breeding. Why has it not persisted?

WHY SEX?

To answer these questions about sex, we must turn from individuals to groups. What happens in families produces the characteristics of the group. It will be clear by now that variation in a group of animals or plants is the normal state of affairs. The last chapter touched on some of the reasons for this; the next few chapters are devoted to expanding some of these reasons. Before we develop the explanations, we must first consider some of the mechanisms for shuffling the existing variation. If each new organism is merely a bud or shoot of an old one, the potentialities for 'trying out' new genetical types are extremely limited. Charles Darwin realized this. In his day, chromosomes and genes had not been discovered and Darwin had to invent particles (*pangenes*) with high mutability to maintain variation. We now know that the events of meiosis and fertilization provide ample opportunities for new combinations of genes to be produced, whilst retaining the chief features which enabled the parents to survive. This is what sexual reproduction is about. It involves:

1. The problem of short-term fitness and long-term flexibility. This is a source of potential tension that we shall return to several times (see especially pages 182–3, 291–6): an organism may be amply equipped for life under one set of circumstances but be unable to adapt to a change. This is probably what happened to the dinosaurs millions of years ago; it is certainly what happened

to three butterfly species, Mazarine Blue (*Cyaniris semiargus*), Black-veined White (*Aporia crataegi*) and Small Ranunculus (*Hecatera dysodea*), which have become extinct in Britain in the last 100 years as a result of changes in agricultural practices (Heath, 1974), and to another butterfly the Large Copper (*Lycaena dispar*) which was confined to the fens, and could not adjust to their drainage as well as the predations of collectors (but see pages 267–8). Equally an organism cannot afford to be so variable that few members of any group are suited to their environment in any generation. There has to be a compromise.

2. The mechanisms by which different species maintain a reasonable balance between self-fertilization or close inbreeding (which from the genetical point of view comes close to splitting off parts of the body, like a potato produces its tuber) and having sufficient mates available to have a reasonable chance of successful reproduction.

3. If two individuals are to reproduce, they must meet – or, more strictly, their gametes must meet. Common species that roam widely (like birds, or many sea fish, or wind-pollinated plants) have a chance to mate with a higher proportion of the rest of the species than a more restricted form (such as a land-snail or an insect pollinated plant). This means that the amount of gene flow is related to the breeding structure. Put the other way round, the sexual organization affects – indeed, controls – the genetical structure of a population. Consequently a knowledge of the method of breeding of a population underlies an understanding of the genetical processes acting on it.

4. Breeding produces new individuals which have to establish themselves in competition with each other and with existing members of the species. In many species, parents produce as many young as possible, but there are species in which the control of reproductive rate is associated with successful breeding. So we must consider this topic under the heading of sex and breeding.

SEPARATE SEXES

Probably most plants but a minority of animals are hermaphrodite – the same individual carries both male and female organs. In such cases (buttercups, snails, earthworms) the sex organs differentiate during development just like (say) liver and kidney, and there are no special problems of sex determination. On the other hand most higher animals and representatives of most of the orders of higher plants, as well as many fungi and bacteria, have separate sexes. In this case by far the most common mechanism of sex determination is chromosomal.

We have already seen that humans have 23 pairs of chromosomes: 22

autosomal pairs and a single sex chromosome pair (page 32). Females have two identical sex chromosomes called X chromosomes; males have one X chromosome and a Y chromosome which does not occur in females. At meiosis in males two sorts of sperm are produced – equal numbers of ones with 22 autosomes plus X, and 22 autosomes plus Y. All human ova carry 22 autosomes plus X. Half the ova will be fertilized by X bearing sperm to give a zygote with 44 autosomes plus XX (*i.e.* females), half by Y bearing sperm to give 44 autosomes plus XY (*i.e.* males). The segregation of the sex chromosomes at meiosis is the basis of the approximately equal numbers of men and women. In practice more men than women are conceived, because Y-bearing sperm seem more efficient at fertilization than X-bearing ones (perhaps they are lighter and swim faster). The sex ratio at conception was once believed to be 1.6 males to 1 female, but studies of the chromosomes of early spontaneous abortions have shown that it is not too different from unity. At birth the sex ratio is about 1.06 : 1. In the days before modern medicine, boys died more rapidly than females so that sex equality was reached around 6 years of age, and the adult population contained an excess of unmarried women. Now that the medical profession is keeping all the weedy males alive, we are faced with a surplus of bachelors.

All the chromosomal methods of determining sex are similar in principle. Some species have multiple sex chromosomes, some have sex chromosomes determined by a cytologically indistinguishable segment of a pair of chromosomes (or even by a single segregating locus), some have the male with different sex chromosomes and others the female (Table 11). The sex with

TABLE 11. Chromosomal sex-determination in selected organisms

	Male heterogametic	*Female heterogametic*
PLANTS		
	Docks (*Rumex*)	Strawberries (*Fragaria*)
	Campions (*Melandrium*)	
	Hops (*Humulus*)	
	Asparagus	
	Spurges (*Mercurialis*)	
ANIMALS		
	Nematodes	Copepods
	Echinoderms	Butterflies and moths
	Spiders	Thrips
	Some mites	Some fishes
	Centipedes	Snakes
	Most insects	Birds
	Some lizards	
	Mammals	

different sex chromosomes is the heterogametic one, the other is the homogametic one. Thus the human male is heterogametic (XY), the female is homogametic (XX).

In plants the condition where both sex organs are on the same plant is called *monoecious*, and where they are on different ones *dioecious*. Only about 2% of the 2200 or so species in the British flora are dioecious. Well-known examples are Stinging Nettles (*Urtica dioica*), Dog's Mercury (*Mercurialis perennis*), and several species of campions (*Silene*) and sorrell (*Rumex*). Another species whose separate sexes have attracted notoriety is the Hop (*Humulus lupulus*). Under the European Common Market regulations, male plants are illegal. Sexual reproduction in hops is said to affect the quality of the beer.

When the sexes are distinct, self-fertilization is impossible; when they are combined in one individual, self-fertilization may be possible or necessary. Almost every permutation of self- and cross-breeding occurs in higher plants. The simplest device for reducing self-fertilization in flowers containing both male and female organs is probably that found in the Umbelliferae (Cow Parsley, etc.) and Caryophyllaceae (Campions, Chickweed, etc.) where the pollen is shed before the stigmata of that flower become receptive. In the Autumn Crocus (*Colchicum autumnale*) and Horse Chestnut (*Aesculus hippocastanum*) the stigmata are receptive first and wither before the pollen-grains are shed. Then there is heterostylism. For example in Primroses (*Primula* spp.) virtually all flowers are either 'pin' or 'thrum' (see figure on page 186). In the former the stigma is at the mouth of the tube formed by the petals, and the 5 stamens are half-way down the tube; in the latter the anthers are at the mouth and the stigma about half-way down. An insect pushing its proboscis down the tube will collect pollen at such a level as to deposit it on the stigma of a flower of the opposite variant. In Purple Loose-strife (*Lythrum salicaria*) there is tristyly – three lengths of stigma and anthers.

However, most dioecious plants also have a self-incompatibility mechanism (page 65) so that pollen grows slowly or not at all on a stigma of the same plant. If there are a large number of self-incompatibility alleles, a given plant will be able to cross with almost any other individual in the population. Only in cultivated strains do problems arise. All plants of one variety may be derived vegetatively from the same stock, and be genetically the same as each other. Consequently, different varieties may have to be planted close to each other to ensure that fertilization takes place, like the Cherry trees described on page 65. Similar restrictions apply to Apple trees – unless flowers of one variety are pollinated from certain other varieties, no fruit will set.

SELF-FERTILIZATION

Bulbs, corms, rhizomes and tillers are common methods of vegetative reproduction in plants. We know of many plants whose spread is exclusively by vegetative means. For example, the Butterbur (*Petasites hybridus*) is a dioecious

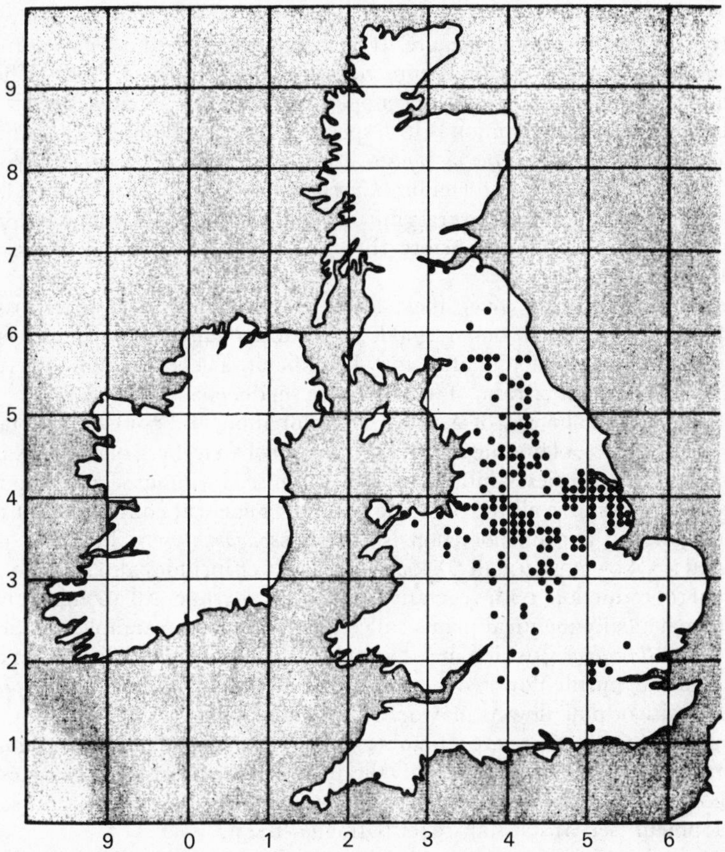

FIG. 25. Distribution of female plants of Butterbur (*Petasites hybridus*); the male has been recorded for all parts of Britain (from Perring and Sell, 1969).

plant locally common throughout Great Britain – but it is only the male plant which has a wide distribution; the female is largely restricted to the Midlands and north of England. Over most of the country males spread entirely by rhizomes. Nobody knows why this is: it is possible that the females once had a wider distribution which has declined recently, but this is guesswork.

However, there are many plants which are pharisaic: they have the outward and visible signs of out-crossing, but little of the reality of it. The simplest version of this is *cleistogamy*, where all the mechanism of out-crossing is present, but where some of the flowers never open and are perforce self-pollinated. For example both the Sweet Violet (*Viola odorata*) and the Dog-Violets (*V. riviniana*, *V. reichenbachiana*, *V. hirta*, and *V. canina*) produce conspicuous coloured flowers well-adapted to insect pollination, but seed-setting from such flowers is irregular and sometimes meagre. If one examines a plant after its normal flowering is complete, one finds numerous short-stalked 'flower-buds'. These are flowers which remain small, never open and are self-pollinated. Seed is set in abundance. Other common British species usually producing cleistogamic flowers are Toad Rush (*Juncus bufonius*), Chickweed (*Stellaria media*), Wood Sorrell (*Oxalis acetosella*) and Henbit (*Lamium amplexicaule*). An animal version of cleistogamy is found in the grass mite *Pediculopsis* where the young become sexually mature and mate before they are born, ensuring brother-sister mating.

Some plant species cover their losses by producing seed from cross-fertilization when conditions are good, but fertilize themselves during adverse conditions when cross-pollination may be difficult. These conditions may be extremes of drought or rain, of temperature, shade, etc.

Cleistogamy is one way of assuring self-pollination, and not having to take the risk, as it were, of finding a mate. It is the tip of an iceberg of non-crossing in plants. Many species habitually produce their seed without sexual fusion of gametes. In this case all the offspring have the genetical constitution of the female parent. The phenomenon is known as *agamospermy* (literally, 'unmarried sex') or more usually, *apomixis*. This latter includes also all forms of asexual reproduction, both vegetative and agamospermic. All apomixis gives rise to genetically identical plants called *clones*. Common examples are Self-heal (*Prunella vulgaris*) which may have clonal patches of pale flowers mixed with normal purple-flowered plants, and Bluebells (*Endymion non-scriptus*) where white or pink flowers may occur in small patches.

Embryologically there are a number of different ways of producing agamospermy. They have in common only the abandonment of fusion of gametes in the normal sexual process as a necessary preliminary to embryo and seed development (see Gustafsson, 1946; Battaglia, 1963).

Apomictic species occur very widely in higher plants, both in ferns and flowering plants (angiosperms), but no apomictic gymnosperms are known.

Some flowering plant families show a great deal of apomixis affecting several genera. The most common well known families of this nature are the Rosaceae and the Compositae. In such families there is a correlation between the occurrence of apomixis and difficulties of classification. Every clone is likely to be genetically different from every other one, and it would be possible to give each one a name as a distinct species. Indeed, this has been done. One Jordan (who believed all species were specially created) described over 200 'species' from the one apomictic species Spring Whitlow (*Erophila verna*). Many of the so-called 'critical' genera which involved disagreement among old-style 'splitter' and 'lumper' taxonomists have turned out to be agamospermic. Familiar examples are Blackberries and Raspberries (*Rubus*), the Whitebeams (*Sorbus*), the Hawkweeds (*Hieracium*) and Dandelions (*Taraxacum*).

One example of obligate apomixis is in the Lady's Mantles, to which Linneaus gave the name *Alchemilla vulgaris*. As far as is known all *Alchemilla* species in Europe are apomictic, with the exception of a very distinct alpine species *A. pentaphyllea* and a few Alpine groups belonging to another sub-section of the species group. The collective Linnean species *Alchemilla vulgaris* and *A. alpina* include about 300 clones, many of which have a wide distribution and are no more difficult to identify than sexual species in many other genera.

Complete agamospermy is easy to detect. For example, a full head of fruit will be set by most plants of the common Dandelion (*Taraxacum officinale*), even if all the stamens and anthers are carefully removed and the capitulum covered to exclude foreign pollen. The trouble is that dandelions sometimes reproduce sexually in the normal way, and such partial apomixis is much more difficult to detect (and much rarer) than obligate apomixis.

Certainly, the most difficult groups for taxonomists are those in which an occasional sexual cross can occur to mix different clones. Dandelions have already been mentioned. Another example is Spring Cinquefoil (*Potentilla tabernaemontani*) which is a mat-forming rhizomatous perennial of chalk and limestone lowlands. Both it and a closely related non-rhizomatous and woody upland species, *P. crantzii*, are apomictic. In some hilly areas, especially in northern England, puzzling intermediate plants occur which obscure the otherwise fairly clear distinction between the two species. The intermediates have higher chromosome numbers than the parental species. In an experimental plot at Cambridge, both species were reared together. Most of the offspring resembled the female parent. However occasional aberrant individuals occurred, differing from both parents (Smith, 1963). One such plant, born to a *crantzii* pollen parent from Ben Lawers in Perthshire and a female *tabernaemontani* from Cambridgeshire had 70 chromosomes, although the male parent had 42 chromosomes and the female 49. The offspring had presumably arisen from the fusion of normal haploid pollen with 21 chromosomes with an egg cell which had not undergone meiosis and accordingly had the full number of 49 chromosomes. Such experimental crosses showed that sexual reproduction does not seem to be very rare, and groups of hybrid plants could arise which might be interfertile.

The common Wall Cress (*Arabidopsis thaliana*) is described in the *Flora of the British Isles* as 'automatically self-pollinated'. Nevertheless by growing together clones differing in alleles determining the presence or absence of hairs on the stems or leaves, 1.73% out-crossing has been found (when over 10,000 progeny were examined) (Jones, 1971). This level of out-crossing may seem trivial but it is sufficient to maintain a not insignificant degree of heterozygosity. In other words, a small amount of sexual reproduction can ensure the benefit of long term flexibility by circulating variation, whilst the great majority of individuals are the unvarying members of successful clones possessing a high fitness. The fitness *versus* flexibility compromise has been resolved in partially apomictic species on the fitness side, but they have not completely burnt their boats and given up all genetical flexibility.

REGULATION OF THE SEX-RATIO AND ANIMAL PARTHENOGENESIS

A monoecious plant with a well-developed system of self-incompatibility alleles can successfully breed with the great majority of other plants in its species. A dioecious plant with its two sexes determined by a segregating system of sex chromosomes producing approximately equal numbers of males and females can breed with only 50% of its fellows. The same applies to those animals which are not hermaphrodite. As in apomictic plants, many animal species have developed ways of improving the efficiency of the process.

One ingenious variation of the hermaphrodite state is practised by a number of shallow sea molluscs, such as oysters, scallops, top-shells and limpets. When these animals enter breeding condition they are males, but they change to females as they age so that older members of a population are almost entirely female. This is entirely different from human 'sex change', which is psychological and hormonal, and can only be continued by plastic surgery. This again is distinct from the rare human 'intersexes' who are usually a mixture of male and female cells. This may happen if, for example, a male zygote loses a Y chromosome in one of its early cell developments, so that it develops as a mosaic of XY and XO cells. An XO individual is a sterile female, so an XY/XO mosaic may have both male and female organs.

Notwithstanding, *parthenogenesis*, the animal equivalent of apomixis, has arisen in most major groups. As in plants, it may be complete or partial. A completely parthenogenetic species will contain only females. This is known in round-worms, earth-worms, sawflies, moths, bony fish, even in a Lizard (*Lacerta saxicola*). An interesting situation exists in a British Sawfly species, *Mesoneura opaca*, which is among the more variable species in the sawfly fauna and yet is entirely parthenogenetic (Benson, 1950). The thorax varies from black through orange to yellow, and different colour patterns can be collected from the same tree on the same day. Another Sawfly, *Eutomostethus epihippium*, shows a progressive change from a red-marked thorax predominating in southern England to an entirely black form in Scotland. In Britain this species consists of females only, although males are produced in southern Europe. This persistence of variation is intriguing, because one of the assumed advantages of the parthenogenic habit is to maximize success in a habitat to which an animal (or plant) is well-adapted. The variation in these two species suggests either that the colour has no importance for the animals, or that there is a stable diversity of different niches into which different clones fit. It is a situation which illustrates the possibilities for specialists to explore the reason for intra-specific variation in their groups (Suomaleinen, 1950).

The amount of parthenogenesis may vary climatically or geographically, just as cleistogamy varies in plants. There is a Millipede, *Polyxemus lagurus*, in which the proportion of males decreases in a northward direction: in France 41.6% of the population are males, in Denmark 8.7%, in Sweden 5.6%, and in Finland 0% (Palmen, 1949).

Parthenogenesis can be produced in some forms by stimulating unfertilized eggs to divide by giving them a physical or chemical shock. This is regularly done in embryological studies of amphibians, but no-one has yet proved parthenogenesis in humans. Some years ago a Sunday newspaper collaborated with medical workers in searching for virgins births in man, *i.e.* parthenogenesis. The newspaper advertised for mothers who had given birth to daughters without having prior sexual intercourse. Virgin births in the non-miraculous sense must be of girls, because there is no way of introducing a Y chromosome if a male parent is not involved. It would theoretically be possible for an ovum with 22 autosomes plus an X chromosome to develop entirely asexually, and double its chromosome number by autopolyploidy or to be 'fertilized' by one of the other daughter cells formed at meiosis. Nineteen women claimed to have produced a child without intercourse having taken place, but eleven of these proved not to know the facts of fertilization. In seven of the remaining eight there were blood group differences between the mother and daughter. The final pair had the same blood groups and were given skin-grafts: mother to daughter, and daughter to mother. These lasted six and four weeks respectively, longer than between random brothers and sisters, but less than between identical twins (who do not react against each other's tissues) (Balfour–Lynn, 1956). So the question of human parthenogenesis remains unproven: it can never be excluded, but no certain example has yet been demonstrated.

Artificial methods of regulating the sex ratio (or of 'choosing the sex of a baby') depends on separating X and Y sperm. If semen is centrifuged, there is some separation but the two sorts of sperm are not completely sorted out. In both plants (notably maize) and animals (*Drosophila*, butterflies and grasshoppers), genes are known which lead to only one sex appearing in the offspring. There is some evidence that the vaginal secretions in mammals may affect X and Y sperm differently. It is said of the Incas that their life high in the Andes tended to make their body fluids slightly alkaline (because of having to breath heavily to obtain sufficient oxygen), and that this slowed down the Y sperm leading to an excess of girl children. The Incas overcame this by having regular mating races, so that when the men caught up the women, their bodies had large amounts of lactic acid produced in the sprint, and the result was an adequate production of boys. This is probably too good to be true: an experiment in which mice were selected for more acidic blood produced a significant change in sex ratio, but when the selection was repeated for more alkaline blood, the sex ratio shifted in the same direction.

The most important and regular method of regulating the sex ratio in animals is found universally in hymenopteran insects (sawflies, ichneumons, wasps, ants, bees, etc.) except for a few species which have become wholly parthenogenetic. This method is called *haplodiploidy*, and in it males arise from unfertilized eggs, females from fertilized ones. It is thus half sexual, half parthenogenetic. The same mechanism exists in some scale insects (Coccoidea) and white flies (Aleurodidae), in a few species of beetles, in thrips (*Thysanoptera*), and in some groups of mites. The males do not have any meiosis since their cells all contain the haploid chromosome number. The great advantage of this system is that males can be produced to order. There is no need to produce vast numbers where a few can do the job.

This can be illustrated by the social organization of Honey Bees (*Apis mellifera*). A colony during the summer months consists of forty or fifty thousand workers (females), a few hundred drones (males), and a single queen bee who is the mother of them all and may lay up to 1500 eggs a day. The workers only lay eggs in exceptional situations. A virgin queen leaves the hive early in her life, mates and is supplied with sperm which will last her all her laying life. The drone who inseminates her is quite likely to have come from another hive, as drones pay no attention to virgin queens in their own hive and do not necessarily accompany one when she leaves on her mating flight. Soon after mating, the queen begins laying eggs: fertilized ones into 'worker cells', and unfertilized ones into 'drone cells'. How she differentiates no-one knows. Young worker larvae fed on the appropriate food will develop into queens. There is no genetical difference between workers and queens. The sole function of a drone is copulation. Individuals live four or five weeks, about the same span as workers: a queen lives three or four years and may lay as many as 600,000 eggs.

The final version of parthenogenesis is where it alternates with sexual reproduction:

1. There are two genera of midges (*Miastor* and *Heteropeza*), neither of which occurs in Britain, in which the larvae reproduce parthenogenetically under the bark of rotting logs, so that a whole cloud of midges can come from a single egg. Adults are only produced when conditions become unfavourable, and act as dispersers.

2. In many gall wasps (the insects which produce oak apples, among other galls) there are two generations a year – one sexual, one parthenogenetic.

3. Most important from the economic point of view, most species of aphids reproduce parthenogenetically throughout the summer months, the females so produced being wingless when conditions are good (so that they can make the most of available food), and winged either every other generation or when they become crowded, so that they can spread to new hosts. Only when conditions become bad in the autumn do the parthenogenetic females give rise to both males and females, allowing sexual reproduction and over-wintering as egg or larva. In many climates, a sexual phase may rarely, if ever, be produced. One of the commonest

aphid pests in Britain, the Peach Aphid (*Myzus persicae*) rarely reproduces sexually in glasshouses.

Cyclical parthenogenesis may involve strong genetical pressures, as shown by the situation in the tiny fresh-water crustacean, *Daphnia magna*. This species breeds parthenogenetically every two weeks when food is abundant. Males are only produced if the animals become crowded, and sexual reproduction then follows. An enzyme (malate dehydrogenase) is commonly represented by two variants in most colonies, determined by two alleles. Heterozygotes at the enzyme locus increase disproportionately during the summer non-sexual breeding season: in one pond near Cambridge containing many thousand individuals, the frequency of heterozygotes was that expected on the Hardy-Weinberg principle in May, but had increased to 23% excess by November (Hebert, Ward & Gibson, 1972). Changes in the frequencies of the three genotypes were strongly correlated with the fecundity of the population in each of the four preceding weeks (Hebert, 1974).

Evidence like this suggests that animals which alternate between parthenogenesis and sexual breeding ordinarily have in their populations a range of genotypes which respond differently to different ecological conditions. Other examples are differences between clones of the Rotifer *Euchlanis dilatata* and populations of the Pea Aphid (*Acrythosiphon pisum*) (King, 1972; Fraser, 1972). We shall return to generalizing these genetical differentiations in Chapter 9.

All the animals which exhibit cyclical parthenogenesis have XX females and XO males, and all individuals are diploids.

SEX LINKAGE

At this point it is necessary to digress from the main subject of this chapter to describe the mode of inheritance of genes carried on sex chromosomes. As we have seen, several of the genes concerned with blood-clotting in man are carried on the X-chromosome (page 43). So is the locus causing the two commonest types of colour blindness, one of the blood groups, Duchenne's muscular dystrophy (the most frequent form), and something like 80 other known loci. This is a far greater number than is known for any other chromosome, and is due to the fact that any allele must manifest in the heterogametic sex because there is no homologous chromosome to affect its expression. It is sometimes said that genes on the X chromosome are only seen in males. This is not true: most X linked variant alleles produce recessive traits, and they will only manifest in females when they are homozygous. For example, red-green colour blindness afflicts only 7% of males. This means that the frequency of X chromosomes carrying the colour blindness allele is 0.07. The chance of a woman being homozygous for the allele is $(0.07)^2$, or 1 in 200. The frequency of haemophilia is 0.00004 in men; the occurrence in women will be not more than $(0.00004)^2$ or 16×10^{-10} (the first known female case occurred in 1951). (Actually female haemophiliacs are much less common, because male carriers of the allele only rarely breed and transmit their allele to their daughters.)

There is nothing peculiar about sex-linked genes. They are transmitted in exactly the same way as autosomal genes, but females will more often be carriers and males sufferers from recessive traits. This is well-illustrated by Queen Victoria who was heterozygous for the haemophilia gene (which

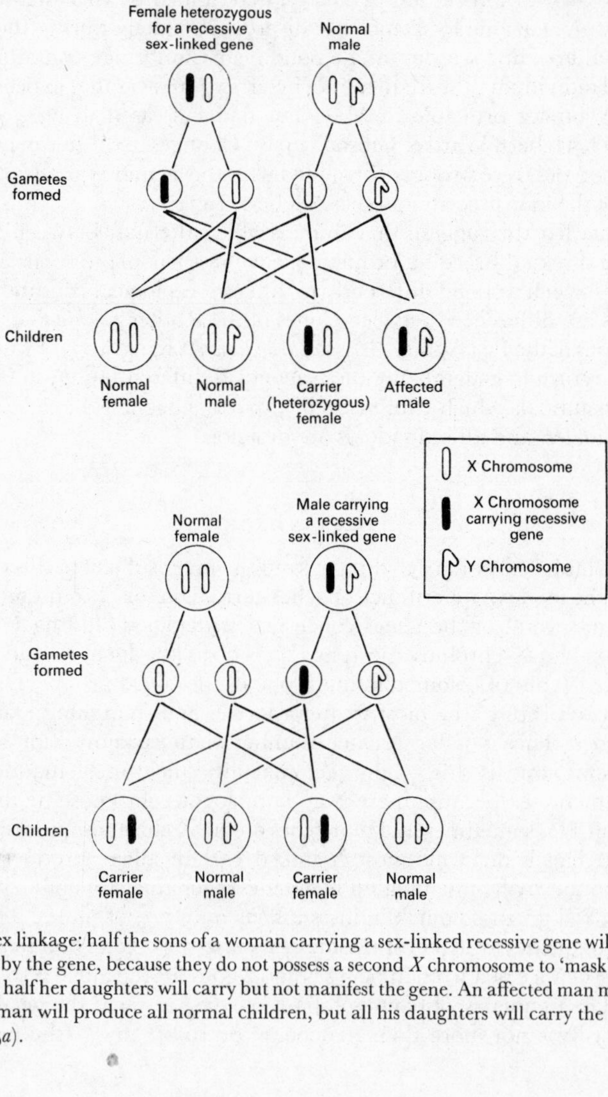

FIG. 26. Sex linkage: half the sons of a woman carrying a sex-linked recessive gene will carry and be affected by the gene, because they do not possess a second *X* chromosome to 'mask' the effects of the gene; half her daughters will carry but not manifest the gene. An affected man married to a normal woman will produce all normal children, but all his daughters will carry the gene (from Berry, 1965*a*).

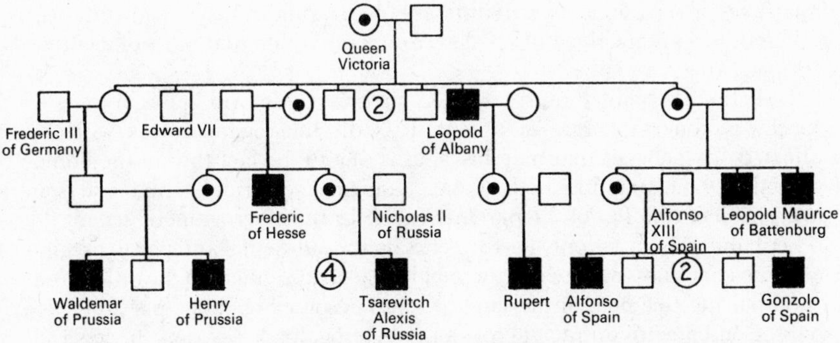

FIG. 27. Pedigree of the descendants of Queen Victoria to show the transmission of haemophilia (a trait determined by a sex-linked recessive gene). All of Queen Victoria's own children are shown, but only some of their children. □ represents a man, ○ a woman; a black square indicates a haemophiliac; ⊙ are women who are heterozygous for the haemophilia gene (and are therefore unaffected by it, although they are carriers). A number in a square or circle means several (two or more) unaffected children of that sex.

presumably arose as a fresh mutation in one of her parents or maternal grandparents). She transmitted this to one of her four sons (Leopold of Albany, who died at the age of 31), and two of her five daughters. These two carrier daughters passed the disease into the Russian, Prussian, and Spanish Royal Families. The British Royal Family is not at risk from the disease, because Queen Victoria's son, Edward VII was not a haemophiliac, and could not be a carrier for it.

No genes are known on the Y chromosome in man (nor in most other animals, save in those which have homologous or pairing segments common to both X and Y chromosomes). Nevertheless the Y chromosome is essential for the production of maleness. Mammals are odd in that only one of the X chromosomes in any cell is functional; X chromosomes in excess of one form a blob (or a series of blobs) on the inside of the nuclear membrane, known as *sex chromatin* or *Barr bodies*. These blobs make it possible easily to determine the number of X chromosomes carried by a person, and is the basis of routine checking for sexually abnormal babies in many maternity centres.

A female may have 2, 3 or 4 X-chromosomes, and still be a relatively normal woman, although multiple X people have a lower intelligence than their brothers and sisters. An individual with a Y chromosome will always be male however many X chromosomes he has as well (although multiple X chromosomes usually lead to infertility in males). Multiple Y chromosomes produce aggressive mental defect in a proportion of their carriers, but some XYY men are apparently completely normal.

In insects sex is determined by a balance between the X chromosomes and the autosomes. The Gipsy Moth (*Lymantria dispar*) produces a variety of intersexes in crosses between different geographical races, showing that different loci are involved in sex determination at different parts of its range (*cf.* page 193).

Haldane has pointed out that when one sex of a hybrid between races or species is abnormal, rare or absent, it is the heterogametic sex which is affected – whichever that may be. This is due to the fact that in the homogametic sex the quantities both of male and female-determining material is an average between that of the parental races. In the heterogametic sex on the other hand, this is true only for the genes on the autosomes and pairing region of the sex chromosomes, whilst those balanced against them in the differential (non-pairing) segment of the single X chromosome are of one parental race only. Consequently in racial crosses the heterogametic sex must be less well adjusted than the homogametic one. For example, crosses between males of the Early Thorn Moth (*Selenia bilunaria*) and females of the Purple Thorn (*S. tetralunaria*) produce either male progeny only or a large excess of males.

A variation of this argument has been used to detect the effect of various agents in producing mutations. If a female is treated with a mutagen, mutations will be induced in any chromosome. Recessive mutations in an X chromosome will be transmitted to sons and daughters in equal numbers, but will be masked in the daughters by the second X chromosome received from the father. The sons will suffer the full consequences of any deleterious effect. Since a new mutation is a random change in development, its effect will be similar to throwing a spanner into the works and thus probably deleterious. Sons may not survive, or may be reduced in vigour. Consequently any deficiency of sons of a treated female will show the mutagenic effect of the treatment. This is the basis of the C l B method in *Drosophila melanogaster* which earned H. J. Muller a Nobel Prize, for with it he was able to measure the genetical effects of X-rays (page 253).

Older books often describe a condition called 'Porcupine Man' as determined by a gene on the human Y chromosome. The origin of this belief was a family which allegedly produced only affected boys and normal girls for 5 generations in the eighteenth and nineteenth centuries. Some of the males made their living by exhibiting themselves as 'a new species of man'. A report of 1832 touches upon the prospects for the future: 'Where the propagation of men brought into the world thus cased is to end – who shall tell? The biped armadillos find no difficulty in wiving. Nay, the present Mr Lambert says with regard to his skin, that his good lady "rather likes it of the two". *De gustibus*, etc.' However research in Parish Registers showed that the apparent Y chromosome inheritance was a by-product of showmanship. A full pedigree of the family revealed that not all the sons of affected males were affected themselves (as they should have been if the condition was carried on the Y chromosome), not was it confined solely to the males (Penrose & Stern, 1958). 'Porcupine Men' are merely possessors of an autosomal dominant gene producing a condition known to doctors as *icthyosis hystrix*.

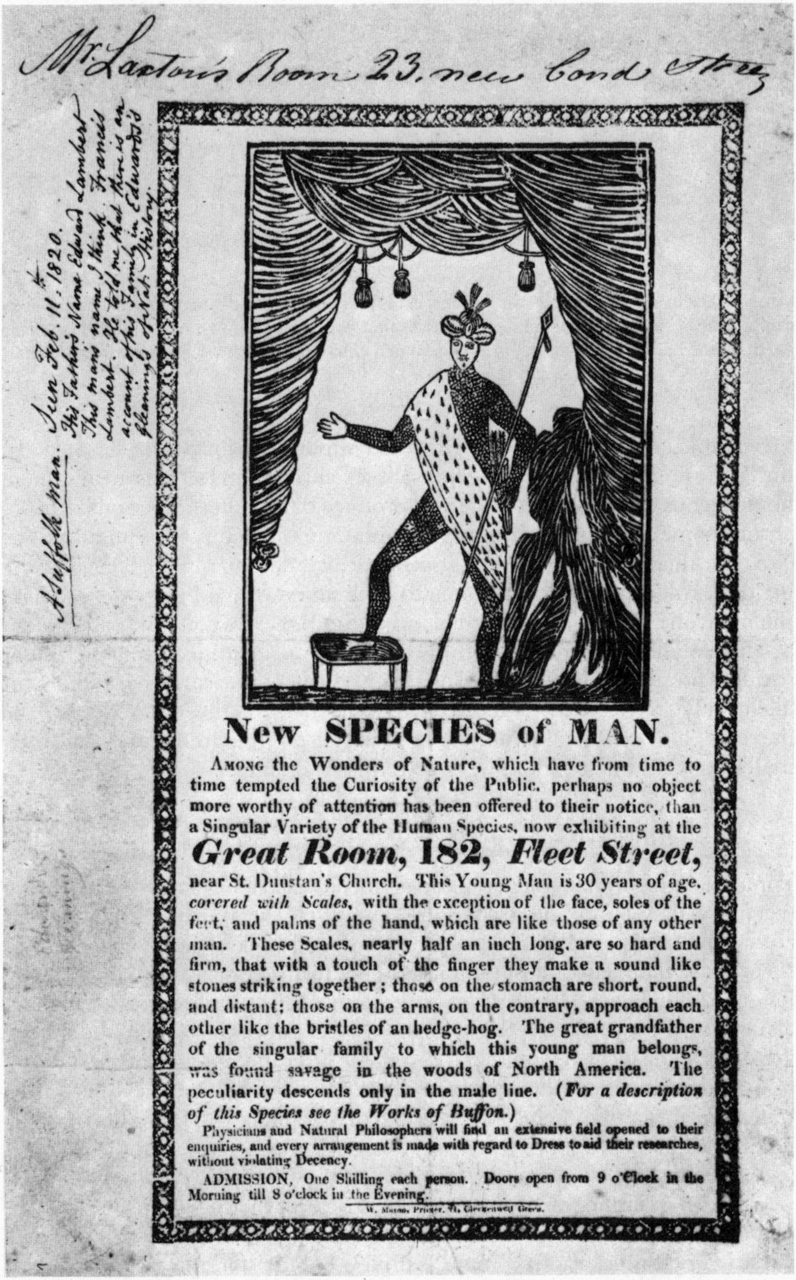

Mr Laxton's Room 23. new Bond Stree

A Suffolk man. Seen Feb. 11th 1820.
His Father's Name Edward Lambert
This man's name I think Francis
Lambert. He told me that there is an
account of his Family in Edwards's
gleanings of Nat: History.

New SPECIES of MAN.

AMONG the Wonders of Nature, which have from time to time tempted the Curiosity of the Public, perhaps no object more worthy of attention has been offered to their notice, than a Singular Variety of the Human Species, now exhibiting at the

Great Room, 182, Fleet Street,

near St. Dunstan's Church. This Young Man is 30 years of age, *covered with Scales,* with the exception of the face, soles of the feet, and palms of the hand, which are like those of any other man. These Scales, nearly half an inch long, are so hard and firm, that with a touch of the finger they make a sound like stones striking together; those on the stomach are short, round, and distant; those on the arms, on the contrary, approach each other like the bristles of an hedge-hog. The great grandfather of the singular family to which this young man belongs, was found savage in the woods of North America. The peculiarity descends only in the male line. (*For a description of this Species see the Works of Buffon.*)

Physicians and Natural Philosophers will find an extensive field opened to their enquiries, and every arrangement is made with regard to Dress to aid their researches, without violating Decency.

ADMISSION, One Shilling each person. Doors open from 9 o'Clock in the Morning till 8 o'clock in the Evening.

FIG. 28. Hand-bill dated 1819, illustrating 'Porcupine Man' from the Royal College of Surgeons collection (reproduced by permission).

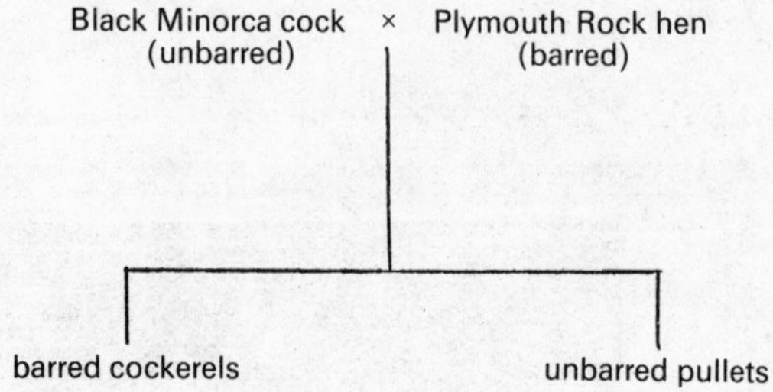

FIG. 29. Use of sex-linked genes for 'instant sexing' of poultry. The distinction between the sexes can be told in the down at hatching: a chicken that will grow up into a barred bird has a light patch on the back of its head; this is not found in a chick that will grow into an unbarred bird.

Sex linkage is of considerable practical importance for poultry farmers. In birds, the female is the sex with different sex chromosomes. This means that a homozygous dark-coloured cock will produce dark pullets, since one of his X chromosomes will be the only sex chromosome carried by the female chicks. Now the allele which produces light colouring is dominant to the one producing dark colouring: if a dark-coloured cock mates with a light-coloured hen, the male offspring will be light-coloured (and heterozygous for colour), and all the females dark-coloured. If the genes can be identified at birth, instant sexing is possible. For example, if a hen's X chromosome carries the dominant allele for barred feathers (as in the Plymouth Rock), all the male chicks will be barred. The early recognition of sex makes it possible to discard unwanted males at once, and avoid unnecessary expense on feed.

POPULATION STRUCTURE AND GENETICAL MIXING

Populations of animals and plants do not behave like molecules in gas, mating whenever they bump into each other. However, the Hardy-Weinberg principle only strictly applies if populations are churned like a gas. Fortunately for theoretical geneticists, most populations get near enough to random mating for gene frequencies to be describable in terms of the simple Hardy-Weinberg assumptions (page 49), but the more intensively any population is studied, the greater are the local deviations. We shall see later the genetical consequences of territoriality through restriction of individual movement (Chapter 5). At this point it may be useful to bring together some of the limitations on the mixing of alleles produced by the breeding systems described in the previous sections.

The two extremes of population structure are, firstly, complete randomization of individuals before mating – perhaps most nearly reached in small aquatic organisms; and secondly, complete subdivision into a series of self-reproducing clones. The problem is that population of most species fall between these two extremes, and it is wrong to make the simple assumptions associated with either. Take Harberd's (1961) investigation into Red Fescue (*Festuca rubra*) growing on the slopes of Scald Law south of Edinburgh. This is a creeping grass, which is wind-pollinated and out-breeding, with a well-developed self-incompatibility mechanism – apparently an excellent subject for illustrating random mating.

Harberd grew in an experimental garden nearly two thousand tillers he had collected from a defined area of Scald Law. Red Fescue is a very variable species, and it was immediately obvious that many of the experimentally grown plants were very similar to each other. By making observations on 20 characters in these plants and comparing them with measurements on randomly collected plants from other areas, Harberd was convinced that a high proportion of his plants were of the same constitution. He confirmed this by a wide series of crosses and self-pollinations. Self-pollinated plants set little seed, as would be expected if their self-incompatibility system was working properly. Crosses between morphologically different plants produced a high yield of seed. But – and this was the important observation – crosses between morphologically similar plants gave as little seed as self-pollinated plants, confirming that morphologically similar plants were also genetically similar (and probably identical). This implied that the similar plants were all part of the same clone: the chance of picking up the same incompatibility alleles in different plants is small; the chance of finding them associated with the same morphological features as well is negligible.

Harberd put his results back onto a map of the original collecting sites, and it became clear that some of the plants were very widespread. One genotype was found scattered over a patch more than 240 yards long, and in much of this area it made up from one to three quarters of all the tillers collected; in an intensely sampled 10 yard square, 755 out of 1481 tillers (51%) were of the same plant. In other words, one plant had spread widely through the area. There seems to be no mechanism in *Festuca rubra* by which a single genotype can spread other than by ordinary tillering. In the intensively studied area, the rate of spread of individual plants was about 9 inches per year, which means that the large plant must have been very old – at least 400 years and perhaps as much as 1000 years. The fact that one genotype had survived for so long in a mixed grassland subjected to a range of management histories has implications which are close to the kernel of adaptation and the whole concern of this book (Jones & Wilkins, 1971). We shall return to this theme several times.

However permanent a few genotypes were on Harberd's hillside, they did not make up the whole of the population. In the 10 yard square referred to, 12 genotypes contributed ten or more tillers each, but 150 apparently different genotypes were represented once only. The picture is thus one of a combination of a few abundant genotypes with a large number of rare ones. There

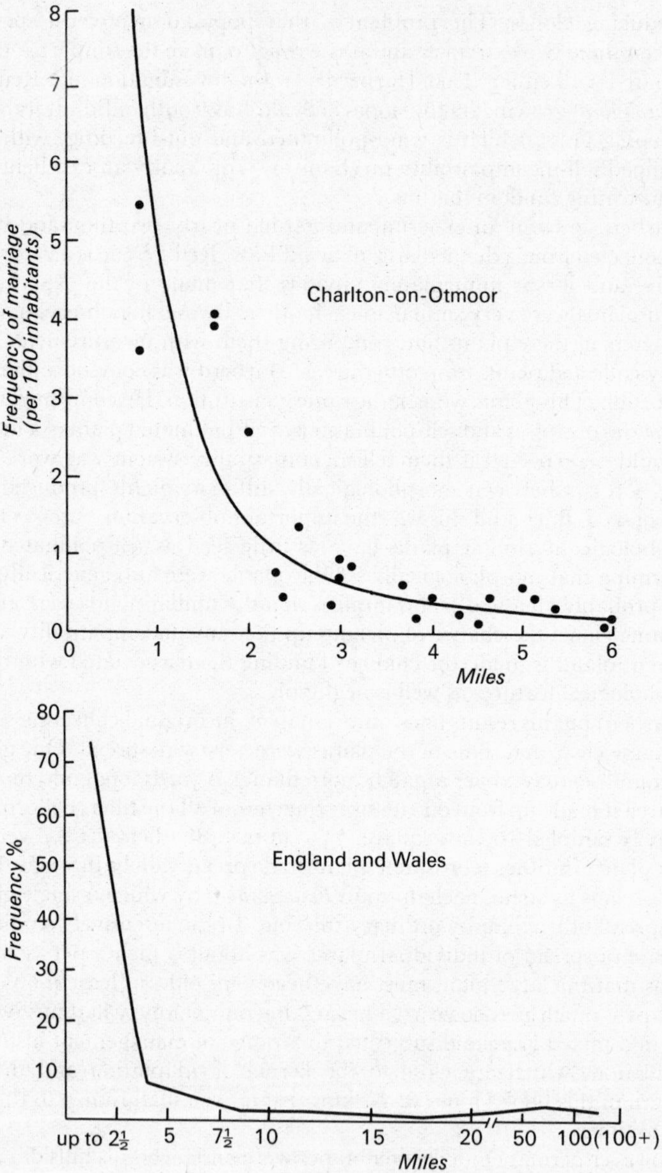

FIG. 30. Average distance apart of marriage partners in (*top*) the Oxfordshire village of Charlton-on-Otmoor in the mid-nineteenth century and (*bottom*) the whole of England and Wales (omitting large towns) between 1920 and 1960 (after Boyce, Kuchemann and Harrison, 1967, and Coleman, 1973).

was no evidence as to whether the rare ones were new arrivals or less successful veterans. The one thing that is clear is that we cannot talk about an average plant in such a community with any confidence at all.

The genes in a population are the result of a whole complex of forces acting and interacting on individuals. What we need to know is how to unravel this complex. Harberd's hillside involved the problem of disentangling vegetative from sexual reproduction. What we can do more directly is to measure the movement of individuals or gametes from their source. Kuchemann, Boyce & Harrison (1967) collected all the information they could on the marriages and movements from 1578 to the present of the inhabitants of a group of seven villages seven miles or so north-east of Oxford. Most of their information came from Parish Registers. Although the total population of the villages varied between 300 and 700 and although more than two-fifths of the marriages were between partners from different parishes, until about 1860 the average distance between the place of birth of bride and groom was six to eight miles. After 1860 (when the railway was built), the average 'marriage distance' increased, but largely as a result of marriages between couples living a long way apart. The bulk of marriages were still between members of the same or nearby villages, so that 95% of individuals after 20 generations still had a common ancestor in the area. Even in Oxford City the marriage distance was small – and has remained small to the present day (Kuchemann, Harrison, Hiorns & Carrivick, 1974).

A similar conclusion was reached by Roberts and Rawling (1974) studying the transmission of surnames through the generations in three rural parishes in Northumberland. The chance of two people having the same surname is related to their having a common ancestor. Consequently it is possible to work out approximately how much inbreeding (and therefore how much immigration) there is in a given area. It turned out that Northumberland people behave much as do Oxford people: each parish was almost an isolate rather than a random section of a large community.

Humans might be expected to be relatively purposive in their mate seeking activity. The same cannot be said for passively transmitted agents like wind-borne pollen. Colwell (1951) measured the range of dispersal of pine pollen. He released one litre lots of pollen from jars 12 feet above the ground, and caught the grains in dishes placed at regular intervals along various radii from the release point. On a mild day with a gentle breeze, the main bulk of pollen was dispersed 10 to 30 feet downwind from the pollen source. Beyond this zone of maximum pollen concentration, the amount of pollen caught fell off rapidly and only a small proportion travelled 50 yards or more. In a pine forest, a female cone will be swamped with pollen from neighbouring trees and will receive only small amounts of pollen from trees a few hundred feet away.

On the other hand, particular individuals sometimes travel considerable distances. House Mice may move only a few yards during their whole life, and mice in different rooms or even the two ends of the same barn may be genetically different, but they are perfectly capable of walking half a mile in a night – and may do so. Another species which has been assumed to move little (and a great deal of genetical inference has been based on the fact) is the Peppered Moth (*Biston betularia*), but Bishop (1972) found that individuals flew up to 3½ miles in a single night on the Wirral Peninsula, although most did not range far. He released 1433 male moths each marked with a spot of cellulose paint, in an area of woodland near Birkenhead during June 1968. Around the woodland he arranged a number of assembling traps in concentric rings 1 km apart. (An assembling trap is built on the same principle as a lobster pot and, contains a fresh virgin female in a small trap within a larger box, so that males attracted by the female's scent are caught.) Sixty-four per cent of moths were caught in the first kilometre from the release point, but 3%

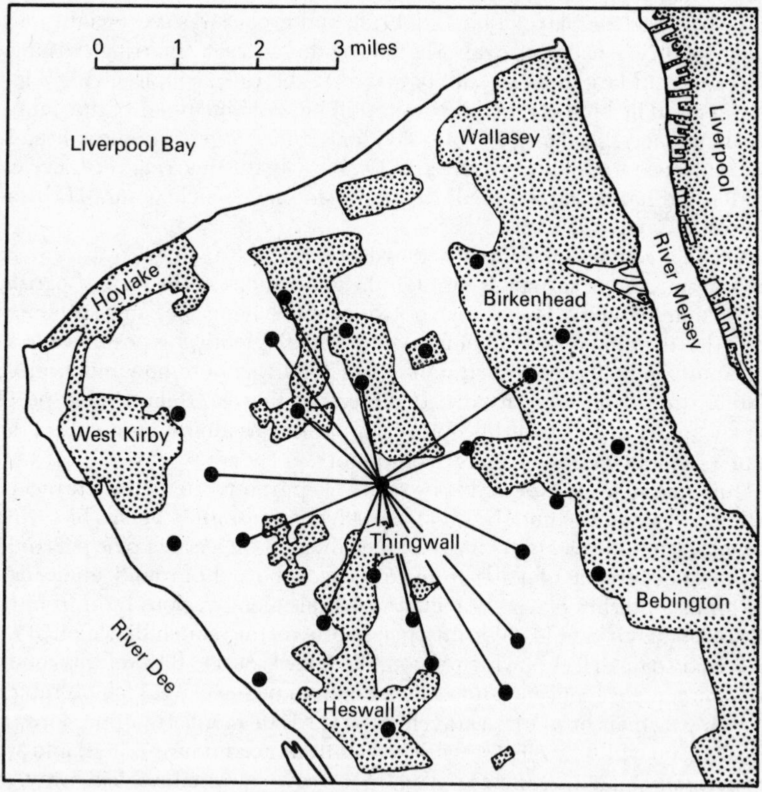

FIG. 31. Flight distances of marked Peppered Moths (*Biston betularia*) released in a park in the centre of the Wirral Peninsula. The black dots are moth traps. Built-up areas are shaded (after Bishop, 1972).

flew more than 6 kilometres. *Biston betularia* has two common forms in Britain – a light one in rural and western areas, and a black one in polluted urban areas. In no urban area is the black form found at 100% despite the fact that the pale form is at a considerable cryptic disadvantage under such conditions, and is quickly eaten by insectivorous birds (page 18). This has been attributed to the existence of a balanced polymorphism for the two forms, with the heterozygote possessing some unknown advantage over the two homozygotes. If pale forms in city centre areas are really migrants and not moths born in the vicinity, the interpretation of the variation has to be changed (but see pages 129, 164, 175).

The picture of gene flow which appears from the distribution of pollen grains by wind, bees, birds, and other agents, and of the dispersal of individual flies, Lepidoptera, rodents, humans and other animals, is a combination of much sedentariness with some long-range movement. The curve representing the pattern of gene dispersion is *leptokurtic*, showing a higher proportion of short and long range than medium dispersal (Bateman, 1950). The normal situation for most species (a dangerous generalization) seems to be that close inbreeding between neighbouring individuals may prevail within a population, but that a certain amount of gene flow over longer distances will be expected to occur.

REPRODUCTIVE RATE

In man identical twins occur in about 3 of every thousand births; non-identical (or two-egg) twins in about 10 per thousand births in England but only one per thousand in Japan. Twinning in sheep is clearly under genetical control, with different breeds producing different frequencies of twin lambs. Domestic poultry have been intensively selected for egg production, so that they carry a large number of alleles contributing to a high rate of laying.

A particularly clear cut example of the genetical control of reproductive rate has been uncovered by Murton working on feral pigeons, which are descendants of domesticated strains of the Rock Dove (*Columba livia*). Feral populations contain a number of colour forms, but those in urban areas tends to have a large proportion of dark or melanic forms. Suggestions as to why this should be have ranged from innuendoes that the birds are dirty, to claims that the dark birds are difficult for cats to see. Murton, Westwood and Thearle (1973) studied the pigeons in the Salford Docks, Manchester, where 68% of the population was melanic and only 21% had the colouring of the wild species, which is now found only on isolated coasts and islands around Britain. They compared the Manchester population with a feral one on Flamborough Head in Yorkshire which, living under conditions similar to those experienced by the wild species, was 70% wild type colouring, and only 24% melanic. Other common phenotypes are pied and grizzled. The New York pigeon populations contain 62% melanic individuals. Melanism is produced by alleles at two loci.

Rock Doves on Flamborough Head breed seasonally, and the testes of the males regress in October. However, about a third of the pigeons in Salford Docks breed throughout the year, the winter breeders being those with a good nest site (Murton, Thearle and Coombs, 1974). In both places, reproductive activity declines with decreasing day length, although the two are only indirectly related *via* a circadian rhythm affecting gonadotrophic hormone secretion. The interesting finding is that melanic males are less influenced by day length than wild coloured birds, and are thus able to produce more young than wild-type during a complete year. If food is available, these young will be able to survive and breed. In parts of the Salford Docks, grain was abundant throughout the year. Males which were dark because of a melanic allele at one locus fledged 4.1 young per year and melanic at the other locus 2.6 per year, while wild-type males only reared 1.3 per year. The effect of plumage phenotype in females was similar but not so marked. Thus in an environment where the young could survive well after leaving the nest, there would be proportionately more melanic alleles than wild-type ones passed on to the next generation, and melanism would increase. However in the Salford population, the melanics had a lower survival rate than wild-type between leaving the nest and becoming sexually mature, and this means that an equilibrium gene frequency would be established with melanics having an advantage at one stage of the life cycle and a disadvantage at another (page 72). The equilibrium will be established with a lower proportion of melanics in populations where food is short during the winter and the melanics at a lesser advantage than in towns. This is the situation in the Flamborough Head and similar non-urban populations.

The factors producing the Rock Dove polymorphism are associated with the resources available to the birds, particularly nest sites and food availability. Another way of expressing the same point is to say that they are *density-dependent*: the more birds there are in the population, the more likely are these factors to be limiting. David Lack, late Director of the Edward Grey Institute for Ornithology in Oxford, assembled a great deal of information about natural selection controlling reproductive rate through density-dependent mechanisms (Lack 1954, 1966). Most of his work concerned birds because of the ease with which data about breeding biology can be collected in contrast to most wild-living organisms, but he applied his arguments to animals generally.

Lack's starting point was the general stability of numbers in most species. For example, one of the species studied most intensively by Lack and his colleagues has been the Great Tit (*Parus major*) population in Wytham Woods near Oxford where the breeding population fluctuated between 21 pairs and 51 pairs between 1948 and 1962, with a single abnormal year in 1961 when 86 pairs bred. Each pair lays an average of 8 eggs in a clutch so the potential for increase was obviously not realized. Invertebrate populations fluctuate more than vertebrate, but return to a similar base number after a peak. This suggests that the numbers in all populations (except those about to become extinct) are regulated.

The next stage in the argument is that the normal clutch size in birds is that which, on average, results in the largest number of surviving young. Survival in the smallest broods is good because the parents can provide for the young, but broods above the average size tend to be undernourished so that fewer young survive per brood than in broods of normal size. These deaths may be in the nest as in the Swift (*Apus apus*) or after the young fly as in the Starling (*Sturnus vulgaris*).

The relationship between clutch size and survival is very clear for the Swift (Table 12). In the Oxford area Swifts normally lay 2 or 3 eggs. Parents are

TABLE 12. Survival in relation to brood-size in the Swift, *Apus apus*, at Oxford

Year	*Brood size*	*No. of broods*	*No. of young dying*	*Per cent lost*	*No. raised per brood*
1958	2	21	2	5	1.9
	3	4	1	8	2.8
	4	2	4	50	2.0
1959	2	15	0	0	2.0
	3	4	0	0	3.0
	4	4	5	31	2.8
1960	2	18	2	6	1.9
	3	6	4	22	2.3
	4	5	14	70	1.2
1961	2	18	1	3	1.9
	3	6	4	22	2.3
	4	5	13	65	1.4

more successful in feeding their young in fine summers than in cold wet ones when airborne insects are scarce. In the unusually fine summer of 1949, 100% of the young hatched were reared; in 1951 and 1955 nine-tenths of those in broods of 3 fledged, and in the fairly fine summer of 1952, three-quarters. Hence in all those four years, more young were raised per brood of 3 than could have been raised per brood of 2. In contrast, in the moderately poor summers of 1953 and 1956, only three-fifths and one half respectively in broods of 3 survived whereas all those in broods of two fledged. This means that the average number raised per brood was higher in broods of 2 (2.0) than in broods of 3 (1.8 and 1.5). Consequently it is an advantage for the Swift to have a variable clutch size, to make the best of good years but not to be penalised in bad years.

Why do Swifts not lay 4 eggs or more so that they can over-compensate in good years? This has been tested by transferring newly-hatched nestlings from one nest to another to make up broods of 4. In all four years that this was done, survival in broods of 4 was lower than that in broods of 3, and in two of the years it was lower than in broods of 2. Lack claimed that this would result in natural selection acting strongly against the laying of clutches larger than 3, since any individuals doing so would raise fewer offspring per brood than those laying clutches of 3.

The presumed genetical control of clutch size may be obscured by phenotypic flexibility, whereby clutches tend to be larger when more young are likely to be raised. For example, the Nutcracker (*Nucifraga caryocatactes*) stores nuts in the autumn to feed its young of the following spring. If it has a poor nut store it lays three eggs; if it has a large store (or extra nuts supplied by an experimenter) it lays four. The Great Tit lays smaller clutches later in the season, at higher densities, in habitats with few large trees, and when breeding for the first time. Perrins (1965) believes that these are all situations where the parents are more likely to raise healthy young which will survive to breed, than if they had laid a full size clutch.

Litter size in mammals has many parallels to clutch-size in birds, with a higher death rate in larger litters and seasonal variations in size correlated with the food situation. Breeding in the House Mouse has been studied by physiologists, geneticists and ecologists, and so is particularly good for testing Lack's ideas. Generally speaking mice continue breeding as long as food is available unless their density rises to a very high level or they are faced with competition from other small mammals (voles or field mice), when the females go out of breeding condition. However on three small islands (Skokholm in Pembrokeshire, the Isle of May in the Firth of Forth, and Foula in the Shetland Islands) evidence of other controlling factors have been found which could be taken as support for Lack. In laboratory mice and those natural populations which have been tested there is a correlation of 53% between litter size and maternal size: the bigger the mother, the more young she produces. This correlation completely disappears on the islands where the mice live under very different conditions to commensal mainland animals, and probably are close to their physiological limits for survival during much of their life. In other words, litter size is not under the chance influence of mother's size in circumstances where density-dependent factors may be presumed to be acting (Batten and Berry, 1967).

General correlations between reproductive rate and food supply exist also in social insects, in certain freshwater copepods, in differences between lizards on islands; Lack has compiled a formidable dossier in favour of his hypothesis. But he has been challenged on two fronts:

1. By biologists who deny that density-dependence is important in regulating population size. This line of thought is particularly associated with the Australian ecologists, Andrewartha and Birch (1954).

2. By Wynne-Edwards (1962), who accepts that density-dependent regulation is important and that food shortage is the ultimate factor limiting numbers, but suggests that animals normally regulate their own density far below the potential upper limit set by food, because through group selection they have evolved both dispersive behaviour and restraints on reproduction. This means numbers are regulated and food supplies conserved.

We shall return to Wynne-Edwards' ideas later (pages 230, 305), because his claim that selection can act on a group as well as on an individual is potentially important, as well as being anathema to some. Andrewartha and Birch need to be discussed briefly in the present context.

Perhaps the simplest way to look at the problem is to recognize two facts:

1. The actual cause of death of an animal or plant may be quite different from the predisposing cause. For example, when an old person dies of heart failure, this may be the result of furring of the arteries, deterioration of the kidneys raising blood pressure, or straight-forward inefficiency of heart muscle; when a young person dies from the same 'cause', this is more likely to be because of a congenital weakness of the heart itself. It is therefore possible to distinguish between the *proximate* cause of death (heart failure), and the *ultimate* cause (which may be one of a number of mechanisms in our example). If all cases of heart failure are lumped together, we should have no useful information about the more important ultimate factors which lead to the heart failing.

The same argument can be used for natural populations: if a starving animal is found in mid-winter, this does not necessarily mean that there is an overall shortage of food; the explanation may be that it is an animal which in the struggle for existence has failed to find a satisfactory niche for the lean season. This has been well-documented by Murton (1965) for Wood Pigeons (*Columba palumbus*). Here the chance of a young bird surviving its first winter depends almost entirely on its incorporation into a successful flock, where it will be better protected against disturbance and protection than in a more marginal group.

2. Over-crowding undoubtedly imposes a stress on an organism. This may be through a straightforward lack of food, or *via* behavioural or hormonal mechanisms. This will lead to competition between individuals and probably an advantage to certain genotypes. In other words, we have the classical situation for natural selection to operate.

When we put these two facts together, we may see the reason for the apparent disagreement between the Andrewartha and Birch school, and those who follow Lack. Proximate causes of death (and hence of number regulation) are likely to change as the environment changes. Lack repeatedly justified his work in the face of apparent exceptions on the grounds that comparatively recent changes in the environment will affect the ecology of a species first, and any adjustment in its reproductive rate may be very slow in following. For example, the Great Tit supports Lack's argument in its traditional deciduous wood habitat, but not in the coniferous forests which it has successfully colonized in the recent past; Gannets (*Sula bassana*) and Fulmars (*Fulmarus glacialis*) lay only one egg despite being able to feed two young, but they are both currently increasing at a rapid rate in the North Atlantic.

This brings us right back to the basic ecological and genetical problems. The questions we have to ask concern the factors which determine genetical variation in populations. Reproductive rate is an inherited character, but it may be misleading to seek an explanation for its level or fluctuation in a particular species from a study of particular populations. What we have got to do is to delve sufficiently deeply into the biology and history of the species we are interested in, and derive an experimental model for testing our conclusions (Birch and Ehrlich, 1967; Lack, 1968). When we do this, the control of reproductive rates from the genetical point of view becomes one particular case of density-dependent selection, to which we return later (pages 145–7, 300–5).

CHAPTER 4

NATURAL SELECTION AND HOW IT WORKS

ADAPTATION of living things to their environment is a fact of natural history that provided one of the mediaeval proofs for the existence of God, excited the speculation of the nineteenth century evolutionists, and is now available for dissection by a combination of genetical and ecological techniques. Adaptation is the result of natural selection. If an animal or plant does not function as well as its peers in a particular environment, it will have a reduced chance of reproducing and perpetuating its malfunction. In other words, the precision of adaptation is measured by the intensity of natural selection.

CUCKOOS, MOSQUITOES, RATS AND MOTHS

Natural selection may act to change or to stabilize gene (allele) frequencies. As we shall see, the detection of natural selection in practice is usually based on estimating frequencies through several generations or (less satisfactorily) in geographically separated populations. If we know the main factors determining population size and individual mortality, and have some knowledge of the method of inheritance of the traits that interest us, we can come to firm conclusions about the genetical pressures at work. However, before considering ways to classify the different actions of natural selection, let us consider some examples. Take the European Cuckoo (*Cuculus canorus*): this relies on birds of other species to hatch its eggs and rear its young (*brood-parasitism*). If a host rejects a cuckoo's egg, that egg (and its genes) is doomed. Victims (both species and individuals) differ in their sensitivity to foreign eggs laid in their nest, but in general the closer the colour, markings and size of an egg to its own, the more likely is a cuckoo's egg to be accepted (Swynnerton, 1918). For example, Baker (1942) found that 8% of cuckoos' eggs laid in the nest of normal fosterers were found deserted, compared with 24% of those laid in the nests of abnormal fosterers.

Now the accuracy of imitation varies, but at its best it may be difficult for the human eye to distinguish the egg of the parasite from its host. This is achieved by 'specialization' on the part of the female cuckoo. Any given bird will only lay in the nest of a particular host species. Chance (1922) studied the movement of some individual female cuckoos through several seasons. Each laid eggs uniform in colour and marking; one of them laid 61 eggs over four seasons and 58 of these were in nests of the Meadow Pipit (*Anthus pratensis*). Young cuckoos will imprint on their host, and when adult, lay their eggs in nests of the same species. There will be different groups or *gens* of cuckoos associated with particular host species.

From the cuckoo's point of view, the problem about specializing in this way is that while a female cuckoo may concentrate on a particular host, she will mate with any male cuckoo that happens to be around. This will tend to prevent any close egg mimicry developing, and reduce the efficiency of the cuckoo's habits. The answer to this is that different gens seemed to have originated in different areas where a particular host species is common, and then spread into areas already inhabited by another gens (Southern, 1954). This explains why only 2, 3 or 4 gens are found together although the cuckoo probably parasitizes 20 or more species in different parts of its range. Each gens behaves as a separate 'race', although crossing will occur with other 'races'. Interestingly enough, poorly matching eggs are commonest in areas where more than one gens is present.

Jensen (1966) has suggested that the genes for egg pattern may be carried on the large Y chromosomes of the cuckoo. This will be transmitted to all daughters (remember the female is the heterogametic sex in birds), and if imprinting is efficient, there will be opportunities for selection of greater resemblance to the host eggs between different female cuckoos.

Mosquitoes are another group where efficiency of an inherited character affects their breeding success. In this case the crucial factor is the speed with which females can bite and suck sufficient blood for their eggs to mature. If they are too slow, the host will be more likely to detect them and react violently. Gillett (1967) found that 90% of Mosquitoes (*Aedes africanus*) caught in the wild could feed and escape before the host was irritated, while only 59% of individuals of a species (*A. aegypti*) which had been bred in the laboratory for three years could do so. Furthermore the feeding time of the 'domesticated' mosquitoes was more variable as well as slower than the wild flies. Gillett concluded that inefficient feeders are eliminated in the wild, but can survive in the laboratory where there is no disadvantage in slowness. In other words, natural selection is constantly maintaining adaptedness under natural conditions.

A cuckoo who lays eggs which do not perfectly match those of her host species or a mosquito which is slow in feeding may 'get away with it', and successfully transmit her genes. There are other examples of natural selection where non-conformity means death. This is particularly clear-cut for genes which give protection against toxic agents or pesticides. For example, humans homozygous for a particular rare gene suffer from agammaglobulinaemia; they cannot make circulating antibodies against any disease, and may die from any mild infection. In this case only a few people are at risk. By contrast, there are an increasingly large number of examples where a changed environment has put the great majority of a population at risk and only a few adapted (one might say 'lucky') individuals survive.

For example, resistance to the widely used anticoagulant poison warfarin is

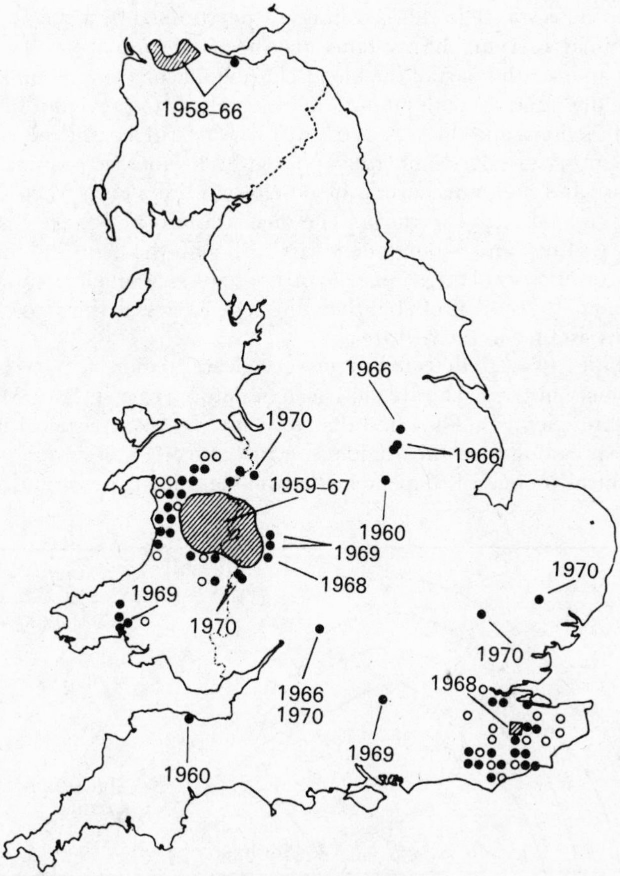

FIG. 32. Areas in Britain where warfarin-resistant populations of the Brown Rat (*Rattus norvegicus*) have been found. Undated samples indicate samples of rats tested in 1972: ● – resistant sample; ○ – non-resistant sample (from Greaves & Rennison, 1973).

now extremely common in rats and mice. Warfarin was first introduced in 1950 and became a poison of choice because of its low toxicity to other animals. The first case of resistance in British rats (*Rattus norvegicus*) was reported from animals caught in 1958 in Lanarkshire, east of Glasgow (Boyle, 1960). By 1972, twelve areas in Britain were known in which warfarin resistance had developed. Nine of these represented isolated outbreaks, but three were well-established resistant populations – in the Central Lowlands of Scotland, in Montgomery and Shropshire, and in south-east England (Greaves and Rennison, 1973).

Warfarin resistance in rats is commonly determined by a single dominant gene. Animals carrying the resistance gene need large quantities of vitamin K, which is an essential part of the blood-clotting mechanism in animals. Warfarin usually interacts with vitamin K in some way to prevent it fulfilling its normal functions, and thus opens the way to warfarin fed animals suffering a fatal haemorrhage. Resistant rats apparently do not use vitamin K in the same way, and maintain normal blood-clotting times even when they have eaten large amounts of warfarin. The mechanism of resistance is similar in both the Scottish and Montgomery rats, although the Scottish animals succumb to lower doses of the poison and do not require as much vitamin K as the Welsh ones. It seems probable that different alleles at the same locus determine resistance in the two areas.

Resistant rats were detected in an agricultural region near Welshpool in Montgomery in 1959. The area has been monitored ever since by Ministry of Agriculture scientists. They find that the places where resistant rats can be found is spreading from the original site at a rate of 3 miles every year. This geographical increase is dependent on the continued use of anticoagulant

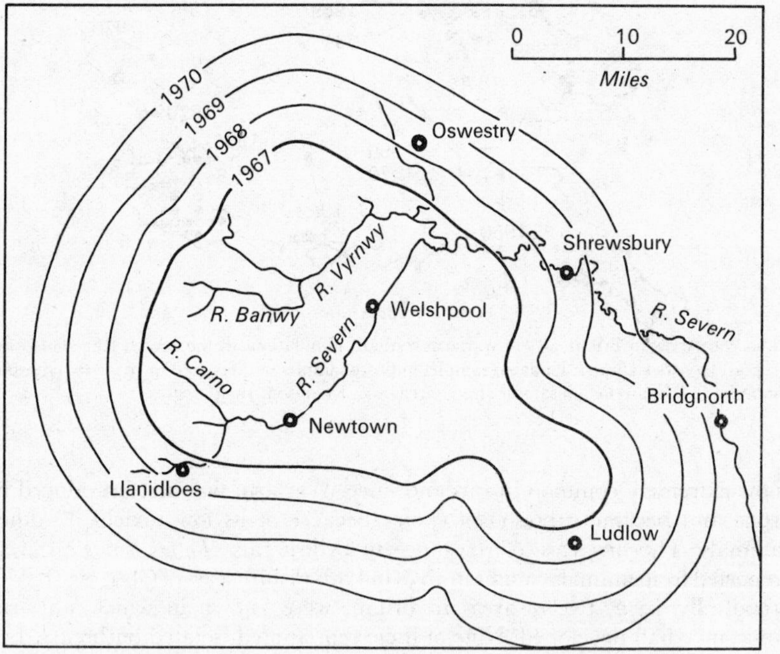

FIG. 33. Map of the Shropshire-Montgomeryshire area where the spread of warfarin resistant rats has been intensively studied. The concentric circles show the limit of the resistant population in each year (after Greaves, 1974).

poisons, and is limited by the migratory ability of rats. It is similar to the rate at which the species colonizes new territory (Drummond, 1970). Now it might have been expected that all the rats at the centre of this area would now be resistant. Intriguingly the incidence of resistant animals rose rapidly to about 50%, and has remained constant at this level, at least until 1972. This stability in the face of the repeated use of anticoagulants (even though their intensive use in the region is now restricted) suggests that a balance may have been reached between the advantage to the resistant rats of being resistant, and the disadvantage of a twenty-fold increase in the requirements of the homozygous resistant rats for vitamin K. If we call the resistance locus *Rw* with a susceptible (*s*) and a resistant (*r*) allele, there will be three genotypes:

$Rw^s Rw^s$ – normal rats at an unconditional disadvantage in the presence of lethal doses of warfarin.

$Rw^s Rw^r$ – rats resistant to warfarin, and needing slightly more vitamin K than normal for full health.

$Rw^r Rw^r$ – rats resistant to warfarin poisoning, but requiring prohibitively large amounts of vitamin K.

This will give a balanced situation very similar to that of sickle-cell anaemia in man in malarial areas (pages 35–6, 69–70). There is no direct evidence of any lowered viability of homozygous resistant animals in the wild, but they are clearly hampered in the laboratory (Greaves and Rennison, 1973). In an artificially enclosed colony in Denmark containing resistant rats which had arisen in that country, Lund (1968) found that the rate of population growth was slower than expected, and that the proportion of resistant individuals dropped over several generations from its initial frequency.

Rats from two other areas where resistance has arisen in the wild (from farms in Gloucestershire and in Nottinghamshire) seem to have their resistance determined in some other way than in the Welsh and Scottish areas: standard toxicological tests that distinguish efficiently between the resistant and non-resistant Welsh phenotypes failed to work as well on individuals from these areas. Certainly in House Mice, resistance is usually produced multigenically (by many gene loci acting together), although a single gene producing resistance in the species has been identified biochemically, and is linked to the same genes as *Rw* in the rat (Wallace and MacSwiney, 1975). This had the result that although a range of susceptibility to warfarin developed in many mouse populations during the period of intense use of the poison, for a long time inherited resistance was 'officially' denied, since it was impossible to distinguish between unequivocally 'susceptible' and 'resistant' individuals. Warfarin continued to be used in attempts to control populations increasingly resistant to it, with the result that mouse infestations received additional food supplied by those trying to kill them. Resistant populations of mice are now widespread throughout Britain.

Warfarin is used in humans in diseases where clots may form and block vital blood vessels. A dominantly-inherited resistance to its action has been described, functioning in a similar way as the resistance in Welsh rats, but this seems to be rare (O'Reilly, Aggeler, Hoag, Leong & Kropotkin, 1964).

It would be easy to fill up the rest of the book with discursive descriptions of selection acting in wild populations, but we must seek some order in the possible examples. However before submitting to the tedium of order, we must explore in more depth than hitherto (page 18) the best known and one of the clearest examples of genetical change in nature, that of industrial melanism in tree trunk-sitting moths.

In the first half of the nineteenth century a new England was built, housed (in Arthur Bryant's words) 'not so much in towns as in barracks . . . Tall chimneys and gaunt mills multiplied a hundredfold, and armies of grimy, grey-slated houses had encamped round them. Overhead hung a perpetual pall of smoke, so that their inhabitants groped to their work as in a fog. There were no parks or trees: nothing to remind men of the green fields from which they came or to break the squalid monotony of the houses and factories' (Bryant, 1940). This pollution was not new. In 1661 John Evelyn was moaning about 'that Hellish and dismall Cloud of SEA-COALE perpetually imminent over London . . . which was so universally mixed with the otherwise wholesome and excellent Aer, that her Inhabitants breathe nothing but an impure and thick Mist, accompanied with a fuliginous and filthy vapour . . .' What was new was the scale of the pollution. For centuries men had been modifying the land surface of Britain for their own ends; suddenly air pollution and its effects became a major ecological factor.

Now air pollution consists of gases and solid 'fall-out'. The bulk of fall out (measured in tons per square mile per month) drops near its source, but there is a lighter portion which may travel a long distance down-wind. It is the light part of the fall-out which has the most widespread biological effects. It is precipitated with rain, and carried onto the boughs and thence the trunks of trees. Vegetative lichens are killed by atmospheric sulphur dioxide, and tree trunks become bare, and often black (although black 'rain runs' are found in non-polluted areas, and come from dark coloured mosses growing on wet barks). A large number of moths (and other insects) have evolved wing and body patterning making them almost invisible on a lichen covered tree trunk. Once the lichens are killed and the bark darkened, a moth which was hidden (to the human eye at least) becomes very conspicuous. Visual predation by birds has lead to the virtual elimination of the previously adapted typical moths in polluted areas, and usually only species which have evolved black (melanic) forms have been able to survive in industrial areas.

The proof that the spread of melanic moths in industrial areas was the result of natural selection through bird predation dates only from the work of H. B. D. Kettlewell in the early 1950s. Before this it was frankly not believed that birds could be discerning enough to select between morphs. For example, White (1876) cited climatic factors as being the 'exciting' cause of melanism in the Scottish Highlands; Merrifield (1894 and earlier) believed that melanism was the result of the direct result of temperature during development; Tutt (1899) argued (but did not prove) that in wet areas, particularly where soot was present, melanic moths were better concealed and 'natural selection' augmented by 'hereditary tendency' favoured them; Ford (1937, 1940) suggested that the increased viability of some melanic mutants would result in their spread in industrial areas once their cryptic disadvantage had disappeared. However the most important red herrings were produced by Heslop Harrison (reviewed 1956). He fed caterpillars of the Early Thorn (*Selenia bilunaria*) and Engrailed (*Ectropis bistortata*) moths on leaves impregnated with lead nitrate and manganese sulphate, and produced a number of melanic offspring from them. He attributed this to the direct results of the salts on the 'soma' (*i.e.* the bodies of the moths), which in turn affected the 'germplasm'. Unfortunately for himself, Heslop Harrison chose species which possessed recessively inherited melanics. Hughes (1932) repeated the experiments and bred 3265 individuals under conditions identical to those of Heslop Harrison, and obtained no melanics. Heslop Harrison proved no more than that heterozygotes are common in wild populations. In neither of the species he studied intensively are industrial melanics found.

In the early 1950s Kettlewell tested and showed that birds could be responsible for the spread of industrial melanics. He chose to work on the Peppered Moth (*Biston betularia*) because it is widely distributed at high densities in both rural and urban areas, and has an easily recognized and well known melanic form (*carbonaria*). Moreover the Peppered Moth produced the first known industrial melanics in Britain: as early as 1848 it was reported as occurring in Manchester (Kettlewell, 1973 – a review of all work on melanism).

Kettlewell began by demonstrating that Great Tits (*Parus major*) found and ate Peppered Moths on trees in an aviary in their order of conspicuousness to human eyes. The same thing happened when moths were released in a heavily polluted oak wood outside Birmingham: 8 out of 366 *carbonaria* and 137 out of 154 typicals were judged conspicuous; direct observation from a hide revealed that Hedge Sparrows (*Prunella modularis*) and Robins (*Erithacus rubecula*) removed 54% of the typicals but only 37% of the *carbonaria*. As already noted (page 18), when Peppered Moths were released at the same site, twice as many *carbonaria* as typicals were recaptured relative to the number released (27.5% of 477 *carbonaria*, and 13.0% of 137 typicals).

The relative survivals of the two forms were reversed when the experiments were repeated in an unpolluted wood in Dorset. The typicals were all inconspicuous, and all the *carbonaria* were conspicuous. Spotted Flycatchers (*Musicapa striata*), Nuthatches (*Sitta europaea*), Yellowhammers (*Emberiza citrinella*), Robins (*Erithacus rubecula*) and Song Thrushes (*Turdus ericetorum*) were seen to eat 164 *carbonaria* but only 26 typicals. Following the release of marked forms, 14% of typicals and 5% of *carbonaria* were recaptured.

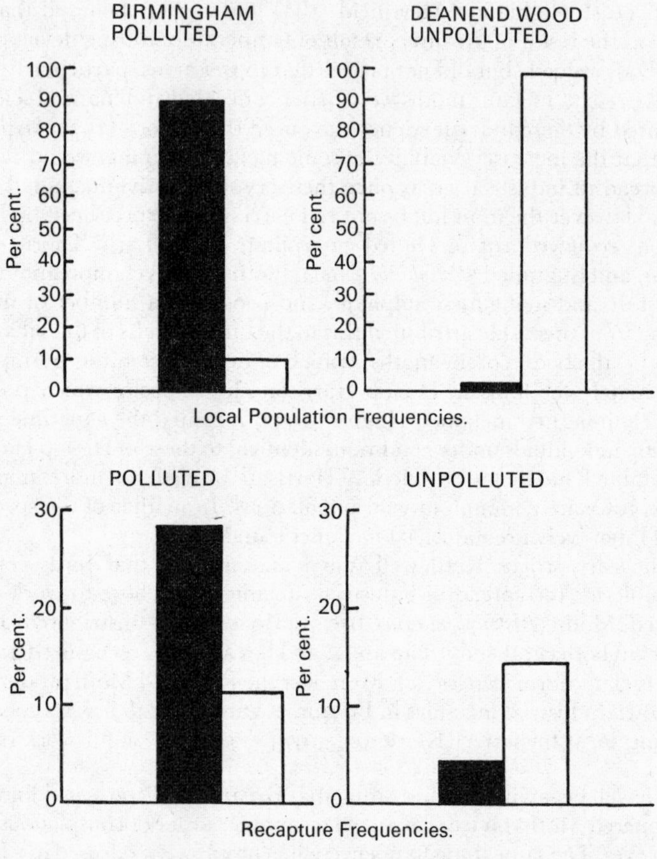

FIG. 34. Relative frequencies and survival of melanic (black columns) and typical (white columns) Peppered Moths (*Biston betularia*) in a polluted area on the edge of Birmingham, and an unpolluted one in Dorset. The lower histograms show the proportion of marked moths of each form recaptured in the two areas (from Kettlewell, 1956).

It follows from the disadvantages that the 'old-fashioned' typicals suffer in polluted areas and the melanics in rural areas, that the *carbonaria* form should be mostly confined to industrial areas. So the next step was to map the distribution of the forms. Kettlewell enrolled the co-operation of more than 150 lepidopterists from all over Britain to supply him with the numbers of Peppered Moths caught by them. The result is shown in the figure on page 123.

Here we must mention a complication: besides the all black *carbonaria*, there are also several less intensely pigmented Peppered Moth melanics

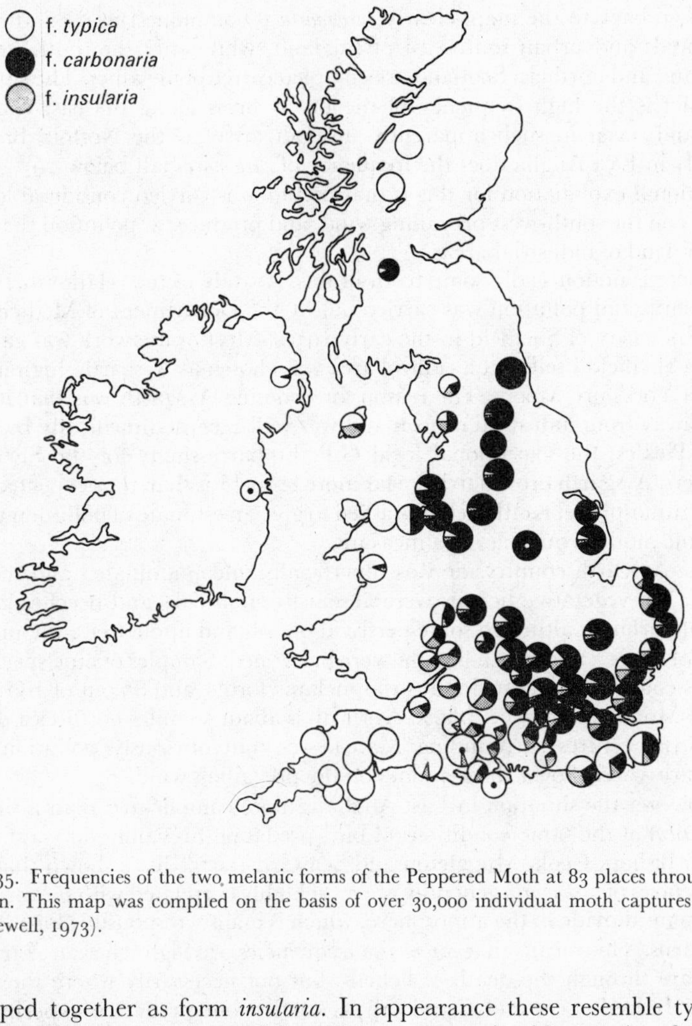

FIG. 35. Frequencies of the two melanic forms of the Peppered Moth at 83 places throughout Britain. This map was compiled on the basis of over 30,000 individual moth captures (from Kettlewell, 1973).

grouped together as form *insularia*. In appearance these resemble typical moths with heavy black markings which are determined by a series of alleles at the *carbonaria* locus (Lees, 1968). *F. insularia* reaches its highest frequencies in industrial areas of South Wales, but also occurs commonly on the western edges of industrial areas. There is some evidence that *insularia* represents an attempt at melanic disguise, to be replaced by a *carbonaria* allele if one is available. However, *insularia* remains in the population even when high frequencies of *carbonaria* occur (Lees and Creed, 1975). In such circumstances it may be maintained by frequency-dependent selection – predators failing to recognize it, so long as it remains fairly uncommon (page 140–2).

To go back to the map: clearly *carbonaria* is commonest in the industrial Midlands and urban south-east of England, while only the south-west of England and northern Scotland are completely free of melanics. The surprising fact is the high frequency of the black forms along the east coast of England, even in such apparently unspoilt areas as the Norfolk Broads. Rarely in East Anglia does the frequency of *carbonaria* fall below 70%. The traditional explanation for this is that pollution is carried considerable distances on the south-west prevailing wind, and produces a 'pollution shadow' down-wind of industrial areas.

There is undoubtedly some truth in this. A study of the relation between bronchitis and pollution was carried out by the Department of Medicine in the University of Sheffield in the early 1950s. Most of the work was carried out in Sheffield itself, but a control area was chosen at Aysgarth, high in the North Yorkshire Moors. The reason for choosing Aysgarth was that it was well away from industrial centres and was well-recorded medically by William Pickles, the exceptional local G.P. Embarrassingly for the Sheffield workers, Aysgarth proved to have far more bronchitis than they expected in a clean atmosphere. Kettlewell was asked to give an estimate of pollution using melanic moth frequencies as a measure.

The Aysgarth countryside was superficially uncontaminated and the air clean. Yet vegetative lichens were absent from all oaks and beeches in the neighbourhood, although some persisted on ash and apple trees; the undersides of the oak and beech boughs were blackened. Samples of nine species of moths collected contained industrial melanic forms; and 85 out of 100 Peppered Moths caught were black. Aysgarth is about 50 miles north-east of the industrial centres surrounding Manchester, but obviously pollution was regularly carried over the Pennines on the prevailing wind.

However the situation in East Anglia is more complicated than a simple extension of the same conditions of bird predation prevailing in (say) Aysgarth. Bishop, Cook, Muggleton and Seaward (1975) have shown that the frequencies of *carbonaria* generally are most highly correlated with the amount of sulphur dioxide in the atmosphere, which is mainly responsible for the loss of lichens. This means that *carbonaria* frequencies are high where tree trunks are bare through the death of lichens, but not necessarily where they are black. Indeed, Lees and Creed (1975) carried out predation experiments similar to Kettlewell's original studies in Norfolk and Suffolk and found that *carbonaria* consistently disappeared more rapidly than the typical form, *i.e.* it was at a cryptic *dis*advantage despite being at a population frequency of 70%. The only way that the melanic could be maintained was for it to have some compensating advantage, almost certainly at another stage of the life-cycle.

This receives support from another consideration: nowhere in Britain does the *carbonaria* allele rise to 100%. If melanic homozygotes are less fit than heterozygotes, this would mean that typical moths would still occur under conditions where they are rapidly eliminated by predation. Indeed, Haldane

(1924, 1956) calculated from assumed equilibrium gene frequencies that *carbonaria* homozygotes had 8% selection against them when compared with heterozygotes, whilst in the Liverpool area Clarke and Sheppard (1966) deduced values of 8% and 15% from predation experiments and rate of change of phenotypic frequencies. The advantage of the heterozygote cannot be a consequence of its colour, since *carbonaria* heterozygotes and homozygotes are normally inseparable in appearance; heterozygotes must have some invisible physiological superiority. This would mean that when the visual advantage of typical moths is less than their physiological disadvantage, *carbonaria* will occur, but slight changes in selection intensity are likely to lead to relatively large changes in phenotype frequency – greater than expected on the grounds of resemblance to resting places alone. In fact *carbonaria* has decreased during the past 20 years north and east of a line from Chester to west of London, and increased to the south and west of it (Lees and Creed, 1975). In some parts of the country, these changes can be related to decreasing smoke pollution (Table 13), but this can only be part of the story. Only where the cryptic advantage of the typical form is so great that it outweighs its physiological disadvantage will *carbonaria* be eliminated.

TABLE 13. Numbers of Peppered Moth morphs (*Biston betularia*) caught in Didsbury, near Manchester (from Cook, Askew and Bishop, 1970)

	carbonaria	*insularia*	*typica*
1952–64	749	11	0
1966–69	935	12	25

These conclusions are supported by detailed studies of Peppered Moth melanism in the North Wales-Liverpool-Manchester area (Clarke and Sheppard, 1966; Bishop, 1972; Bishop, Cook, Muggleton and Seaward, 1975). The frequency of *carbonaria* in central Liverpool is 97.3% falling to 11.8% at Colwyn 50 miles from Liverpool, and 3.1% at Bangor 70 miles away (see figure on page 126. Mark-recapture experiments have shown that the typical form has a disadvantage compared with *carbonaria* of 60% in Liverpool, and 20% in Caldy on the Wirral Peninsula. Bishop (1972) used these figures, together with estimates of migration rate and heterozygous advantage to work out the expected rate of change of gene frequency. His calculated graph showed a much more rapid decline in typical frequencies than occurred in actual populations, even where estimates of migration rate and heterozygous advantage were made as large as possible: in the simulations, the *carbonaria* allele would have been expected to have disappeared from the population in parts of Wales where it still reached 30–60%.

In the early days of speculation about the causes of spread of industrial

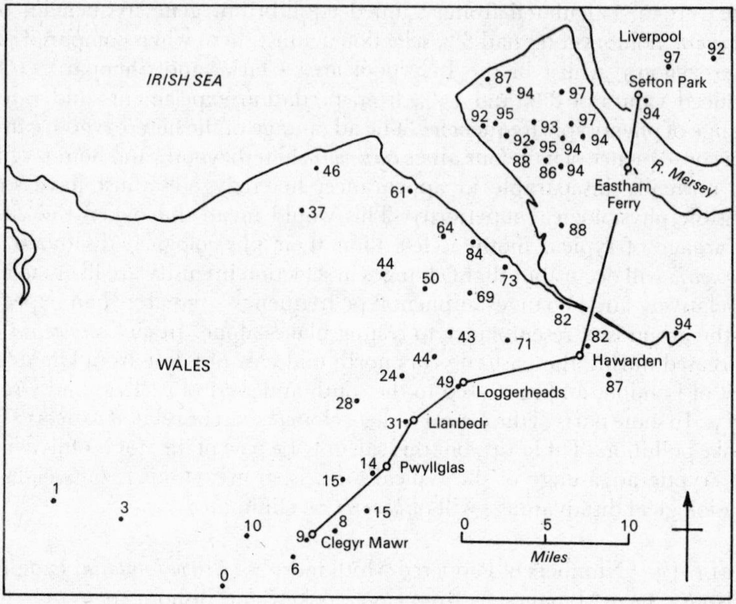

FIG. 36. Frequencies of the *carbonaria* melanic of the Peppered Moth on Merseyside and in North Wales. The line joins sites for which estimates of the selection pressures were made by Bishop (see figure 37) (after Bishop, 1972).

melanics, E. B. Ford (1937, 1940) placed considerable weight on the higher viability of dominantly inherited melanics over typicals. (Recessively inherited melanics almost always have a reduced viability: the Brindled Beauty (*Lycia hirtaria*) and the Oak Eggar (*Lasiocampa quercus*) are two of the extremely few species which have recessive industrial melanics.) Ford fed the caterpillars of some broods of the Mottled Beauty (*Cleora repandata*) adequately and starved others. In all broods he expected equal numbers of typicals and melanics, but he had a marked deficiency of typicals in his starved broods (Table 14). If pollution leads to feeding difficulties, the melanics would have a clear advantage. Kettlewell (1973) has reviewed the attempts to show different viabilities of the typical and melanic forms of the Peppered Moth, and concluded that there is no convincing evidence of any such difference.

These results show that industrial melanic morphs have more traits than a dark wing pattern. But even when colour alone is considered, melanics still have the behavioural problem of actively hiding themselves: cryptic moths must take up resting positions on backgrounds they match. In other words they must be able to choose their background as well as resemble it. This was originally investigated by Kettlewell (1955*b*), who released Peppered Moths into an old cider barrel lined with vertical black and white stripes. Sixty-five

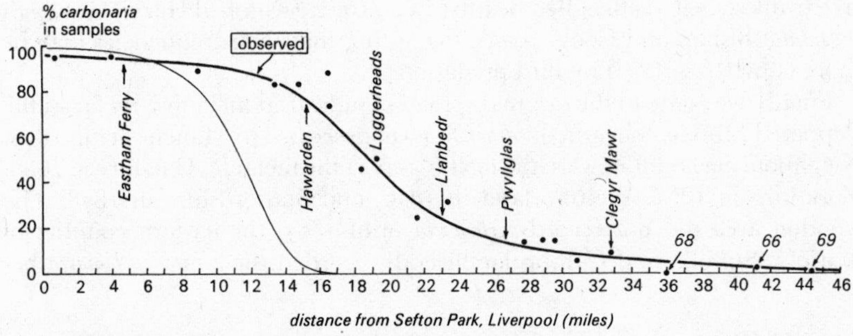

FIG. 37. Changes in the frequencies of the *carbonaria* melanic of the Peppered Moth between Liverpool and North Wales. The heavy line shows the observed frequencies; the lighter line the expected frequencies based on the best available estimates of migration rate and heterozygous advantage (after Bishop, 1972).

TABLE 14. Survival of larvae of typical and melanic forms of the Mottled Beauty (*Cleora repandata*) in starved and non-starved back-cross families (after Ford, 1945)

	Melanic	*Typicals*	*% of typicals*
Non-starved	101	91	47·4
Starved	52	31	37·3

In each case equal numbers of melanics and non-melanics are expected.

per cent of the moths took up position on the background they matched (*i.e.* typicals on white stripes, *carbonaria* on black). This behaviour was confirmed and the experiment extended by Boardman, Askew and Cook (1974), who released 31 species of moths in containers like Kettlewell's, and also in a 'garden tent' containing grass and trees. They found that light-coloured species usually rested on fresh vegetation, and dark ones on bark or bare ground. *F. insularia* took up a resting place on a background intermediate between the choice of *carbonaria* and typical.

Another behavioural difference in a species showing industrial melanism occurs in a North American relative (*Phigalia titea*) of the British Pale Brindled Beauty (*Phigalia pilosaria*), where melanic males fly earlier in the evening than typicals (Sargent, 1971; Lees, 1971). They therefore have a greater chance of copulating with the wingless females of the species, which could explain why melanic frequencies of up to 20% are found in unpolluted areas where trees are covered in lichens (Lees and Creed, 1975). The frequency of melanics in other species is closer to that expected from the blackness (or 'reflectance') of tree trunks (*e.g.* Pale Brindled Beauty: Lees, 1971; Scalloped Hazel, *Gonodontis bidentata*: Bishop and Cook, 1975), suggesting that their frequencies may be more closely regulated by bird predation.

Finally we come to the original spread of industrial melanics. As far as the Peppered Moth is concerned, *carbonaria* was described in Manchester in 1848. Neighbouring counties were the next to report the melanic: Cheshire in 1860, Yorkshire in 1861, Westmorland in 1870, and Staffordshire in 1878. The London area did not record *carbonaria* until 1897; the eastern counties of Norfolk, Suffolk, and Cambridgeshire all recorded their first *carbonaria* be-

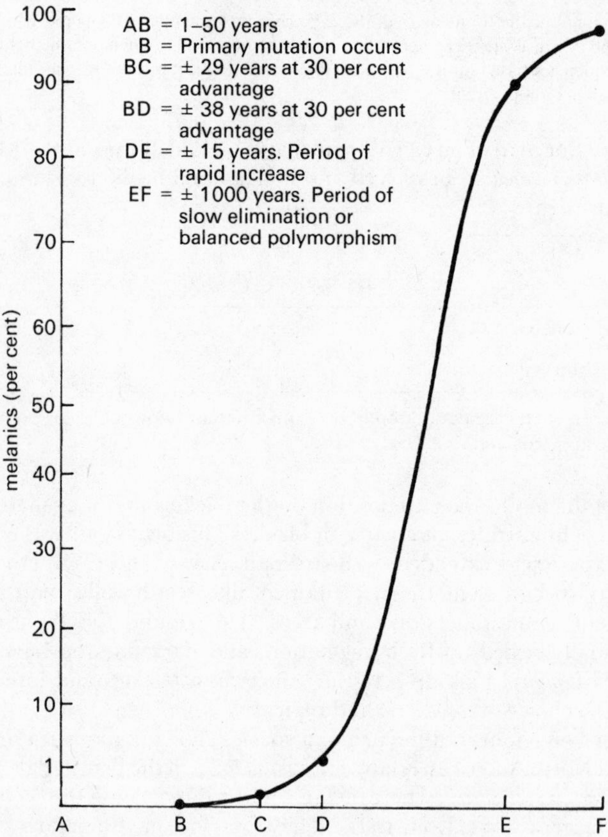

FIG. 38. Rate of spread of a dominantly-inherited melanic mutant (such as in the Peppered Moth) with a mutation rate from the normal allele of one in a million, assuming a constant advantage of the heterozygote throughout and a 50% selective advantage of the melanic over the typical form (from Kettlewell, 1973).

PLATE 5. *Mimicry in British insects (page 134)*

Above left, Italian Honey Bee *(Apis mellifera)*; *right*, a Hover Fly *(Eristalis tenax)*.

Mimicry of Bumble Bees (*Bombus* spp.) by the Narcissus Bulb Fly *(Merodon equestris)*: top row, *Bombus hortorum*, *B. terrestris*, *B. agrorus*; bottom row, inherited variants of *Merodon equestris*.

PLATE 6. *Banding and colour in molluscs*

Dog-whelks *(Nucella lapillus)* from the north coast of Cornwall (page 222).

The yellow five-banded (12345) form of the land snail, *Cepaea nemoralis* (page 148).

A thrush 'anvil' with the remains of *Cepaea* shells collected and broken by a local thrush *(Turdus ericetorum)* (page 149).

A bank providing an apparently uniform habitat for *Cepaea nemoralis* between the New Bedford River and a road near Ely.

tween 1892 and 1895. The same picture seems to have occurred on the Continent: *carbonaria* was recorded first in Hanover in 1884, Holland and Thuringia in 1888, and in various parts of the Rhine Valley during the next few years.

These fragmentary reports are consistent with the spread of a successful form from its centre of origin, in the same way that warfarin resistant rats spread in the west Midlands. Haldane (1924) pointed out that if the Peppered Moth population of Manchester in 1848 contained 1% *carbonaria* and by 1898 it contained 95%, this represented an average 33% disadvantage of the typical form in comparison with *carbonaria*. The increase of melanics may then be regarded as involving a period of adjustment cryptically (*q.v.* pages 190–2) and physiologically as one allele largely replaced another; a period of rapid spread; and a period of slow elimination of the typical form or establishment of a stable situation. However, Kettlewell (1973) believes an additional factor has been repeated mutations to the *carbonaria* allele. He cites the origin of the form in Belfast in 1894, the Isle of Man in 1904, Dublin about 1950 and Torquay in 1956, each locality being 50 miles or more from the nearest contacts. Notwithstanding, the main causes of the spread of melanic Peppered Moths must have been selective predation.

When all the information is put together, it is clear that the differences between Peppered morphs are much more than a matter of colour: the morphs are distinguished by adult background choice, non-adult viability, speed of larval feeding, the attractiveness of females and probably other traits (Kettlewell, *loc. cit.*; see also Lees, Creed and Duckett, 1973). These complications do not contradict nor lessen the importance of Kettlewell's original discoveries on visual predation in the species, but they highlight the danger of glib extrapolations that some scientists have made about particular pressures in natural populations. To summarize the Peppered Moth story, there seem to be three situations:

i. When *carbonaria* frequencies are high, the form has a clear visual advantage against predation.

ii. When *carbonaria* is rare or absent, the typical form is maintained by its cryptic advantage.

iii. In areas where *carbonaria* occurs at up to 80%, the typical form may not suffer heavy visual predation, but nevertheless be discriminated against on other grounds.

Nevertheless, however much of the effectiveness of visual predation in the Peppered Moth is qualified, there is clear evidence of cryptic adjustment in the species. In the earliest examples of *carbonaria* caught a century and more ago, which are prized specimens in entomological collections, there were white markings on the upper sides of the wings. These never occur nowadays, although the majority of *carbonaria* have four white dots round the head and at the base of the wings. In the Sheffield area even these are tending to disappear, leaving a completely black insect.

The story of industrial melanism in the Peppered Moth is important both because its investigation marked a turning point in our understanding of selection pressures and also because of the clarity with which it illustrates some of the genetical forces acting on populations in nature. We shall have to return to the Peppered Moth again(pages 164, 190–1, 294), but it is proper now to attempt to put some order into our understanding of selective processes.

TYPES OF SELECTION

Natural selection may change, stay or split phenotype frequencies at any locus. In more conventional language, selection may be:

directed – favouring one extreme of the distribution of inherited variability
stabilizing – favouring the mean at the expense of the extremes
disruptive – favouring the extremes at the expense of the mean.

TABLE 15. Estimate strengths of directed selection (extreme phenotype favoured) (after Antonovics, 1969; Berry, 1971)

Selection for:	*% strength of selection*
Heavy metal tolerance of grasses of mine soil:	
(a) *Agrostis tenuis* on a copper mine	54–65
(b) *Holcus lanatus* on a lead/zinc mine	46
Heavy metal susceptibility on pasture:	
Agrostis tenuis downwind from a copper mine	27–62
Non-banded *Cepaea nemoralis* in woodlands	19
Light-coloured *C. nemoralis* on dunes (*v* brown)	6+
Single-banded *C. nemoralis* on dunes	5+
Female Meadow Brown butterflies (*Maniola jurtina*) with low hind wing spot numbers	69–74
Melanic (*carbonaria*) form of Peppered Moths (*Biston betularia*) in various regions of Great Britain	5–35
Spotted form in overwintered Leopard frog, *Rana pipiens* (*v* unspotted)	23–28
Unbanded water snakes (*Natrix sipedon*) (*v* heavily banded)	77
Tooth-size in a fossil horse (*Merychippus primus*)	27–61
Normality in man (*v* many biochemical and chromosomal disorders giving low-grade mental defect)	100

TABLE 16. Estimated strengths of stabilizing selection (intermediate phenotype favoured) (after Berry, 1971)

Selection for:	*% strength of selection*
Survivors among Sparrows (*Passer domesticus*) stunned in a storm	—
Coiling in snails (*Clausilia laminata*)	8
Variation in over-wintering wasps (*Vespa vulgaris*)	10
Size and hatchability in duck eggs	10
Birth weight and survival in human babies	2.7
Clutch size in Swifts	—
Inversion heterokaryosis in *Drosophila pseudoobscrua*	up to 50
Mating efficiency and morphological variation in the beetle *Tetraopes tetraophthalmus*	—
Tooth variability (i.e. *de*-stabilizing selection) in *Mus musculus*	21–26
Tooth size in *Rattus rattus*	4
Non-metrical skeletal variation in over-wintered *Mus musculus*	0–27
Shell variability in Dog-whelks (*Nucella lapillus*) on shores exposed to different strengths of wave action	0–91
Colour morphs in *Sphaeroma rugicauda*	50+

Directed and stabilizing selection are the more familiar modes. We have already delved into the theory underlying them (pages 67–72). The spread of melanic Peppered Moths and warfarin resistant rats are examples of directed selection; the maintenance of egg mimicry in Cuckoos and feeding speed in mosquitoes represent examples of stabilizing selection.* Generally speaking stabilizing selection will be commoner but its intensities less than that of directed selection since it is concerned with preserving the *status quo* and eliminating newly arising deviations, while directed forces will be related to rates of change in the environment (Tables 15, 16).

Birth weight in man is affected by stabilizing selection. Although there are many factors that contribute to the size of a baby at birth (for example, mothers who smoke have smaller babies than they would if they did not smoke), about half the variation is inherited (Penrose, 1961; Polani, 1974). Consequently if more babies than average of a particular weight die, this could affect the genetical constitution of the babies' generation. Karn and Penrose (1952) plotted the birth weights of all babies and of those who died, in the 13,730 babies born in a London hospital between 1935 and 1946. There was a higher death rate among very light and very heavy babies – in other words there was stabilizing selection for birth weight.

*As do the *maintenance* of melanic Peppered Moth and warfarin resistant rat frequencies.

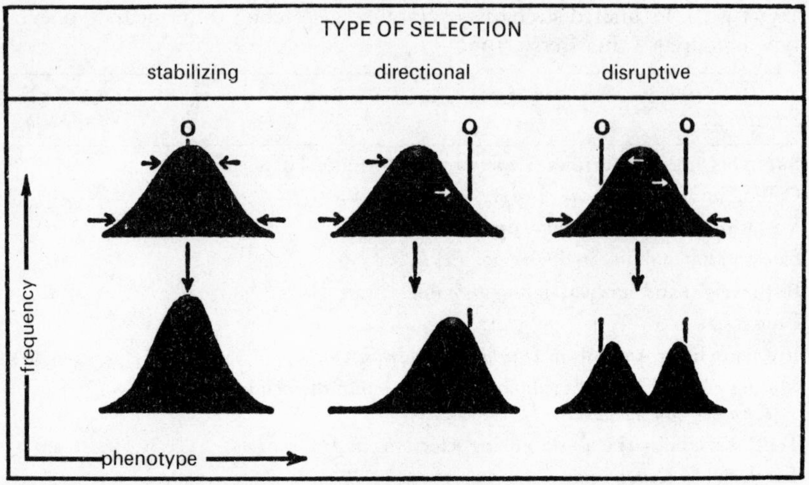

FIG. 39. The three basic types of selection. The optimal phenotype towards which selection is acting is shown as 'O'; the direction and force of selection is shown by the arrows in the upper row of phenotypic distributions. The lower row shows the consequence of the different selections (from Mather, 1953).

In fact the situation was more complicated. The lowest mortality was in babies of 7.97 lbs. This was therefore the optimum weight. But the mean weight was one per cent less (7.89 lbs). The effect of this would be of directional selection for heavier babies as well as stabilizing selection for the optimum weight.

The classical example of stabilizing selection is a description by an American who collected a number of Sparrows (*Passer domesticus*) cold and exhausted after a severe snow-storm in New England (Bumpus, 1899; O'Donald, 1973). Out of a total of 136, 72 survived and the rest died. Bumpus made 9 measurements on all the sparrows (body weight, total length, wing span, etc.), and concluded that it was the more variable birds which failed to survive. In other words, the extremes perished, and the average lived to pass on their genes.

Historically little attention was paid to *disruptive selection* until the publication of a classical paper by Kenneth Mather (1953), although the idea was familiar to palaeontologists concerned with diverging lines of fossils. G. G. Simpson, for example, speaks of the same process as centripetal selection. From the point of view of evolution, disruptive selection produces two different phenotypes from a single one and therefore leads to divergence, the beginning of the origin of new species.

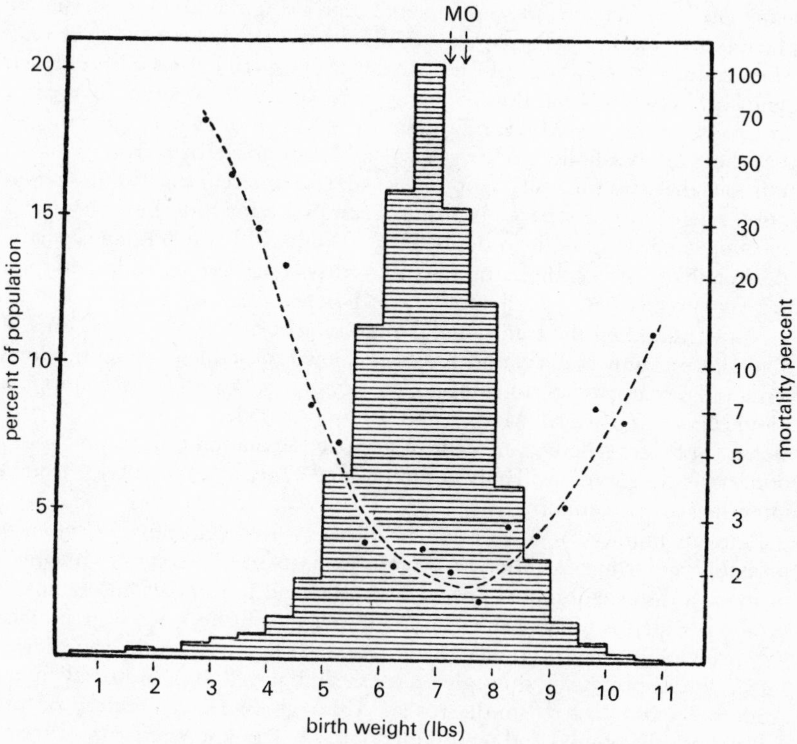

FIG. 40. The distribution of birth weight among 13,730 children and the mortality of each birth weight class. The broken line is the death rate in relation to birth weight. *M* is the mean birth weight, *O* is the birth weight associated with the lowest mortality, and hence the optimum weight (from Mather, 1964).

The theory of disruptive selection is simple: in the calculation for heterozygous advantage (page 71) substitute a reduced instead of an increased fitness for the heterozygote when compared with the homozygotes. Where there has been considerable argument is the relation between gene flow and the establishment of stable equilibria by random mating. If the population is a single breeding unit, selection against heterozygotes (because that is what disruptive selection is) gives a single unstable equilibrium, and any displacement from that unstable point will lead to the eventual loss of one of the alleles from the population. Disruptive selection *per se* is not a sufficient condition for the production or maintenance of a stable polymorphism or – more important in the long run – the development of reproductive isolation between the divergent forms (Smith, 1962).

What is much more important is to decide on the conditions where divergence *can* arise between parts of a population being selected for alternative phenotypes. Traditionally population divergence was believed to arise only when a single population split into separate sub-populations with reduced gene flow between them, thus allowing each population to adapt by normal directed selection (*e.g.* Mayr, 1963). Smith (1966) has shown mathematically that there are possibilities of achieving stable equilibria by disruptive selection, but that it is difficult – population sizes in different parts of the population range must be separately adjusted, and selection must be strong (with selection coefficients of the order of 30%). While neither condition is particularly unlikely, the combination is sufficiently restrictive for differentiation to be commonly based on disruptive selection *plus* other forces.

Notwithstanding the traditional demand for isolation as a prerequisite of divergence is unnecessary, and has been shown to be incomplete in a long series of laboratory experiments on a variety of organisms (*Drosophila* spp., a Flour Beetle *Tribolium*, Maize, and *Brassica*) (reviewed Thoday, 1972). Genetical heterogeneity can occur in a single population if disruptive selection is operating, but it will only be maintained if the environment or selection intensities are not uniform.

Batesian mimicry is a situation in nature where disruptive selection is probably important (Mather, 1955). Here a tasty species comes to resemble or *mimic* a distasteful model and is thus protected to some extent against a predator which is wary of the distasteful species (on the 'once bitten, twice shy' principle). It has been suggested that the inoffensive Slow Worm (*Anguis fragilis*) gains protection through a pattern similar to that possessed by many Adders (*Viperus berus*) (Smith, 1974). Although Batesian mimicry occurs throughout the animal and plant kingdoms, the best known examples are in butterflies, particularly the beautiful *Papilio* species of the tropics. In Europe, bees and wasps are mimicked by the Broad-bordered and Narrow-bordered Bee Hawk Moths (*Hemaris fuciformis* and *H. tityus*) and several of the Clearwing moths (such as the Hornet Clearwing, *Sesia apiformis*). The various White butterflies are probably mimics of each other (Mullerian mimics), so that the distasteful Large Whites (*Pieris brassicae*) 'put off' bird predators, and make it less likely that the less uncommon Small and Green-veined Whites (*Pieris rapae* and *P. napi*) and Orange Tip (*Anthocaris cardmines*) will be eaten (Marsh and Rothschild, 1974).

Some orchids (notably those of the genus *Ophrys*, which includes the Bee and Fly Orchids, *O. apifera* and *O. insectifera*) have flowers mimicking the appearance of females of some Hymenopteran species in shape, colouring, and hairiness. Male bees and wasps are attracted to the flowers, and attempt to copulate with them. In doing this, they dislodge pollen which in their lustful frenzy they then carry to another flower and so secure cross-pollination. The Bee Orchid in Britain is not visited regularly by any insects, and is nearly always self-pollinated. Selection for exact mimicry is therefore relaxed, and this species tends to be much more variable than the Fly Orchid.

In Batesian mimicry, the greater the relative number of mimics resembling a particular model, the less successfully can they shelter behind his distastefulness, for the less likely will predators associate the appearance (of model plus mimics) with distastefulness. The best strategy for a mimic species is therefore to develop a number of forms, each resembling a different model. This has happened over and over again (*q.v.* Wickler, 1968; Rothschild, 1971). Selection acting disruptively in a single species has produced resemblance to different models. Once the disrupting situation has occurred, frequency and density-dependent selection will take over and increase the resemblance of model and mimic (see page 141).

Another predator-prey situation involving disruptive selection occurs when a predator hunts by forming a 'searching image' so that a prey is at an advantage if it exists in more than one form. This means that the predator may form an 'image' and hunt for one phenotype and miss an alternative one. The experiment of putting out different coloured food pellets for Blackbirds (page 73) was an attempt to test this idea (Clarke, 1962).

NATURAL SELECTION IS NOT A CONSTANT FORCE

Any individual at any moment has a measurable chance of survival and of reproducing. In other words, he has a 'fitness' in the sense in which the word is used in biology (page 61). Similarly a population will contain animals or plants with different fitnesses, and hence one of the three modes of selection – direct, stabilizing or disruptive may be acting. This is fine and true but wholly unnatural. The fact is that the conditions experienced by an individual at any moment in his life are almost certainly not going to remain constant. Even a fish living in the most constant of physical environments, the ocean depths, is going to grow up, age, escape or not from predators, and have to find food and mates. If we translate this into genetical terms, we are saying that selection coefficients are not constant. It is convenient to speak as if they were, but we must be careful not to involve ourselves in a circular argument which runs: 'The strength of selection acting on (say) birth weight variability in human babies is 2.7%; so the action of selection on humans is 2.7%; so we can work out the advantage and disadvantage of particular genes; and so . . .' It is obvious that the moment infant mortality changes, the selection coefficient will change. Indeed the data on birth weight and mortality collected by Karn and Penrose can only give information about the situation in particular hospitals at particular times (Jayant, 1966). To generalize blithely from such premises is as stupid as the weather forecaster who may be very good with isobars but never looks out of the window to see if it is raining. A selection coefficient has a meaning if we can freeze a population in a particular environment, but only refers to the effects of that environment on the phenotypes in that population.

The forces of natural selection vary in both space and time, through both physical and biological conditions. This is easier to see with physical conditions, since they are less variable than biological properties.

One of the more satisfying environments to 'measure' is that of the seashore, particularly on rocky coasts. Every child can recognize differences between exposed wave-battered shores and sheltered ones with heavy growths of sea-weed almost to high tide mark. There have been various attempts to give numerical values to the amount of wave action experienced by a shore, most of them in physical terms such as the distance of open sea in front of the beach, direction of the prevailing wind, and so on. Ballantine (1961) has devised a biological 'exposure scale' in terms of the relative abundance of plants and animals, which is of particular relevance in attempts to estimate genetical forces. The scale grades shores from No. 1 ('very exposed') to No. 8 ('very sheltered'). It has been criticized on the grounds that it is only strictly accurate in the Milford Haven area where Ballantine worked (Lewis, 1964). This is a counsel of perfection: in fact slight modifications of the Ballantine scale enable it to be applied usefully on most shores throughout Britain.

There are some species which occur on shores over the whole gamut of exposures. It is obviously of interest to compare the adaptations they achieve to do this. Some work along these lines is involved in the magnificent series of studies carried out by Ebling, Kitching, Muntz and their co-workers at Lough Ine in Ireland (Kitching and Ebling, 1967). However the first attempt to detect and measure selection pressures on shores of different exposure has been on the Dog-whelk (*Nucella lapillus*) (Berry & Crothers, 1968).

The inspiration for the work on whelks came from a classical study on a land snail (*Clausilia laminata*) made by Weldon (1901) in the palmy days of biometrics.* He collected samples of snails and ground their shells away, progressively exposing the shell whorls laid down earlier in the snail's life. When he measured the radius of the whorls, he found the earlier ones of adult shells (mean length 6 inches) were less variable than the same whorls in young snails (less than $3\frac{1}{2}$ inches long). From this he concluded that only the less variable snails survived to adulthood and breeding. Haldane (1924, 1959) showed that the intensity of natural selection on a metrical character can easily be calculated if we know the mean and standard deviation of the character before and after selection. He estimated that a selection coefficient of 8% would explain the differences between Weldon's old and young shells.

The great advantage of Weldon's work was that he could calculate the difference in variance between survivors and non-survivors in shells of the

**Footnote*. In the early years of the century before the general acceptance of the chromosomal nature of heredity there was a virulent controversy between the biometricians led by Weldon and Karl Pearson, and the 'Mendelians' under Bateson. The biometricians were unwilling to accept the effectiveness of unit genetical factors and supported their case by a version of blending inheritance involving correlations between relatives. The furore lasted until the death of Weldon (Professor of Zoology at University College London and then Oxford) in 1906. It was a fascinating example of how the so-called 'scientific method' does not work in practice (Froggatt and Nevin, 1971; Provine, 1971).

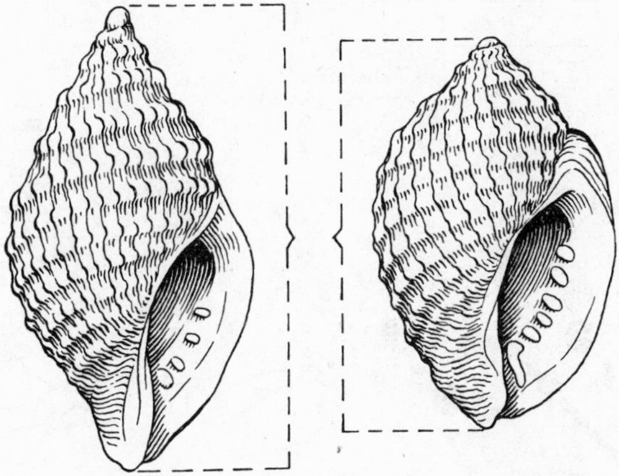

FIG. 41. Dog Whelk (*Nucella lapillus*) shells from a sheltered shore (*left*) and an exposed shore (*right*) (from Berry and Crothers, 1968).

same age without the necessity of a longitudinal study; the disadvantage was that he broke a high proportion of his samples when grinding them down, and the procedure was extremely tedious.

To avoid this problem, Dog-whelk samples collected from a range of exposures around the mouth and along the shores of Milford Haven were divided into groups of different age on the basis of size and thickening of the 'lip' at the growing edge of the shell. Here a difficulty presented itself. Dog-whelks differ in shape on sheltered and exposed shores. On sheltered shores they have a high, thick shell, and a relatively small aperture for attachment to rocks; on exposed shores the shell is thinner and stubbier, so that they are better able to hold on to the rocks they inhabit when subjected to the pounding of waves. The main predators of whelks are probably crabs, which will be active almost entirely on sheltered shores. Shell thickness is clearly protective here.

To compare old and young individuals in differently shaped samples meant discovering a measurement which changed neither with size nor shape. The easiest one turned out to be the ratio of maximum shell length divided by the cube root of dry shell weight. This is superficially an odd function, but both length and weight can be measured easily, and the standard deviation of the ratio gives an easily understood measure of morphological variation.

Berry and Crothers (1968) collected shells from 23 areas ranging from the exposed rocks below St Ann's Head at the entrance of Milford Haven where the Atlantic rollers from the south-west meet their first obstruction for many thousand miles, to sheltered havens in the estuary where large seas can never develop. In all the most exposed populations there was a reduction in

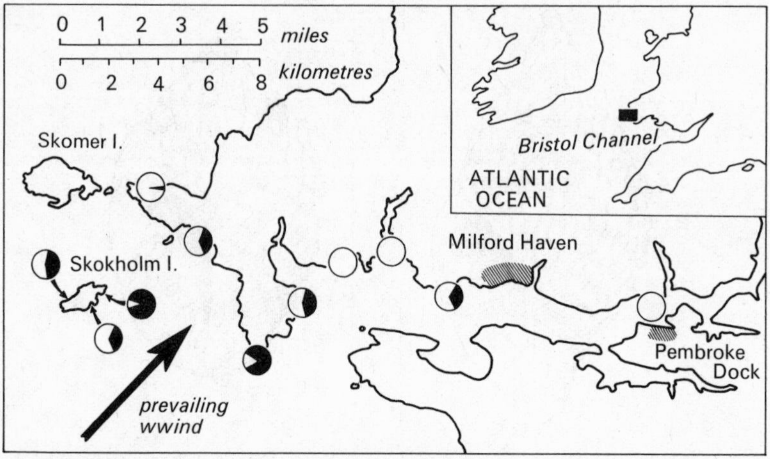

FIG. 42. Stabilizing selection in different populations of the Dog Whelk (*Nucella lapillus*). The proportion of black in each circle represents the difference in variance of a shell character between young and old members of the same population: this equals the strength of selection acting on that population (based on Berry and Crothers, 1968).

variance between the youngest and oldest groups in the population, whilst in the most sheltered there was no significant change during life. Using Haldane's argument about the relation between variance reduction and natural selection, these data show that the strength of selection ranged from over 90% in the population of St Ann's Head, to no selection acting on the particular trait measured on more sheltered shores. The selection concerned was, of course, stabilizing selection; the conclusion was that only individuals near the mean shape (*i.e.* highly adapted) could survive the physical stresses of exposure to wave action, whereas the niceties of shell shape were comparatively irrelevant on sheltered shores where wave action was negligible.

The correlation between degree of exposure and strength of stabilizing selection (= amount of variance reduction) was 63%, or 76% if one anomalous sheltered population was omitted. The simplest genetical model for the situation is that the more variable individuals are homozygous at various loci and the mean ones are multiple heterozygotes. The elimination of the extremes in exposed situations has the effect of removing the homozygotes without changing allele frequencies. In other words, we have an equilibrium situation. We know that equilibrium exists although whether this is the correct genetical explanation cannot be tested until we know more about the action of different genes on shell shape. Somewhat surprisingly an exactly comparable situation exists on Antarctic shores, where morphological variation of limpet shells (*Patinigera polaris*) is affected by similar wave conditions to those selecting Dog-whelks in Wales (Berry and Rudge, 1973).

Whatever the finer details of gene action in Dog-whelks, the point is that selection intensities vary over quite short distances in space. It is inaccurate and misleading, to talk about *the* 'selective intensity' acting on a species. Anyone who has walked along the sea-shore during a storm can testify that environmental pressures change enormously with quite small variations in coastal topography, and should not be surprised if similar pressures are acting in non-littoral situations.

The inconstancy of selection pressures is so important, that it is worth considering another example – with the variation in this case occurring in both time and space.

Sphaeroma rugicauda is a small isopod of salt marshes, with adults about a quarter of an inch long. It can be bred in the laboratory easily, so the inheritance of different variants can be worked out. There are two patterned and four readily distinguishable colour forms: red, yellow, red-yellow and grey. Red and yellow are produced by dominant alleles at two closely linked loci; red-yellow is the combination of the two dominants; and grey is recessive. Yellow homozygotes have a greatly reduced viability in laboratory conditions so that when two yellow heterozygotes are crossed the ratio of yellow : grey in their progeny is nearer 2 : 1 than the expected 3 : 1. Not surprisingly the yellow morph is at a low frequency in most natural populations although it is rarely absent (West, 1964).

Bishop (1969) found that the proportion of yellows in a salt marsh at Neston, Cheshire on the Dee Estuary doubled during the two winters he studied the population (from about 1% to 2%), and decreased during the summers; and that yellows survived low temperatures much better than greys. A similar effect was observed over 3 years by Heath (1974) in a population from the River Tyne near Edinburgh. Here yellow varied from about 4% to 8%. In two years yellow decreased in frequency from October to January as expected on the basis of the laboratory experiments. In only one year did it increase during the January–March period when the lowest temperatures are expected to occur. In that particular year the January–March period was very mild. The probable reason for this effect is that yellows have a faster growth rate than greys under winter conditions and that large animals survive better than small animals. Yellows initially have a lower survival than greys but later, when yellow individuals have grown larger than grey ones they have a higher survival. Furthermore, the populations in different estuaries have different genetical constitutions, suggesting that local habitat factors (such as salinity and vegetation) are more important in controlling frequencies than large-scale climatic conditions such as temperature (Heath, 1975*a*). In other words, the colours of the different forms are merely indicating different physiological tolerances – and complex ones at that. But whatever is the full explanation, the relative advantage of the different forms is constantly changing through the year and from place to place.

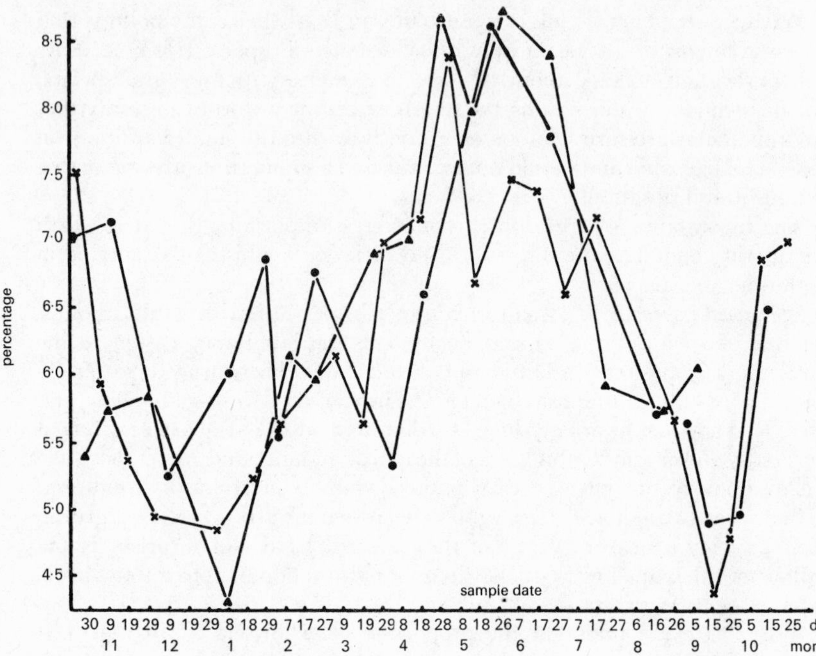

FIG. 43. Changes in the frequency of the yellow morph of *Sphaeroma rugicauda* in the estuary of the River Tyne, East Lothian: X-X, samples collected in 1968–9; ● – ●, 1969–70; ▲ – ▲, 1970–71 (from Heath, 1974).

FREQUENCY AND DENSITY-DEPENDENT SELECTION

Once the inconstancy of selection pressures was realized, theoretical geneticists had a heyday. We are not concerned here with their ratiocinations, however elegant. What we have got to do is to relate the fact of a variable and uneven environment to real populations. There are two sides to this: the changes in populations themselves, particularly in the relative frequencies of different morphs and their densities; and the existence of different niches within the range of a single population (Clarke, 1964).

Let us look first at the consequences of variations with the frequencies and densities of different phenotypes within a population. We can generalize the theorems that we derived when originally considering fitness (pages 69–72):

	A_1A_1	A_1A_2	A_2A_2
Frequency before selection	p^2	$2pq$	q^2
Fitness	a	b	c
Frequency after selection	ap^2	$2bpq$	cq^2

The quantities *a*, *b* and *c* may encompass mutation rates, migration rates, inbreeding, and both frequency-independent and frequency-dependent effects.

Think for a moment about the situation of Batesian mimicry already described. A good mimic of a distasteful model will be well-protected *as long as it is rare*. The commoner it becomes, the more often will its pattern be associated by a predator with good food, and the lower will be its fitness. In other words, selection will be *frequency-dependent* (Murray, 1972).

A direct study of predator-prey relationships was carried out by Popham (1941, 1942). He investigated the efficiency with which Rudd (*Scardinius erythrophthalmus*) (a small fish similar to a Roach) caught Corixid Bugs (*Sigara distincta*) in an aquarium tank lined with brown sand.

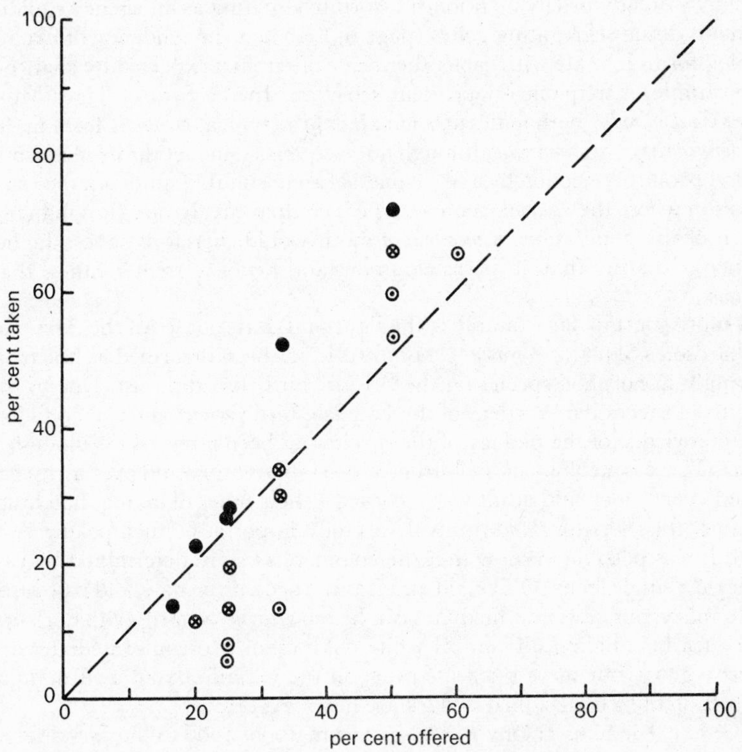

FIG. 44. Predation by the Rudd (*Scardinius eryopthalmus*) on three colour forms of a corixid bug (*Sigara distincta*). Solid circles represent the least cryptic type; circles with dots the most cryptic; circles with a cross are intermediate forms. Each type was taken more frequently than expected when common, and less frequently than expected when rare (after Clarke, 1962, based on data of Popham, 1942).

Sigara bugs vary in colour from light to dark brown. The particular shade is determined during development, and depends on their background. Popham offered fish different colours of prey in equal numbers: the Rudd always took a higher proportion of the more conspicuous insects. When different proportions of matching and conspicuous bugs were offered, the more conspicuous ones were always taken preferentially but the difference in survival between conspicuous and matching changed: every phenotype was more favoured than expected at low frequencies, and less so at high frequencies, *i.e.* the predation was dependent on the relative frequencies of the prey phenotypes.

Assortative mating and frequency dependence

We have already briefly mentioned assortative mating as an agency capable of maintaining segregating genes (page 65). In fact the tendency of like (or unlike) forms to mate with each other more often than expected by chance is an example of frequency-dependent selection. In the Scarlet Tiger Moth (*Panaxia dominula*) both males and females of the typical *dominula* form prefer *medionigra* mates to *dominula* although not vice versa – in fact the mechanism is probably one of rejection by a like female of an unsuitable suitor. In the usual situation where the *medionigra* phenotype is comparatively rare (less than one in ten of the population) a *medionigra* moth would therefore have a higher chance of mating than if it was common, and hence a greater fitness than *dominula*.

A more spectacular example is the light and dark phases of the Arctic or Richardson's Skua (*Stercorarius parasiticus*). It has been described as 'the most strikingly dimorphic' species on the British (bird) breeding list. Due to the labours of successive Wardens of the Fair Isle Bird Observatory, a considerable knowledge of the biology of the species has been amassed (Williamson, 1965). There is a colony of the bird near the Observatory and over a 15-year period every chick and adult was recorded – their times of laying, hatching, fledging, their territory and survival, but most important – their colour. As a result it was possible to show that the colour phases are determined by two alleles at a single locus (O'Donald and Davis, 1959). True dark and pale birds are homozygous, and intermediates can be recognized as birds with dark tips to the feathers hiding all ventral white coloration. Most intermediates are heterozygotes, but there is a wide range in the variation, and a substantial proportion of birds classified as dark are heterozygous.

The Fair Isle Skua colony was extinct from about 1906 to the early 1920s. From 1935 and 1947 8 to 12 pairs bred each year, and then in the period of intensive study it increased from 40 breeding birds in 1949 to 142 in 1962. Throughout the time the bird was studied, pale birds comprised 21% of the total.

What is the reason for the pale-dark dimorphism? The frequency of pales

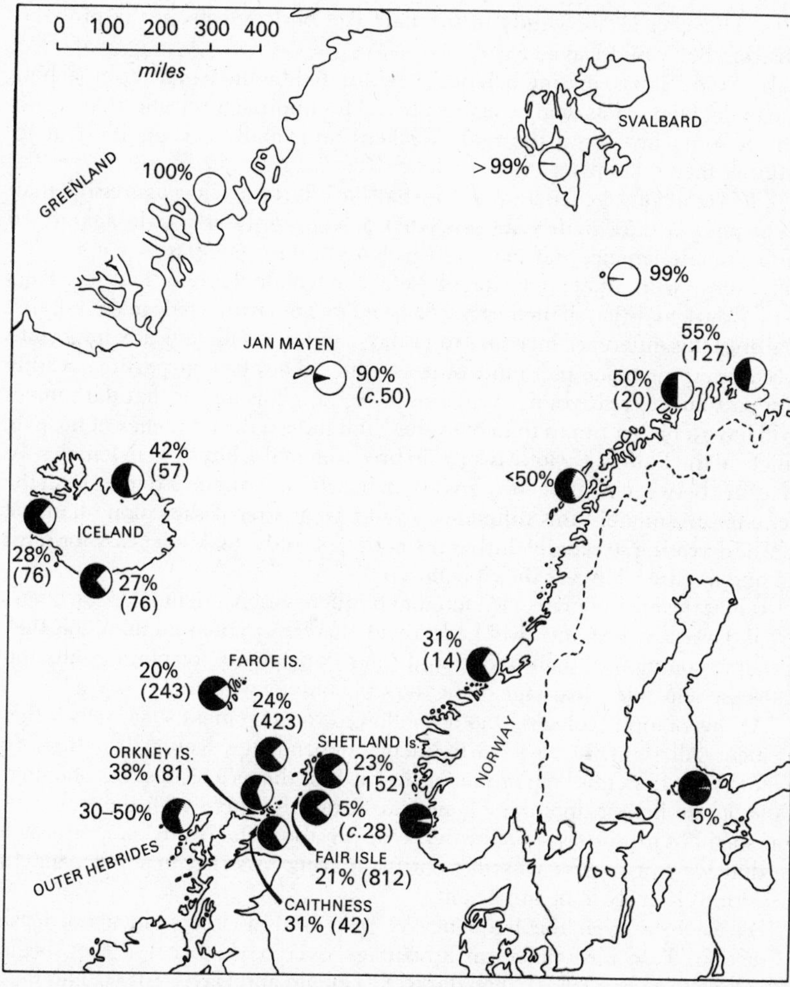

FIG. 45. Map of the North Atlantic to show part of the cline in colour phases in the Arctic Skua (*Stercorarius parasiticus*). For each sample the proportion of pale birds is represented by the white portion of the circle. The sample size (where known) is indicated alongside the circle (from Berry and Davis, 1970).

increases in a northerly direction, and northernmost colonies have few or no darks. It has been suggested that the pales may be at an advantage in hunting lemmings over a snowy tundra, whilst dark and dark-intermediate might be more cryptic over the ocean where they parasitize terns and smaller gulls by harrying them until they yield up their food. There is no evidence for either idea. However a close study of the Fair Isle birds has given the probable answer (Berry and Davis, 1970).

In Arctic Skuas, mating behaviour is initiated by the female approaching the male, who reacts angrily against her. This approach-rejection cycle goes on for some time until the male weakens and finally accepts the female. Mating then takes place.

The significant genetical point is that dark birds are less aggressive than light ones: a dark male does not react as vigorously as a pale against an importunate female, and matings involving a dark male are set up earlier than those with a pale one (the phase of the female does not matter). Pairs with a dark male hatch their eggs 3 days before ones with a pale male. For first pairings this difference increases to 11 days, and this is biologically important because Arctic Skuas pair more or less for life. (They change partners readily if a pair fails to rear young: Williamson, 1959). This means that dark males will have a higher fitness than pale ones, and indeed the frequency of the pale allele in the Fair Isle colony is 43% in breeding males but 51% in females. In matings between phases, 60% involve pairs where the male is darker than the female. Presumably this difference would be greater if the colony had not been increasing so rapidly during the period of study, making it easy for birds to find a mate whatever their phenotype.

It is at the level of mate selection that frequency-dependent selection comes in: if darks are rare, they will have a wide choice of potential mates; as they become commoner, their choice and their opportunity for mating fall, and thus the selective advantage of the dark phenotype will vary.

At this point a biologist who has followed the argument so far gets rather restless. All right, he says, so frequency-dependent selection is acting. So what? How does this help or hinder the birds in the serious business of living? And, to be honest, too many genetical stories stop here with the geneticist satisfied but no one else much wiser. One joy about the Arctic Skua situation is that we can suggest why disassortive mating (and frequency-dependent selection) is likely to be important.

As we have seen the frequency of pale birds increases in a northerly direction. Pale birds have an advantage over darks because they begin breeding at a younger age than darks (O'Donald and Davis, 1975). This may be another consequence of their importunate aggressiveness. It means that a pale bird has a higher chance of surviving to breed than a dark. However another factor comes in. The higher the proportion of pales in a colony, the later will be the breeding season. Now in the north of its range, Arctic Skuas feed largely on lemmings and voles which become common late in the summer; further south it has to adjust its time of breeding to that of the gulls and terns on which it preys, and they in turn will breed at a time when food and climate are best for raising their own broods. The dark/light segregation is really a mechanism for varying the time of peak breeding which happens to result in a colour difference useful for field workers.

PLATE 7. *Local forms of two moth species*

Above, Oak Eggar *(Lasiocampa quercus)* morphs from Caithness: *left*, typical male; *right*, *f. olivacea* (page 164). *Below left*, inherited variation in the Scarlet Tiger moth *(Panaxia dominula)*. Top, typical form; bottom, *f. bimacula* (homozygote); the other specimens are less and more extreme forms of the heterozygote *(f. medionigra)* (page 225).

PLATE 8. Variation in specimens of the Marsh Fritillary *(Euphydras aurinia)* from a Cumberland colony. Top row, original form; specimens caught in 1899. Second and third rows, extreme variants during a time of increase and high numbers; specimens caught in 1922–6. Bottom row, form subsequent to period of commonness; specimens caught in 1933.

Effect of inherited modifiers on dominance in the Currant Moth *(Abraxas grossulariata)*. 1, 2, homozygous typical form *(f. typica)*; 3–6, heterozygotes, with extreme specimens overlapping both homozygotes in appearance; 7–9 homozygous *f. lutea* (page 189).

One of the foibles of geneticists is that they will persist in talking about probabilities and second-order effects (like frequency-dependent selection), and this turns off simple minded naturalists who merely want to know what is going on in natural populations. The Arctic Skua situation has given rise to a mass of valuable theoretical studies (*e.g.* O'Donald, Wedd and Davis, 1974) which may provide the jumping off points for understanding other polymorphisms. Notwithstanding, geneticists have been far too prone in the past merely to identify inherited variations and describe their properties, and not go the extra step of discovering the *biological* significance of the phenomena they are discussing. Dear reader, err not in this way.

A botanical example of assortative mating which need not concern us at this point is heterostyly and related to it, the whole system of self-incompatibility alleles (page 186). Another situation we have already discussed (page 110–12) is selection for reproductive rate in birds and other organisms (Gadgil and Bossert, 1970; O'Donald, 1972). Batesian mimicry depends on frequency-dependent selection, as do other types of predator-prey systems such as the relation between hosts and parasites, and organisms and pathogens (Haldane, 1949; Damian, 1964).

Density-dependent selection

When a germ or a parasite spreads through a population, different individuals will respond differently to it because of their different metabolic and immunological capacities (*i.e.* their inherited variation). The germ will be successful against common types, but less so against variants which it rarely encounters – in other words, we are likely to succumb to an infection unless we have some rare quirk which protects us. For example, the 'Dutch Elm' (*Ulmus* × *hollandica*) is a hybrid with several parental species, and as such exists in a wide variety of forms. Specimens similar to the 'English Elm' (*U. procera*) had a much higher mortality to the elm disease of the early 1970s than typical 'Dutch Elms'. In epidemic conditions rare host variants will be at an advantage because of their rarity, and will increase in frequency. When this happens, a germ will have more chance of changing its weaponry so as to pierce the host's defences, *i.e.* it will be more successful as the host variant becomes commoner. Consequently the variant's selective advantage will decrease as its frequency increases. Haldane suggested that much of the biochemical diversity of a population may be the result of differing resistance to strains of disease organisms.

Myxomatosis resistance in rabbits is probably an example of frequency-dependent selection but the evidence is unclear because of changes in rabbit density, the virulence of the virus, and the difficulty of distinguishing between inherited and acquired immunity. When the virus first spread across Britain in 1953, 99% of an estimated 60 to 100 million animals in the country were killed. Most of the survivors contracted the disease but recovered. There have

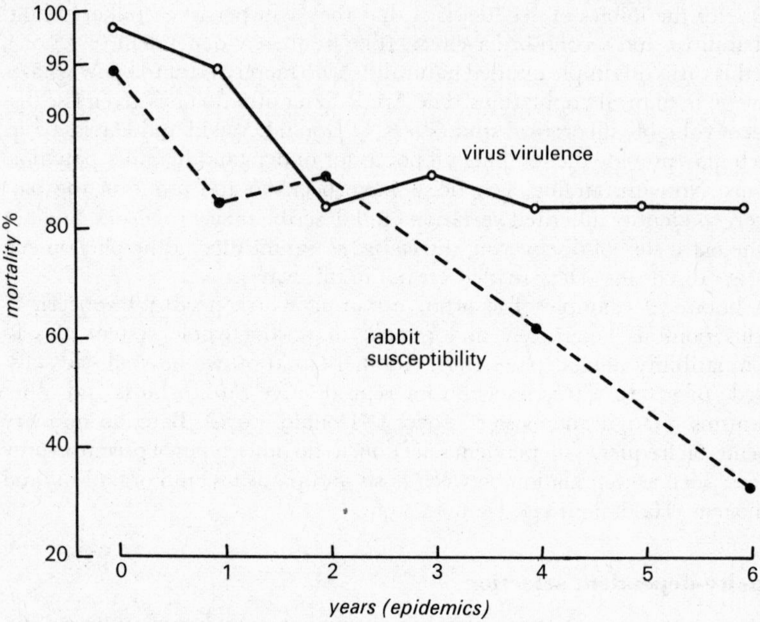

FIG. 46. Changes in the virulence of myxoma virus and in the genetical susceptibility of rabbits in Australia since 1950. Virus virulence is measured by the mortality produced by strains collected in the field on different occasions, and tested on standard laboratory-bred rabbits; rabbit susceptibility is that of young rabbits caught in areas which had experienced the number of epidemics shown and infected (after any maternal immunity had disappeared) with a virus strain isolated about a year after the first epidemic, and known to kill 90–95% of normal laboratory rabbits (after Macfarlane Burnet and White, 1972).

been several further epidemics but on a much more local scale, and the proportion of survivors has increased every time – partially due to a change in the virus, and partially due to the decreasing susceptibility of infected rabbits.

There are many fewer rabbits now than there were in pre-myxomatosis days. Consequently the disease does not spread as easily as it did in 1953, and there is not the same selection pressure for resistance as there was in the early days – a rabbit may escape infection as a consequence of low population density. In other words we have another way in which selection coefficients may vary: there is *density-dependent* selection as well as *frequency-dependent* selection (Turner and Williamson, 1968; Clarke, 1972).

Perhaps it is easier to think of the environment than the population varying. When we think about density-dependent selection, we are concerned with the interaction between the individuals in a population and its environment. If this interaction is lessened, the selection pressures may change. This is a difficult situation to investigate in nature because it depends usually upon the unpredictable – an unexpected environmental change. Ford and Ford (1930) studied a colony of the Marsh Fritillary butterfly (*Euphydras aurinia*) for 19 years. Records and specimens from the previous 36 years had been left by collectors. The butterflies lived in poorly-drained fields outside Carlisle, and were separated from other colonies by several miles of woodland, heather moor, and agricultural land. From 1881 (the earliest records) to 1897 the butterfly was exceedingly common and occurred 'in clouds'. From 1906 to 1912 the colony was reduced in numbers, and became rare until 1920 (probably as a result of parasitism by a Braconid Fly, *Apanteles spurius*). From 1920 to 1924 a great increase of numbers (and density) took place, so that several could often be caught by a single stroke of a butterfly net (Ford and Ford, *loc. cit.*). From 1925 to 1935 (when observations ceased) the numbers remained high.

The amount of variation in wing pattern was small during both the first period of abundance and the phase of rarity. In contrast when the numbers were rapidly increasing after 1920 an extraordinary burst of variability took place: marked departures from the normal form of the species in colour-pattern, size, and shape were so common that hardly any two specimens were alike. A considerable proportion of these were deformed, with the amount of deformity being associated with the degree of variation. The more extreme departures from normality were clumsy on the wing, or even unable to fly. By 1925 the rapid increase had ceased: the undesirable types practically disappeared and the population settled down to an approximately constant form – which was recognizably distinct from that which had prevailed during the earlier period of abundance. As the environmental constraints on numbers were reduced (*i.e.* as selection slackened), new variants were able to persist and spread to the extent that the genetical composition of the population was different when the environment again became limiting. The selective coefficients at different loci changed with the density of the population.

NATURAL SELECTION IN SNAILS

We cannot end our description of the action of selection on natural populations without describing the work that has been done on species of the Land Snail *Cepaea*. We have already seen that one species (*Cepaea nemoralis*) used to be thought to show random effects in the distribution of its gene frequencies (page 55). It is now time to consider the evidence for the operation of selection on *C. nemoralis* and its close relative *C. hortensis*.

The shell of *Cepaea nemoralis* may be brown, pink or yellow. The colours are determined by a locus which is closely linked to one at which alleles control the presence of up to five bands (rarely six) running round each whorl of the shell. These are numbered from the top downwards, with the absence of a

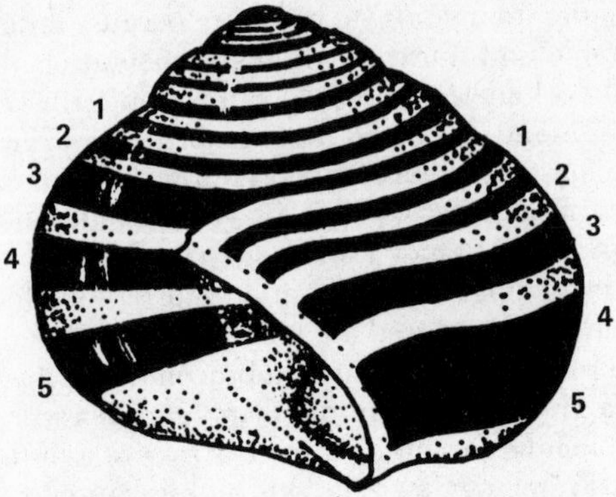

FIG. 47. *Cepaea* shell, showing the convention for numbering the bands.

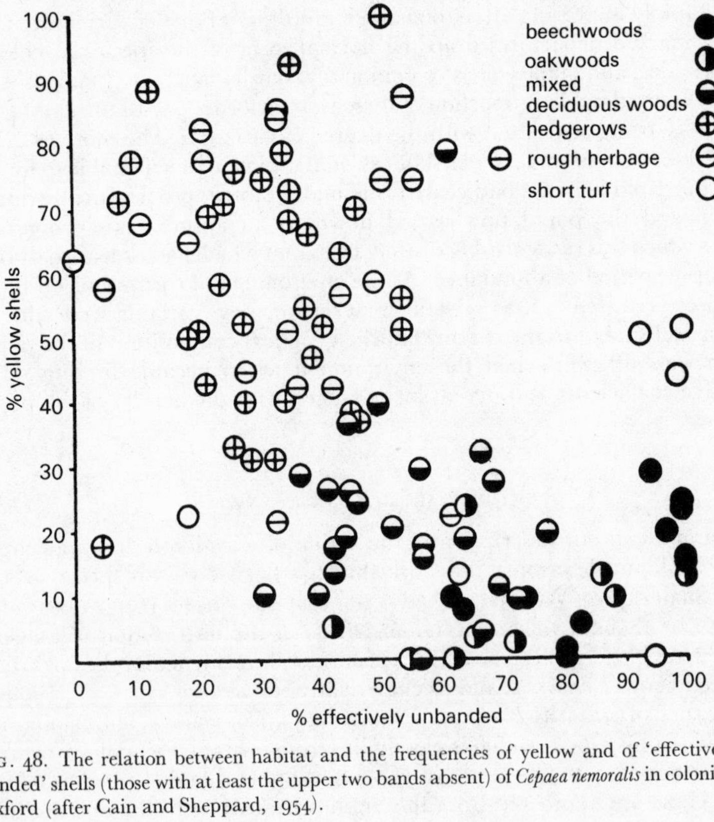

FIG. 48. The relation between habitat and the frequencies of yellow and of 'effectively unbanded' shells (those with at least the upper two bands absent) of *Cepaea nemoralis* in colonies near Oxford (after Cain and Sheppard, 1954).

band recorded as 0. Thus the condition with all five bands present is described as 1 2 3 4 5; the bandless (or unbanded) form as 0 0 0 0 0; and the mid-banded as 0 0 3 0 0. Unbanded is dominant to banded, and alleles at another locus control the number of bands.

At about the same time that Kettlewell was beginning his work on natural selection in the Peppered Moth, Cain and Sheppard (1950, 1954) were starting their studies on *Cepaea*. They found that colonies with diverse proportions of shell colours and banding occurred within a few square miles in the Oxford district, in an area where patches of woodland and rough herbage are scattered in a countryside often crossed by hedgerows. More important, the phenotypes of the snails broadly matched the backgrounds where they were living. On a uniform background, the majority of shells was unbanded, whilst on a variegated one the majority was banded; in woodland most specimens were brown and banded, in hedges most were yellow and unbanded.

Cepaea is eaten by small mammals and birds. Where it is present the chief predator seems to be the Song Thrush (*Turdus ericetorum*). This has the valuable habit for biologists that it picks up snails and carries them to a convenient stone ('thrush anvil') to break open the shells. Here the fragments accumulate, so that it is possible to record the snails which the thrushes in the neighbourhood are eating. Sheppard (1951) collected these snail relicts regularly over a 3 month period (April to June) in two Oxford woods. At first a high proportion (42%) of the broken shells were yellow, but as the grass grew and the leaf litter of the previous year was covered, fewer yellows were found so that by the end of the collecting only 22% of the predated snails were yellow. To eliminate the possibility that yellow was becoming rare at the same time in the natural population (*i.e.* before the thrushes acted), Sheppard liberated in one of the woods 1358 snails (747 yellows, 611 pinks) marked with a small dot of paint. He found that the proportion of colours from the marked animals which turned up on the anvils were precisely the same as those from the unmarked population. There could be no doubt that the thrushes were removing the more conspicuous snails from the population in the same way as birds removed more conspicuous moths in Kettlewell's experiments. Greenwood (1974) showed similar effects in a region of wood and farmland south-west of Worcester (see also Carter, 1968; Arnold, 1970). Cain (1953) has pointed out that mammals which hunt by sight may have a similar selective effect, although they will select by tone rather than colour since they are colour blind. O'Donald (1968*a*) found Glow-worms (*Lampyris noctiluca*) on sand-dunes in Anglesey preferred browns to yellows, to the extent that browns only occurred in areas where there were no glow-worms.

The resemblance of snails to their background is best documented for *Cepaea nemoralis*, but it has been shown for a number of other species. For example, there is a small snail *Hygromia striolata* which is a frequent pest of strawberry beds, and which Jones, Briscoe and Clarke (1974) found to match its backgrounds in woods, fields and hillsides around Malham in Yorkshire.

H. striolata, like *C. nemoralis*, is eaten by thrushes, and these presumably exercise a selective influence. A rather different example is the Rough Periwinkle (*Littorina saxatilis*) which abounds on rocky-shores. This is commonly said to vary randomly in colour, but careful study has shown that this is not so. The key to the Rough Periwinkle situation was the recognition that it consists of not one but four different species (*L. rudis*, *L. patula*, *L. nigrolineata*, and *L. neglecta*), each with different habitat preferences and breeding characteristics (Heller, 1975*a*). When this distinction was made, it was realized that in *L. rudis* red morphs are found predominantly on shores composed of red sandstone rocks (where it is eaten by Rock Pipits, *Anthus spinoletta*), while both *L. nigrolineata* (which occurs further down the shore) and *L. rudis* have high frequencies of yellow morphs on sheltered shores where they are inconspicuous on the fronds of sea-weed. White forms of *L. nigrolineata* increase relative to red on exposed shores, where barnacles are common and give a superficial white appearance whatever the underlying rock (Heller, 1975*b*).

So far so good: gene frequencies are clearly adjusted by visual predation according to a simple selective mechanism. Furthermore specimens of subfossil *Cepaea nemoralis* collected from archaeological sites show roughly similar frequencies to modern snails found in the same place, suggesting that comparable selection pressures have operated for thousands of years (Cain, 1971). Even the opposite situation may help. For example, marked changes in frequencies were found by Clarke and Murray (1962) when they collected snails from sand-dunes from Berrow in Somerset in 1960 and compared them with samples collected by the pioneer British conservationist, Captain Diver,

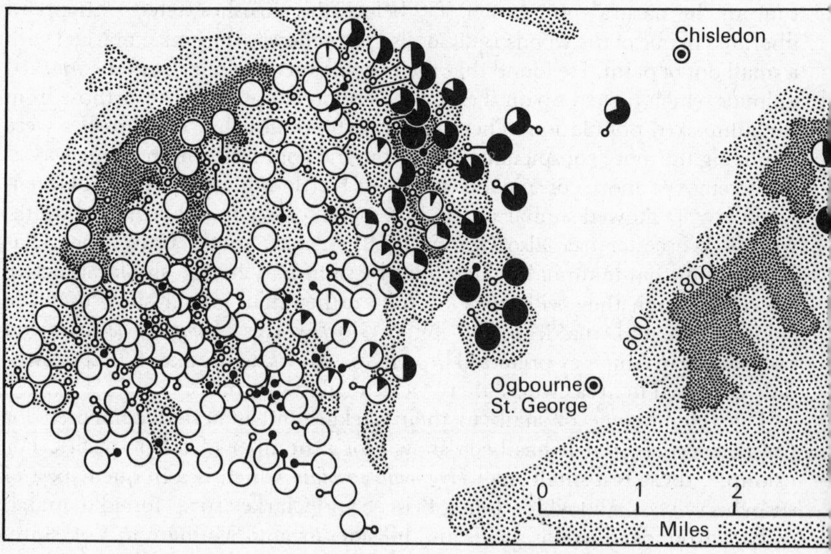

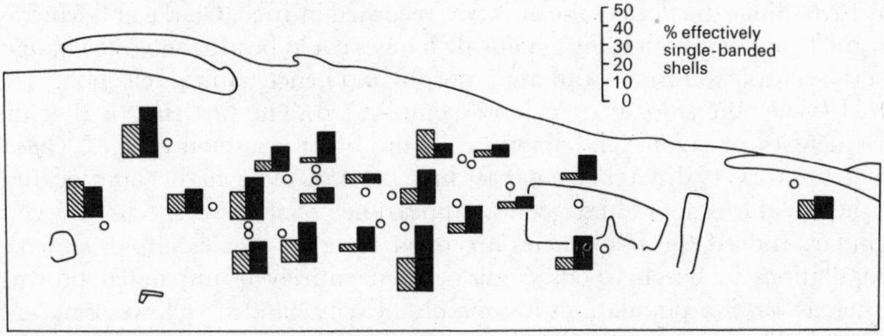

FIG. 49. Changes in gene frequency of shell banding in *Cepaea nemoralis* on sand dunes at Braunton Berrow in Devon. The striped column indicates the frequency of single-banded snails in samples collected by Cyril Diver in 1926; the black column the frequency of the same type in 1959 (after Clarke and Murray, 1962).

FIG. 50. 'Area effects' in *Cepaea nemoralis* on the Marlborough and Berkshire Downs. The black sector in each large circle represents the frequency of five-banded shells. The small circles show where each sample was collected – black circles, woodland populations; white circles, populations living in open habitats. There is no association between morph frequency and either habitat or topography (after Jones, 1973*a*).

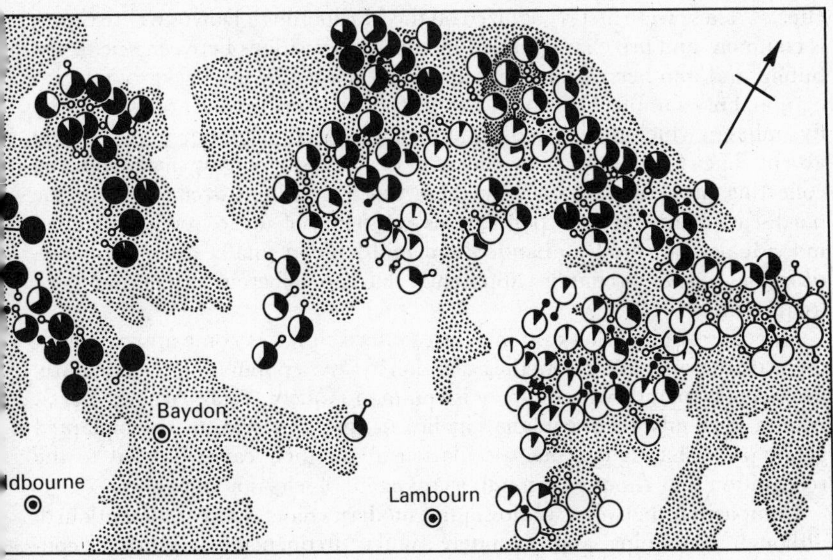

in 1926. Since the flora of the area was recorded in 1926, Clarke and Murray argued confidently that the genetical changes could be attributed to changes in the conspicuousness of different morphs and hence natural selection.

However the story is even more complicated. The first snag is that the frequencies of colour and banding in the other common species, *Cepaea hortensis*, are very different from those in *C. nemoralis* living in the same locality, although the genetics and general appearance of the two species are very similar. Indeed the frequencies are often inverted – for example, *nemoralis* populations in beech woods consist almost entirely of unbanded browns, whereas *hortensis* populations are mainly heavily banded yellows. Pink and brown *hortensis* are usually comparatively rare when compared with *nemoralis*, and this rarity has little or no relation to the conspicuousness of the shells (Clarke, 1960). Notwithstanding there is a fair degree of correlation in *hortensis* between vegetation and shell variation.

In neither British species of *Cepaea* is the matching between snails and background particularly good: pink shells stand out in virtually any habitat; five-banded forms would usually be better hidden if the bands were not so different from each other; while mid-banded shells (0 0 3 0 0) often have a white stripe below the brown band, which serves to accentuate the band. These considerations led Clarke (1960, 1962) to suggest that selection might be operating to make the varieties look distinct from each other. With two species like *nemoralis* and *hortensis* hunted by the same predators, selection will act in one species against any form which is common in the other. In other words selection will be frequency-dependent as well as directed.

The other main complications about *Cepaea* morph distribution are 'area effects'. These were first recognized on the Marlborough Downs where *Cepaea* is common, and are characterized by abrupt transitions between one distribution and another without any apparent change in the background vegetation. For example, Cain and Currey (1963*a*) described one area of three by five miles in which the normally common 5-banded phenotype was virtually absent. They found only 11 specimens of this form out of 5767 snails from 103 collecting sites. In stark contrast there was a transition over a mile at the north-east of this area to populations consisting of up to 100% 5-banded individuals, with both unbanded and mid-banded totally absent. In such situations visual predation apparently has no effect in regulating gene frequencies.

Since the original description of 'area effects' it has become apparent that these are the rule rather than the exception in *Cepaea* populations, particularly from chalk uplands where *Cepaea* is common (Carter, 1968; Arnold, 1971). Almost certainly their explanation lies in the conservatism of coadapted genotypes (Chapter 5), but speculation about their cause has led to the recognition that climate is an important agent of selection in *Cepaea*.

Comparisons between sub-fossil and modern colonies already mentioned, although containing approximately similar frequencies of morphs, con-

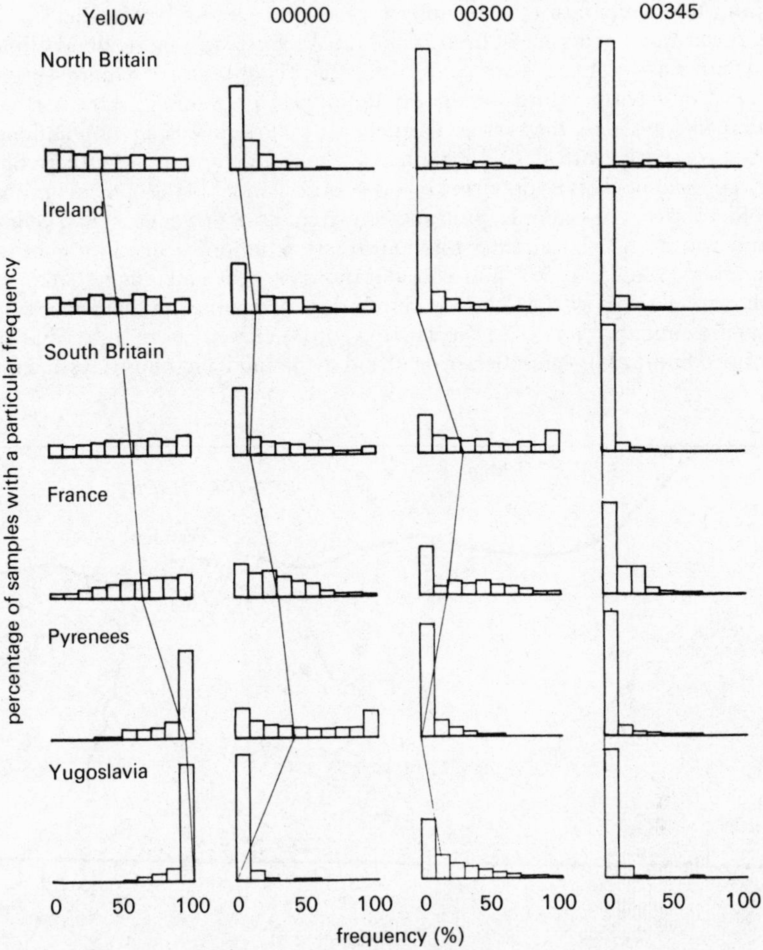

FIG. 51. Morph frequency histograms for populations of *Cepaea nemoralis* throughout its range. Vertical lines connect the median frequencies for each region. Only the yellow morph shows a significant change with latitude (from Jones, 1973*a*).

sistently show that yellow and unbanded was more common 7000 years ago than nowadays (Cain, 1971). Pollen analysis has shown that there was a 'climatic optimum' in the late Boreal and early Atlantic periods (5000–4000 BC), since when there has been a deterioration to our present damp and cool 'Sub-atlantic' period. Partly on these grounds and partly from local distribution of different morphs (in hollows and on slopes), Cain and Currey (1963*b*) suggested that the brown phenotype might be better able to withstand low temperatures than other forms. However no brown shells at all occur at the northernmost limit of the range of *nemoralis* in Scotland (Jones, 1973*b*). This led Jones (1973*a*, *c*) to argue that yellows may be a more sensitive indicator of temperature tolerance than browns: for example, at the northern limit of *C. nemoralis*, the average frequency of yellow in *nemoralis* populations is 16% compared with 53% in Britain as a whole, and 95% at its southern limit in Yugoslavia. This is supported by other observations of differences between colour forms. For example, pink and brown snails lay fewer eggs than yellows in a warm and dry summer although there is no difference in a cool wet summer (Wolda, 1963), and considerations of heat gain suggest that unpigmented shells will be favoured in situations where the snails are exposed to intense sunlight (Jones, 1973*a*; Heath, 1975*b*). Richardson (1974) has described finding large numbers of dead and dying snails on sunny days on sand-

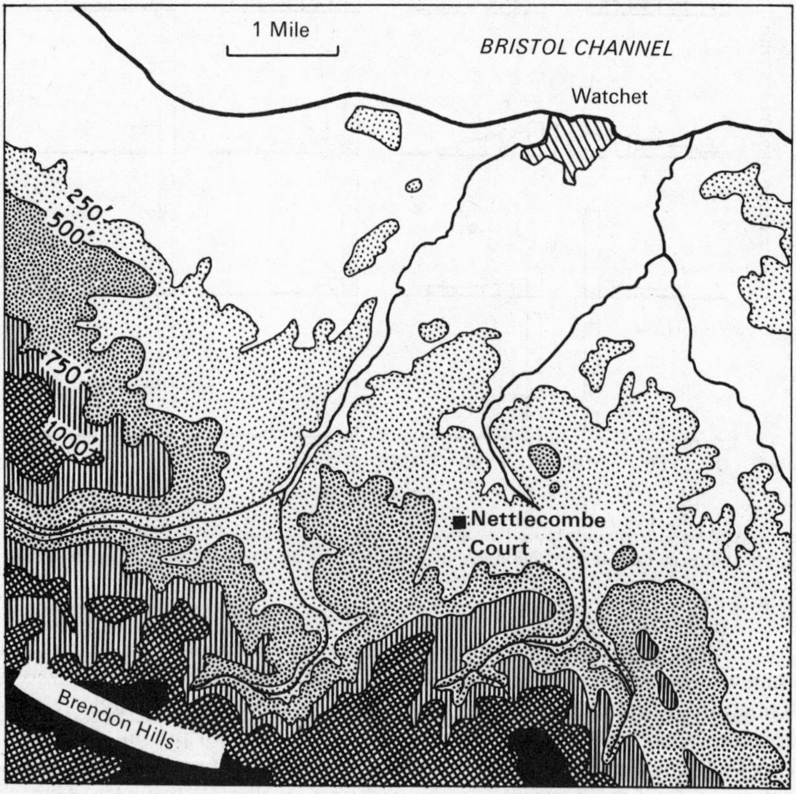

FIG. 52. Map of the area round Nettlecombe Court (after Bantock and Price, 1975).

dunes at Braunton Burrows in Devon. Apparently the upper limit of temperature tolerance can be fairly easily reached. The interesting observation in this unfortunate population was that yellow and unbanded were always less common in the dead snails than in the living ones in the area.

Bantock has sought more direct evidence of differences in climatic tolerances between morphs in a series of experiments at the Leonard Wills Field Centre at Nettlecombe Court in north Somerset (Bantock and Noble, 1973; Bantock, 1974*a*, *b*; Bantock and Price, 1975, and unpublished). He began by making an intensive survey of the phenotype frequencies of both *Cepaea nemoralis* and *C. hortensis* in an area two by four miles between the Brendon Hills and the Bristol Channel. *Cepaea hortensis* was common throughout the area, and there were marked differences in morph frequency in different colonies which were correlated with altitude and could be most easily explained by the different climatic conditions of the valley bottom and sides, together with the change from the maritime climate near the coast to the more variable one only a short distance inland. For example, the growing season is two months longer at Old Cleeve than in the Nettlecombe Valley less than three miles away; there are frequent temperature inversions in the steep valleys of the Brendons, and frost may occur in any month on the valley floors (Ratsey, 1973).

Cepaea nemoralis reaches the limits of its distribution along the sides of the valleys and presents an unusual distribution of phenotypes. The pink morph is by far the commonest (88%), 75% of these being mid-banded. The common forms in Britain are yellow and five-banded: only six of each were found in a total of 1328 individuals collected, although both these morphs are common in regions to the north-west where *nemoralis* is abundant. These peculiarities suggested that if selection is important in regulating gene frequencies in *Cepaea*, they should be readily detectable under these situations.

Bantock (1974*a*) began by setting up cages for snails at the bottom of the valley near Nettlecombe Court. The cages were made of a mesh sufficiently small to keep out vertebrate predators, but open enough to allow wind and rain to act on the snails. During the summer of 1973, survival in the cages of different morphs of *nemoralis* taken from the maritime region a few miles to the north-west was very uneven. Brown unbanded survived best (25% survival), and five-bandeds (13%) and yellow unbandeds (0%) worst. In further experiments with cages on the valley slopes up to 200 feet above the lower ones, Bantock found that the direction of selection differed between sites but not between cages at the same site, and also that it differed at different times of year. Furthermore when he put *C. hortensis* and *nemoralis* together, inter- and intra-specific differences in survival occurred with the density of the experimental populations.

A similar type of result was obtained using *nemoralis* taken from the island of Steepholm in the Bristol Channel (Bantock 1974*b*). There were no survival differences between morphs maintained for 3 years in a large enclosure, but great mortality occurred when the snails were crowded into a small cage, with yellows surviving better than pinks (particularly during the very hot summer of 1975). While the selective agent or agents has not so far been identified in any of these experiments, the results indicate that, in addition to any visual selection exerted by predators, other and complicated selection pressures must operate on the different morphs (Bantock & Bayley, 1973).

The literature and speculation about genetical influences on *Cepaea* is vast. Perhaps the most important lesson from the research has been the complexity of the forces that control gene frequencies. It is still sometimes said that natural selection is 'merely' a sieving procedure for eliminating disadvantageous mutants. Anyone with a knowledge about the situation in *Cepaea* has to recognize that different genetical forces can operate at the same time, and that they may change over a short span of time or space, so producing a fine adjustment between population and environment(s). There is still more to the *Cepaea* story and we shall have to return to it in the next chapter. Suffice for the moment an acknowledgement that natural selection cannot be conveniently slotted into a series of neat pigeon-holes labelled directed, stabilizing, disruptive; frequency and density-dependent; balanced and transient; constant and fluctuating; and probably a few others. When we talk about natural selection we are describing a vast range of interactions between a population's potentialities and an environment's effects which occurs in the present but is limited by the experience and history of the population. Natural selection is only one of the five forces that may change the genetical constitution of a population (page 49), but is so much the most important that it merits the extended discussion that it has had in this chapter.

ARTIFICIAL SELECTION

Our main concern is to learn what goes on in natural populations, but we must note the effects of selection in agricultural and laboratory practice, because of its economic and practical effects. The 'green revolution' is only one example of the application of genetical techniques to commercial practice, and political consequence.

There is no essential distinction between artificial selection and the examples of natural selection we have already described. The main differences are in the techniques and statistics which have been developed for animal and plant breeders but which have not been fully applied to investigations of natural situations. The reason for this is that commercially and economically important traits tend to be determined multigenically (*i.e.* by a number of genes acting together) and to have a substantial environmental component in their causation. In contrast most of the characters that have been studied in natural populations have been ones controlled by single genes, for the simple reason that they are the easiest to follow under conditions where so many data (such as population sizes, growth rates, numbers of births and deaths, and so on) have to be estimated or guessed. This means that artificial selectionists tend to work with the mean and variance of the characters they are interested in, rather than the gene frequency itself. Their techniques are biometrical rather than Mendelian.

Once this distinction is realized, the complicated procedures of animal and plant breeders (to the extent that they are based on rational procedure rather

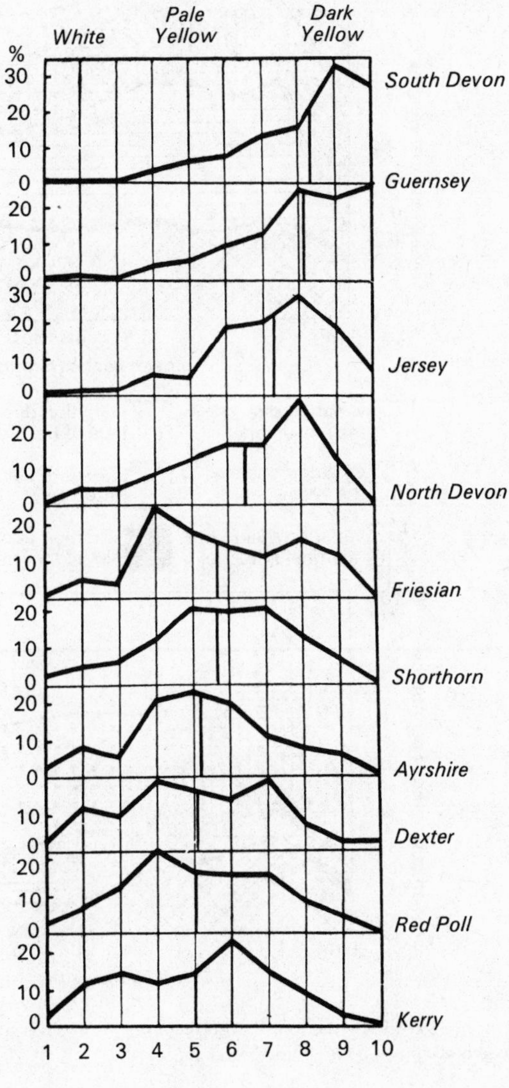

FIG. 53. Variation of butterfat colour from individual cows in breeds exhibited at the London Dairy Show. The colour is affected by both diet and genetical constitution (from Hammond, 1971).

than antique lore) are forced on them by the difficulties of their material rather than a perverted sense of complexity (Fisher, 1918).

The stock in trade of the practical breeder is the distribution of his character in his population. This character will have a mean value, and a spread or variance on each side of the mean. Now the phenotypic variance (V_P) is the

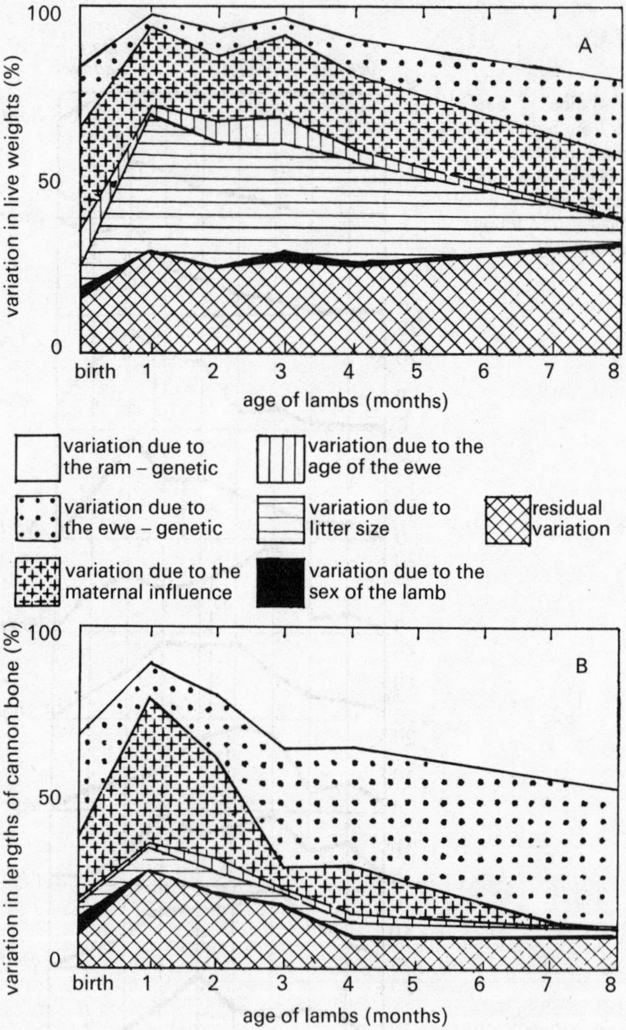

FIG. 54. Partition of the contributing factors to the total variation in lamb size: A, live weight; B, length of cannon bone (from Hammond, 1971).

sum of two independent variances, one genetic (V_G) and the other environmental (V_E):

$$V_P = V_G + V_E$$

Selection for either extreme will be more effective if the environmental variance is small and if there is sufficient variation for the individuals selected to form the parents of the succeeding generation to be distinct from the mean of the population into which they were born. This gives us another measure – selection differential (S), which is the difference between the mean of the population and the mean of the selected individuals.

In any selection, it will be possible to measure the response (R) as the difference between the offspring of the selected individuals and the population from which they were drawn. Different breeds or varieites have different distributions of valuable (or detrimental) traits, and will therefore respond differently to selection. The ratio R/S defines the heritability (h^2) of a trait. This is an unfortunate name because it does not measure the extent to which a trait is inherited, but only its characteristics in a particular environment. It is failure to recognize this that has led to a tremendous amount of sterile argument about the size and significance of heritabilities of intelligence in different human races.

During selection, extreme phenotypes are continuously favoured. This is likely to lead to an increasing proportion of individuals homozygous at the selected loci. Ultimately, therefore, the rate of response to selection will be expected to diminish, and there may be generations where there is no response. However what normally happens is that fecundity and viability fall as selection proceeds, until no further selection is possible. This is due to the accumulation of an unsatisfactory (or *unbalanced*) set of genes brought together because of their linkage to genes controlling traits being selected. These interactions are paradoxically responsible for both hybrid vigour (when it is not determined by a single locus, as in the case of sickle-cell haemoglobin) and inbreeding depression. We shall consider these topics further in Chapter 5.

It is not the function of this account to describe the ramifications and complexities of artificial selection (*q.v.* Lerner, 1950; Falconer, 1960; Mather and Jinks, 1971). Selection programmes have to be linked to performance and progeny testing, and to planned crossing between strains. No new principles are involved but it is relevant for those who have managed to follow the arguments of this book so far to connect the mass of phenotypic variation which keeps the applied geneticist in business with its genetical counterpart in the array of genic variation revealed by electrophoresis which has only recently come to light (page 20). The inherited variation which allows an agriculturist to produce new strains of domestic animals and plants was really rather mysterious until methods of classifying genic variation became available; prior to this we were forced to assume that variation existed, but we had no means of demonstrating it.

CHAPTER 5

GENETICAL PLUMBING AND ARCHITECTURE

WE have already seen that the distribution of the alleles of a gene is rarely uniform throughout the range of a species. When a change of gene frequency occurs, what Julian Huxley has named a *cline* of change in morph frequencies is said to exist. For example, the frequency of black Peppered Moths decreases from the Manchester–Liverpool area into North Wales; the frequency of pale Arctic Skuas increases from Scotland to the Arctic Circle; the occurrence of warfarin-resistant rats declines from a focus near Montgomery; and so on.

CLINES

A series of clines occurs in an interesting situation existing in White Clover (*Trifolium repens*) and Birdsfoot-trefoil (*Lotus corniculatus*). Some plants contain hydrogen cyanide, produced by the complementary action of alleles at two loci, and most populations are polymorphic for cyanogenic (cyanide-containing) and acyanogenic individuals. Daday (1954) obtained wild clover seeds from many areas of Europe, and showed that the frequencies of the cyanogenic phenotype depend on the mean temperature in January: a fall of 1°F in mean temperature leads to a 4% reduction in cyanogenic plants. As far as Birdsfoot-trefoil is concerned, slugs, snails and voles prefer to eat acyanogenic individuals, whilst cyanogenic ones are more susceptible to frost damage (Jones, 1967). Some slugs distinguish between the morphs in clover, others apparently do not (Angseesing, 1974). Since the predators will be commoner in warmer climates, this balance between palatability and environmental resistance might be the main cause for the clines in the two species.

We must look at the features and properties of clines in more detail. A cline may arise from several basic situations (Endler, 1973):

i. random genetic drift producing a frequency difference in one part of a population's range
ii. establishment of contact between two genetically distinct populations
iii. spatially discontinuous changes of environment
iv. continuous environmental gradients.

A cline produced by drift will disappear rapidly as gene flow takes place: for any stable differentiation to evolve as a result of drift the number of dispersing individuals between different breeding groups must be less than one per generation.

These considerations lead us to a distinction between clines: they may be stable or transient. If a new advantageous variant arises in a population, it

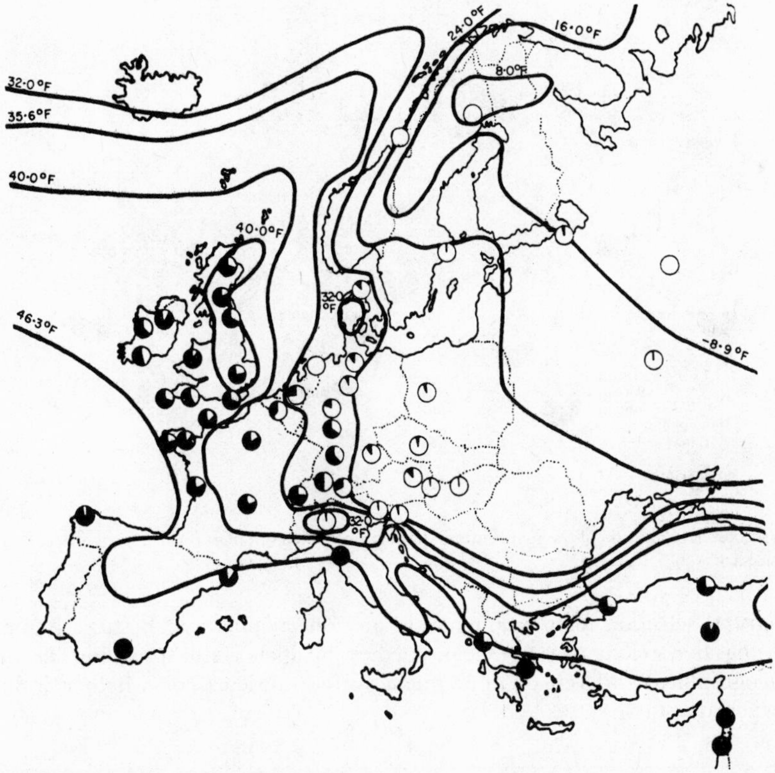

FIG. 55. Distribution of the cyanogenic (cyanide-containing) form of White Clover (*Trifolium repens*) in Europe. The black sector of each circle indicates the frequency of the cyanogenic form. The isotherms are mean January temperature (from Jones, 1967, modified from Daday, 1954).

will spread. The highest frequencies of the variant will be close to its origin, and it will become rarer with distance away from its origin. We have already described the spread of melanic Peppered Moths from the Manchester area in 1848 onwards. There is in man a haptoglobin allele (one of the α-globulin proteins in blood serum) (Hp^2) which produces a protein 142 amino acids long. This contrasts with the haptoglobin produced by the great apes and the other common allele in man (Hp^1), which is only 83 amino acids long. Clearly Hp^2 arose through a mutation involving a chromosomal rearrangement in an Hp^1 individual, and this mutation must have occurred after the separation of the human stock from the apes. It is now found in all human populations, with high frequencies in Asians and Europeans, lower ones in Negroes and American Indians. The mutation must have taken place early in the history of *Homo sapiens* and spread with the species to all parts, although possibly affected by

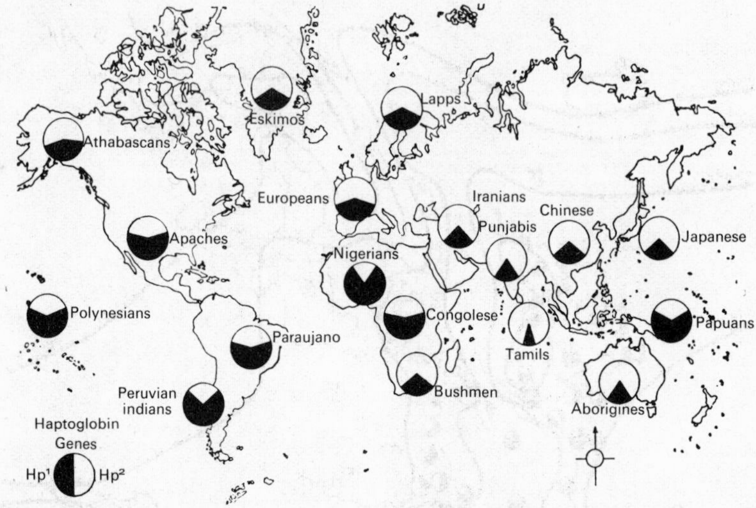

FIG. 56. Distribution of the two common haptoglobin alleles in man (from Harris, 1970, based on Kirk, 1968).

natural selection in its frequency in any one population. Because human beings live so long, it is not clear whether the allele is still spreading (*i.e.* the geographical differences in frequency are transient), or whether it has reached a stable state.

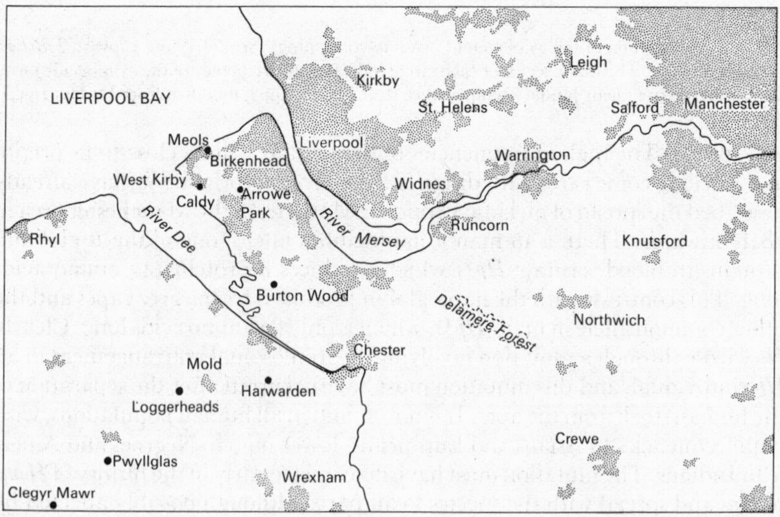

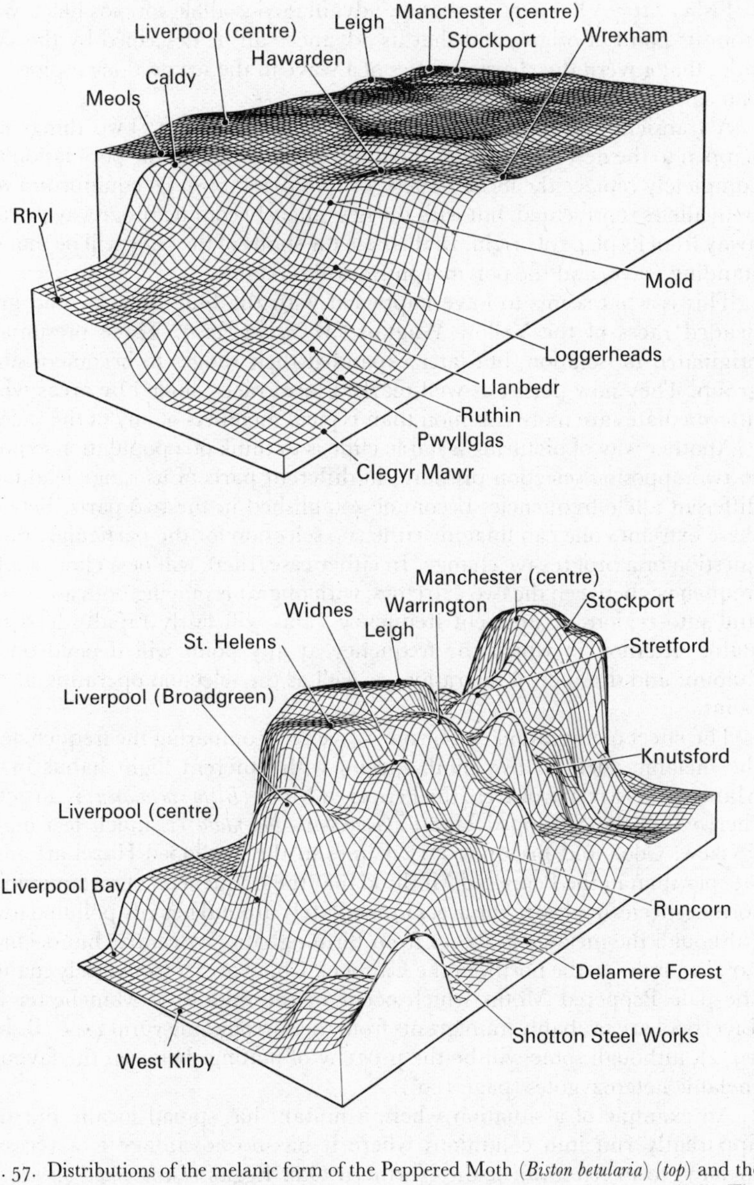

FIG. 57. Distributions of the melanic form of the Peppered Moth (*Biston betularia*) (*top*) and the Scalloped Hazel (*Gonodontis bidentata*) (*middle*) in the Manchester–Liverpool area (*left*). The 'contours' of the distribution graphs give the frequencies of the melanics as a proportion of the total. The graphs were drawn by computer, fed with information about frequencies at more than 100 sites (redrawn from original diagrams supplied by Dr J. A. Bishop).

Fisher (1950) has argued that an advantageous allele spreads like a wave from its point of origin, and that its advance can be described by the same rules that govern the characteristics of a wave in the sea or along a pipe – we can call such a situation genetical plumbing.

A transient cline will have a comparatively short life. Two things may happen to the new allele: either it will spread throughout the population and completely replace the former allele; or it may establish an equilibrium with both alleles represented, but with the new allele becoming progressively rarer away from its place of origin. In this case the shape of the cline will be that of a standing wave, and the polymorphism will be stable.

This is what seems to have happened with the blue-, black-, and grey-headed races of the Yellow Wagtail (*Motacilla flava*). These presumably originated in isolation, but later expanded their ranges to become a single group. They now persist as well-marked local races, linked by areas where intermediates are more common than typical members of any of the races.

Another way of picturing a stable cline is to think of a population exposed to two opposing selection pressures in different parts of its range leading to different allele frequencies becoming established in the two parts. Between these extremes one can imagine either no selection for the particular trait in question or a progressive change. In either case, there will be a cline of allele frequencies between the two extremes, with migrants moving both away from and into regions of different frequency. This will fairly rapidly lead to a stable situation in which the frequency at any point will depend on the amount and distance of migration, as well as the selection operating at that point.

The effect of migration can be nicely seen by comparing the frequencies of the melanic allele in two moth species with different flight habits in the Manchester–Liverpool area: the Peppered Moth (*Biston betularia*) is an active flier, while the Scalloped Hazel (*Gonodontis bidentata*) is much less mobile (Askew, Cook and Bishop, 1971). Clines in the Scalloped Hazel are much steeper than in the Peppered Moth, so that the frequencies of melanics in the former species are less in the areas immediately around heavily polluted towns (although the melanic is absent from Birmingham city centre, but occurs at 30% 15 miles to the north, in the Cannock Chase area). Conversely many of the pale Peppered Moths which occur in the middle of Manchester and Liverpool are probably immigrants from the surrounding rural zone (Bishop, 1972), although some will be the progeny of matings between the favoured melanic heterozygotes (page 109).

An example of a situation where a mutant has spread locally but then apparently run into conditions where it has no advantage is a recessive melanic form (*olivacea*) of the Northern Oak Eggar (*Lasiocampa quercus*) in Caithness at the north-eastern tip of mainland Scotland (Kettlewell, Cadbury and Lees, 1971). A similar form exists on moorland in the Huddersfield–Sheffield–Manchester triangle, and on sand dunes in south Lancashire and

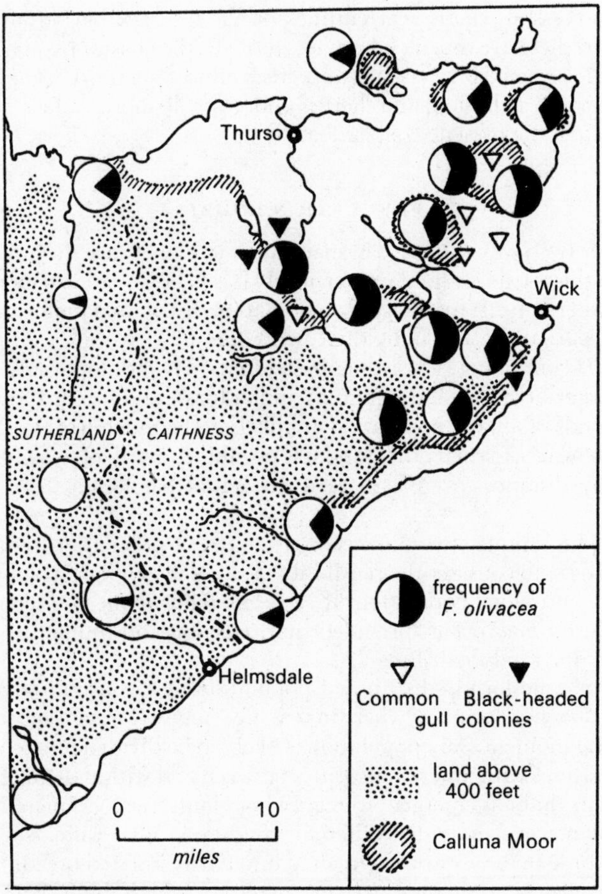

FIG. 58. Phenotypic frequencies of the melanic form (*olivacea*) of the Oak Eggar moth (*Lasiocampa quercus*) in north-east Scotland (after Kettlewell, Cadbury and Lees, 1971).

Cheshire, in both of which areas an industrial melanic might be expected to occur.

There can be no question of pollution in Caithness, yet the *olivacea* form occurs at frequencies of over 50% in several places on the coastal plain beyond the high hills of Caithness and Sutherland. Although Oak Eggars occur on heather moor to the south and east of the area, no melanics have been found. Black-backed and Common Gulls (*Larus ridibundus* and *L. canus*) are common in both the melanic and surrounding areas, and have been shown to catch more typical than melanic moths. Kettlewell has suggested that the melanic may be successful where peat cuttings provide a dark background for the moths, and these are much commoner in the fertile coastal plain area where the melanic occurs. However there are steep clines over short distances within the melanic's distribution, and there is undoubtedly more to the restriction of the form than the presence of peat banks.

POPULATION MODELS

The nearer two individuals are in space, the more likely are they to mate. We have seen this is true for humans: obviously it is likely to be even more true for animals and plants. If we imagine a population distributed uniformly over an area, the parents of any individual will come from within a circle with a diameter D containing N breeding individuals. The grandparents will come from a larger area $2D$ containing $2N$ individuals, and so on. Organisms at opposite ends of a large population are likely to be much less closely related than those nearer to each other, and it is possible to speak of two organisms as isolated by distance, even if there is no actual barrier to them breeding together.

It was this simple, two-dimensional idea of a population that first led Wright (1931) to consider the possibility of random genetical changes occurring in a continuous population: if N is less than about 100, considerable differentiation may occur, provided that selection and immigration do not overpower the random drift.

However a uniformly distributed population in a constant environment probably does not occur anywhere in real life – anyway not for more than the odd fleeting moment. Any population is likely to be divided into a number of sub-populations with different densities between and within them. This is one of the points that has emerged from growing plants in experimental gardens: related plants grown under different conditions may differ so much in flowering time that they are fragmented into partly isolated breeding groups. The same effect occurs if they grow in a heterogeneous habitat. Plants in each type of habitat will breed with each other, but will be genetically isolated from those in different habitats (Heslop-Harrison, 1964; Bradshaw, 1972). The converse of this micro-differentiation is that many plants in northern latitudes where there is a very short growing season, are 'forced' into flowering at the same time as species from which they are normally distinct. The result is that a host of 'illegitimate' crosses occur. For example, the Common and Marsh Ragworts (*Senecio jacobaea* and *S. aquaticus*) are virtually indistinguishable in the Orkney Islands (59°N), and hybrids between them are fully fertile.

Examples of fragmented species ranges abound, both on the large and small scale. A particularly elegant example is provided by the Seaweed Flies *Coelopa frigida* and *C. pilipes*. Breeding of these flies is confined to piles of seaweed cast up at high tide mark, which persist for about four weeks between spring tides. The females choose warm places in the decaying weed to lay their eggs, and the larvae feed on the bacteria producing the decomposition. Different wrack beds function as isolates with little communication between them (Dobson, 1974). Substantial genetical pressures have been demonstrated in the flies of individual beds, as shown by both biochemical and cytogenetic markers.

The effect of population sub-division has long been recognized by theoretical geneticists who have developed a series of models to describe its properties and consequences in detail (Wright, 1949; Li, 1955; Crow and Kimura, 1970).

Perhaps the most important model is the *island* one. This assumes a population sub-divided into islands within which there is random breeding, and between which migration occurs at random. A development from this is the *stepping-stone* model, where migration is only between colonies that are immediate neighbours. This is perhaps the best way to look at most clinal situations. For example, Self Heal (*Prunella vulgaris*) possesses a large number of locally adapted populations (or *ecotypes*), but also shows continuous variation in size, growth-habit and flowering time throughout its range from southern Europe to Iceland (Bocher, 1949). In principle we can describe the conditions affecting gene frequencies in any local population (or island), and the effect of immigration from adjacent islands. This means that we have to make estimates of: the selection pressures operating at any point in the cline;

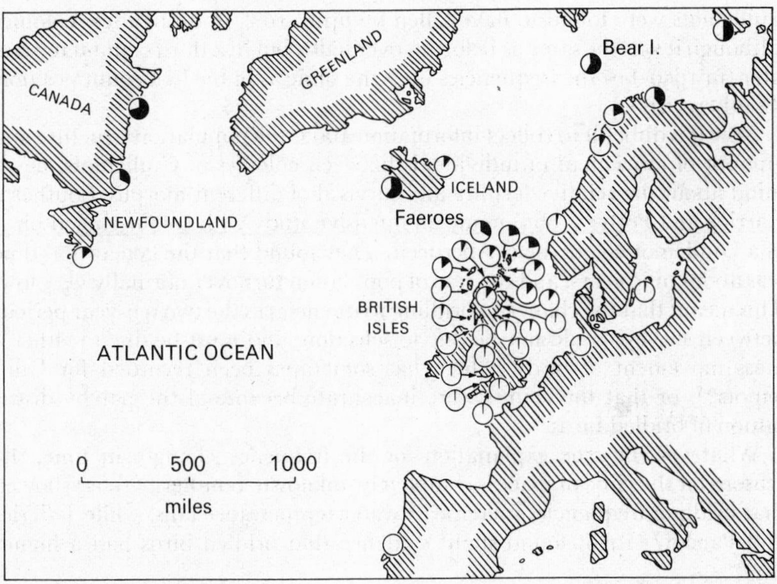

FIG. 59. Frequencies of bridling (shown in black) in the Common Guillemot (*Uria aalge*) (from data given by Southern, 1962).

the effective sizes of local breeding populations to determine whether drift might significantly alter gene frequencies; and the extent of individual movement which might blur any local adaptation.

Such information has been collected in surprisingly few cases, largely because for many organisms it is difficult to get accurate estimates of all the necessary factors. For example, there is a 'bridled' condition in Common Guillemots (*Uria aalge*), where a white stripe forms a ring round the eye, and then runs to the nape of the neck. At one time these spectacled birds were thought to be a distinct species, but recently it has been possible to raise a family in captivity and show that bridling is a recessively inherited condition, produced by alleles at a single locus (Jefferies and Parslow, 1976).

In 1938 and 1939 members of the British Trust for Ornithology were enlisted by H. N. Southern to count the numbers of bridled and non-bridled Guillemots in as many colonies as possible. With a few irregularities, the frequency of bridled birds increased with latitude: at the southern end of the range in Portugal not a single bridled bird was seen, but northward the proportion of bridled birds increased until it reached over 50% in southern Iceland. Bridling does not occur in Pacific populations of the species. In Iceland, the Common Guillemot is replaced by Brunnich's Guillemot (*Uria lomvia*), and the frequency of bridling apparently falls with the species density, so that only 10% of Guillemots are bridled on the island of Grimsey, north of Iceland. The counts were repeated in 1948–50 and the proportions of bridled Guillemots were found to have fallen by up to 10% in quite a few colonies (although it was the same as before in over half); but in a third count ten years later (in 1959–60) the frequencies were the same as in the first count position (Southern, 1962).

Now it is difficult to collect information about the population structure and amount of movement of individuals between colonies of Guillemots, never mind about the relative fertility and survival of different morphs. Southern, Carrick and Potter (1965) made an intensive study with colour-ringed birds of a Guillemot colony near Aberdeen. They found that the generation time was about nine years, and the rate of population turnover normally very low. This meant that the change in bridling frequencies in the two ten-year periods between surveys could not be due to selection, and must be due to either a mass movement of birds (which has sometimes been recorded for Guillemots*), or that the counts were inaccurate because of the patchy distribution of bridled birds.

Whatever the true explanation for the frequency changes in time, the reasons for the cline in space is completely unknown. Southern (1962) showed that bridling frequencies increase as water temperature falls, while Jefferies and Parslow (1976) found slight evidence that bridled birds had a higher

*For example, Guillemots spread rapidly northwards up Labrador to west Greenland, and the highest recorded frequency of bridling (71%) occurs at Nunarsuk, Labrador (Tuck, 1960).

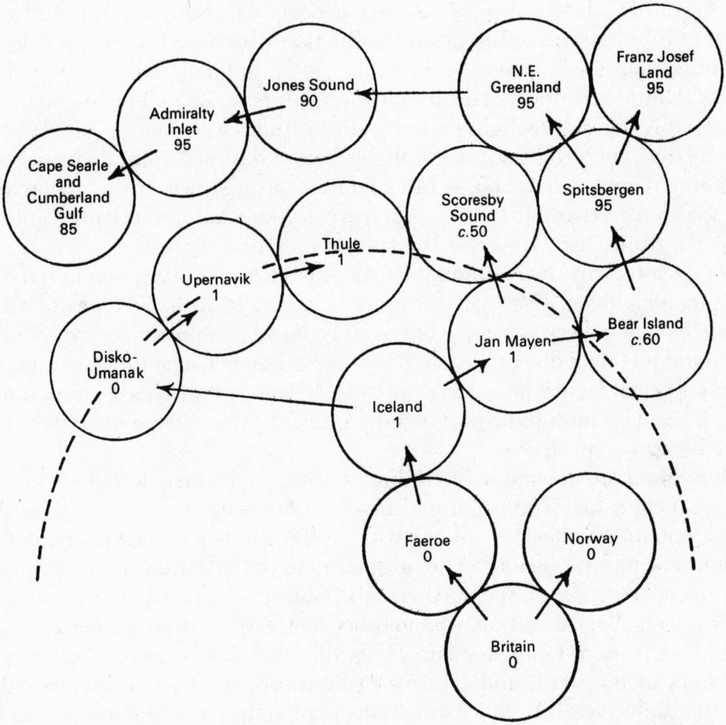

FIG. 60. Percentages of dark ('blue') birds in breeding populations of the Atlantic Fulmar (*Fulmarus glacialis*). The broken line represents the approximate boundary of the 'high arctic'. The arrows indicate the most likely movement of birds between colonies (from Fisher, 1952).

basic metabolic rate than non-bridled ones. If so, this could be the explanation for the northern extension of the species (mainly bridled birds) in this century (Southern, 1962). Certainly, the increase in proportion of bridled birds with latitude is so consistent that it must have some causal explanation, and presumably this will be similar to that in the Arctic Skua, where the proportions of colour phases merely act as indicators for a behavioural adaptation (page 144). It is easy to speculate about bridled Guillemots, but very difficult to find out why they occur.

A parallel cline to that in Common Guillemots is found in the Fulmar (*Fulmarus glacialis*), where high frequencies of a dark or 'blue' colour phase occur in the 'high arctic', *i.e.* northern latitudes where the sea temperature is at or near freezing point even during mid-summer. Nothing whatsoever seems to be known about the inheritance, characteristics, or stability with time of this northern morph (Fisher, 1952).

One of the fullest studies of a cline has been of melanism in the Autumnal Rustic Moth (*Amathes glareosa*) in the Shetland Islands. Twenty-three of the 64 indigenous moth species in Shetland (36%) are melanic or have melanic forms (Kettlewell, 1961*a*; Kettlewell and Cadbury, 1963). The reason for this is probably the long twilight period in the latitude of the islands (62°N) which means that nocturnal moths are flying before darkness fully descends, thus exposing them to predation by insectivorous birds. However this explanation is a guess: the attempt to test it by research on the Autumnal Rustic failed to prove or disprove it. However this species may be a special case.

The reason why the Autumnal Rustic was chosen for study was that it has a distinct melanic (*f. edda*) which cannot be confused with the typical form (*f. typica*) except in very worn specimens. It is unusual in this respect: most of the Shetland melanics do not have a sharp distinction between the dark and light forms. Special factors must have operated to bring about this clearcut dimorphism. As with most industrial melanics, the melanic phenotype is dominant over the grey typical one.

The Shetland melanics (including *f. edda*) were first described by professional collectors who began visiting the islands in the 1880s, and for long they remained collectors' prizes with nothing further known about their genetics or natural history. The original plan of the Autumnal Rustic study was to collect *f. edda* on the northernmost island of Unst (where it was known to fly during August and early September) and *f. typica* from the southern part of the main island (always known as the Mainland); and to release the melanics in the south and typicals in the north in order to measure their differential survival. In other words, the plan of the experiment was basically the same as in Kettlewell's Peppered Moth work. These recapture experiments showed that in Unst the dark form had an advantage of 7.0 ± 6.5% over the lighter one, while in the south Mainland there was no advantage of either form.

The mark-release-recapture experiments were later extended to two more sites (Hillswick and Tingwall). In no place was there much difference in the proportion of dark and light moths recaptured, although moths released at their site of capture always survived longer on average than moths from elsewhere (even though both 'home-caught' and 'introduced' moths were treated similarly to make sure that no inadvertent damage was caused by transporting the moths between sites). For example, at Hillswick marked *f. typica* and *f. edda* of local origin both survived 3.4 days on average between release and recapture, while morphs originating from the south Mainland survived 2.6 and 2.5 days respectively and Tingwall ones 2.8 and 2.6 days. *F. edda* tended to have a slightly higher survival than *f. typica* at its home site, but always survived less well than *f. typica* when moved. As far as the Autumnal Rustic is concerned, home is best. Being interpreted, this suggests that there is a degree of local adaptation which cannot be recognized from a simple inspection of the two morphs.

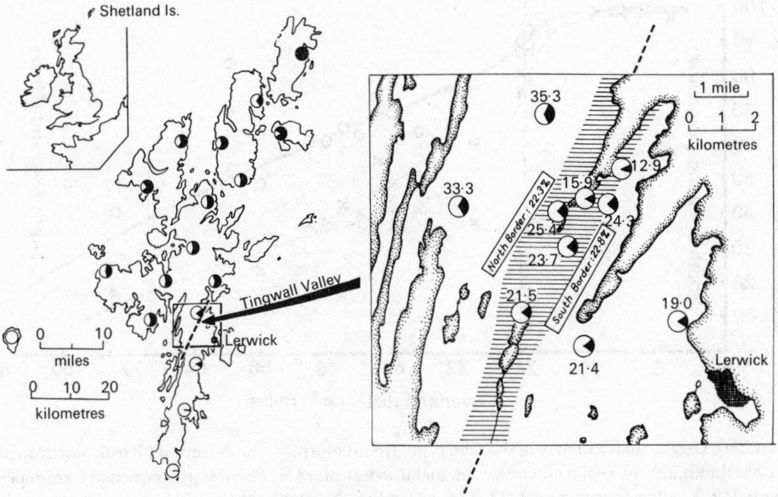

FIG. 61. Frequencies of the melanic form (*edda*) of the Autumnal Rustic Moth (*Amathes glareosa*) in Shetland, showing the situation in the Tingwall Valley in detail. Numbers in the Tingwall Valley are % frequencies of *f. edda*. In this region frequency differences are maintained despite virtually no movement of moths across the valley (based on Kettlewell and Berry, 1961, 1969).

When the work started, *f. edda* was not known to exist outside Unst, but sampling throughout Shetland revealed that it occurred in all parts of the islands with decreasing frequencies towards the south of the archipelago. It has now been found also in Orkney and Caithness (Berry, 1974*a*). When the shortest distance between sampling sites over land ('moth distance') in Shetland is plotted against the frequency of melanics, *f. edda* falls smoothly from 97% at Baltasound on Unst to less than 1% in the south Mainland. The frequency of *f. typica* on the opposite sides of the channels separating different islands is similar, *i.e.* 38.2% at West Sandwick on Yell and 37.7% at North Roe on the Mainland; 42.6% at Burravoe in south Yell, and 48.7% at Mossbank on the Mainland opposite. This again suggests local factors adjusting phenotype frequencies, with similar pressures operating in similar sites separated by water.

Notwithstanding, the rate of change of allele frequencies is so smooth that it invites an explanation of common but uniformly changing factors acting throughout the Shetland range of the moth. Indeed the observed frequencies fit a theoretical model of a cline proposed by J. B. S. Haldane (1948), which was based on a population divided into two parts by a barrier such that the recessive homozygote was at a disadvantage on one side of the barrier but had an overall advantage on the other. In any cline, the rate of change of gene frequency is the result of migration decreasing the slope of change, while

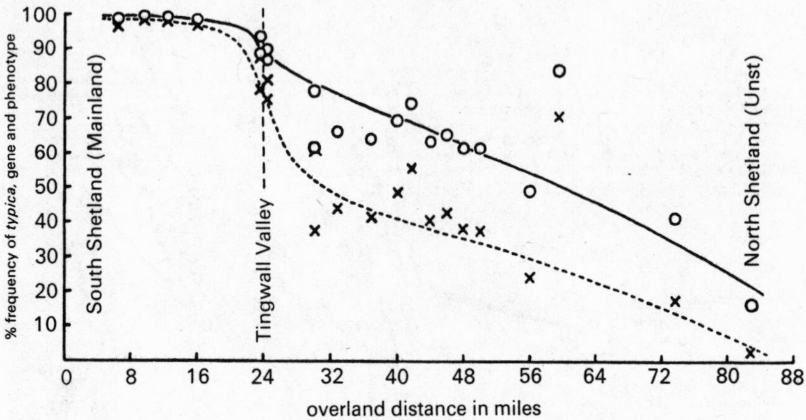

FIG. 62. Decline in frequency of the wild type (*typica*) morph of the Autumnal Rustic northwards in Shetland, and its replacement by the melanic morph *edda*. Phenotype frequencies are represented by X, allele frequencies by O. The dotted line shows the cline in phenotype frequency, the continuous line that of allele frequency (from Kettlewell and Berry, 1961).

selection acting differently on the two phenotypes increases the slope. Since these two forces must be equal in an equilibrium situation, Haldane was able to derive an expression for the relation between the intensity of selection, the mean distance migrated per generation, and the slope of the cline.

On the evidence of the phenotype frequencies alone, there is a good case for a barrier between the population in two parts of the Shetland population (Kettlewell and Berry, 1961). The incidence of *f. edda* remains high over the whole of the northern half of Shetland, falling only gradually (40% in 45 miles) in a southerly direction; in the south Mainland *f. edda* is much less common and the rate of change of frequency even less. However there is an intermediate region where *f. edda* frequency drops 35% in about eight miles. This region is centred upon the Tingwall Valley, a fertile limestone band running south-west to north-east across the Mainland, and varying from one-half to two miles in width. The valley itself contains little of the short heather in which the Autumnal Rustic lives in Shetland. If Haldane's arguments are accepted, the *net* advantage of *f. edda* over *f. typica* north of the valley is less than 1%, which seems very small even though it refers to the equilibrium situation at any point in the cline.

The applicability of the Haldane model depends on a reversal of advantage of the two morphs at the Tingwall Valley. Kettlewell and Berry (1969) carried out an experiment in which moths of both morphs were released on the two sides of the valley with different marks and their movement measured by recording the recaptures in traps situated along the two sides of the valley, and in the middle. Three conclusions emerged:

1. Individual moths frequently flew over a mile, judging by the number of individuals recaptured some way from the site of release.

2. Despite the facts that 65 moths were recaptured (out of 1682 released), that the wind during the experiment would at times have tended to blow moths from one side of the valley to the other, and that many moths moved much further than the distance between the two sides of the valley, only a single moth was recaptured on the opposite side of the valley to the one on which it was released.

3. The frequency of *f. edda* on the northern side of the valley was 22.3%, whilst on the southern slope it was effectively the same (22.8%), although the frequencies in the sparse moths in the middle of the valley ranged from 15.9% to 25.4% over a distance of less than a mile, and these values did not change from year to year. However the frequency of *f. edda* increased up to 35% over two or three miles immediately to the north of the valley in an area of apparently uniform heather moor.

If the Tingwall Valley was really a barrier between a northern and a southern population of the Autumnal Rustic, *either* there should be considerable movement across the valley associated with a rapid elimination of the unfavoured form on each side; *or* the frequencies on the two sides of the valley should be markedly different. As it is, there can be no doubt that the valley does operate as a considerable barrier to moths, but that this barrier merely shows up the distinct populations existing over short distances in an apparently continuous area on the two sides. Far from a situation of a cline maintained by diffusion of migrant moths, the *f. edda* cline is much closer to a network of partially isolated islands with local factors controlling the genetical constitution in each.

Once we have moved away from a picture of the cline controlled by mere physical factors in the way that oil leaks from a punctured tanker, we can appreciate the biological complexity of the situation. Even in an organism like a moth which has no territorial behaviour to complicate its movement, there are differences between populations which can only be determined by the operation of factors peculiar to each population. Home survival is part of the biological picture (Slatkin, 1973).

This is not the same as claiming that identical factors operate on the moths over a wide area. For example, Kettlewell (1961*b*) shot birds (mainly gulls) searching the ground in Unst where the moths were very dense, and found that their crops had 20.7% of *f. typica* in them although the local frequency of typicals was only 2.7%. Selective predation by birds is clearly an important factor for the Autumnal Rustic in Shetland. However there are at least three other pieces of evidence for selective factors acting independently of the main cline in melanism:

1. Most important is the typical pattern of recaptures in north Shetland. With recapture experiments in general there is an exponential decay of recoveries with time – the longer the period since release, the fewer the proportion of recaptures. Autumnal Rustics from south Shetland behave in this way, but in the north there were consistently more moths recaptured on the second night after release than on the first. This was clearer for *f. edda* than *f. typica* because the numbers involved were greater, but it seems to be true for both forms if they originated north of the Tingwall Valley. The simplest explanation of this peculiar observation is that only a proportion of the population flies every night, which means that there will be a residue of the population surviving if catastrophe strikes on a particular night (Kettlewell, Berry, Cadbury and Phillips, 1969). Since a flying creature like a moth is particularly vulnerable on a small island in a high wind, this restriction of flight would obviously be advantageous. Indeed flightlessness has evolved among both insects and birds on many oceanic islands. The point in the present context is that this reduction in flight occurs in some populations of both *f. edda* and *f. typica*.

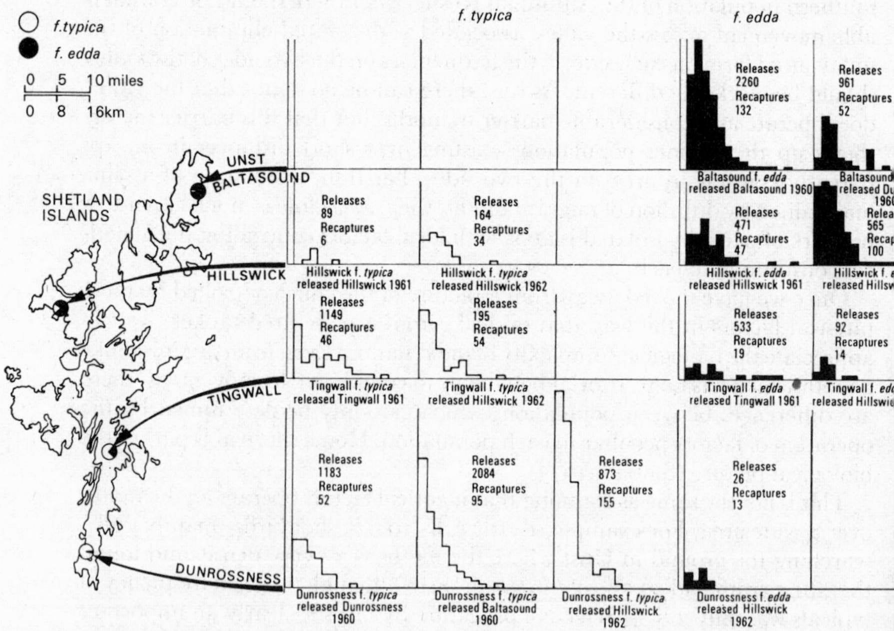

FIG. 63. Histograms of recaptures of marked Autumnal Rustic Moths at four sites in Shetland. The horizontal axes represents days between release and recapture (from Kettlewell, Berry, Cadbury and Phillips, 1969).

2. The *edda* character seems to be incompletely dominant on the Shetland Mainland, and 'dark' and 'light' *edda* can be recognized in the same proportions as those expected of homozygotes and heterozygotes. At the limit of its range on Unst, there was only a third of the expected number of light *edda*. The simplest explanation of this was that there were genes modifying the colour of the moths, so that the character was more completely dominant (*q.v.* pages 188–93). Once again we have a local genetical heterogeneity.

3. Finally, Pillay (unpublished) measured the lengths of a number of elements (legs, wings, antennae, etc.) in hundreds of moths from all over Shetland and found that there was no simple distribution of these size characters such as would be expected if the only genetical variable was the *edda* allele. Moreover correlations between different measurements varied considerably between different localities.

The genetical situation in the Autumnal Rustic in Shetland is clearly not a simple story of changing advantage of the melanic morph. Ironically, we still do not know the reason for the occurrence of *f. edda* in Shetland. Attempts to find the answer to this apparently simple question have revealed a situation of considerable complexity (Manley, 1975). It is frustrating not knowing the answer to the questions which originally inspired a piece of research, but as so often happens, the facts that have emerged may be of greater general significance. It is not irrelevant to repeat that the melanic polymorphism in the Peppered Moth which is the classical example of a genetical difference maintained by a simple selective force (visual predation by birds) also involves: viability and growth rate differences among the caterpillars; resting position behaviour differences; dominance modifiers; possible density dependent involvement of the *insularia* and *carbonaria* morphs; and some evidence of heterozygous superiority (page 109) (Kettlewell, 1973; Lees and Creed, 1975).

POPULATION ARCHITECTURE

We have progressed in our discussion on clines from thinking of them merely as the direct result of gene flow, to picturing them as assemblages of only partially communicating sub-populations. This is highlighted in the case of stable hybrid zones between adjoining sub-species: why should two forms be able to breed together and form completely viable offspring, but not become completely blended into each other?

In extreme cases there is no problem about this. For example when a horse and a donkey cross to produce a mule, the mule is sterile because differences between the chromosome sets from the two parents prevent normal meiosis. In many instances the distinction between the parents is much less.

A good example of the meeting, but only partial mixing of two closely

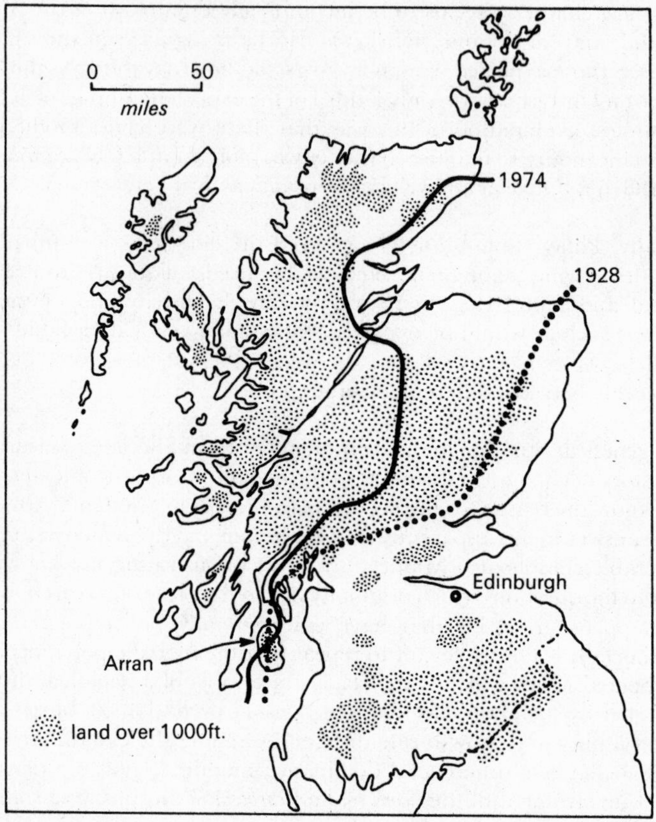

FIG. 64. The shift in the centre of the Carrion/Hooded Crow (*Corvus corone corone* and *C. c. cornix*) hybrid zone in Scotland between 1928 and 1974 (based on Cook, 1975).

related forms is that of the Carrion and Hooded Crows (*Corvus corone corone* and *C. c. cornix*). The Carrion Crow is a western European species; in Italy, the Balkans and Scandinavia it is replaced by the Hooded Crow or Hoodie. Where the two meet there is a narrow hybrid zone in which pairing seems to be at random. The offspring of mixed unions can be recognized by the amount of grey and black colouring on the back and undersides, and it is possible to devise a scale to measure the contributions of each parental form in a local area.

Hoodies occur in Britain in north and west Scotland. Considerable mixture with Carrions occurs in the Cairngorms and Grampians, but north of the Great Glen, there is a steep cline to 'pure' Hoodie. The range of the Hoodie seems to have contracted in recent years: in 1928 the centre of the hybrid zone

PLATE 9. Heterostyly in Primroses *(Primula vulgaris)*. Left, pin with long stigma and short anthers; right, thrum with short stigma and long anthers; centre, rare long homostyle form with long stigma and anthers (page 186).

Heavy metal tolerance in the grass *Festuca rubra*. The area is mine waste at Trelogan Mine, Clwyd which contains high concentrations of zinc and lead. The left plot was sown with normal (sensitive) seeds which have virtually all died; the right plot was sown with seeds from an adapted mine population (page 265).

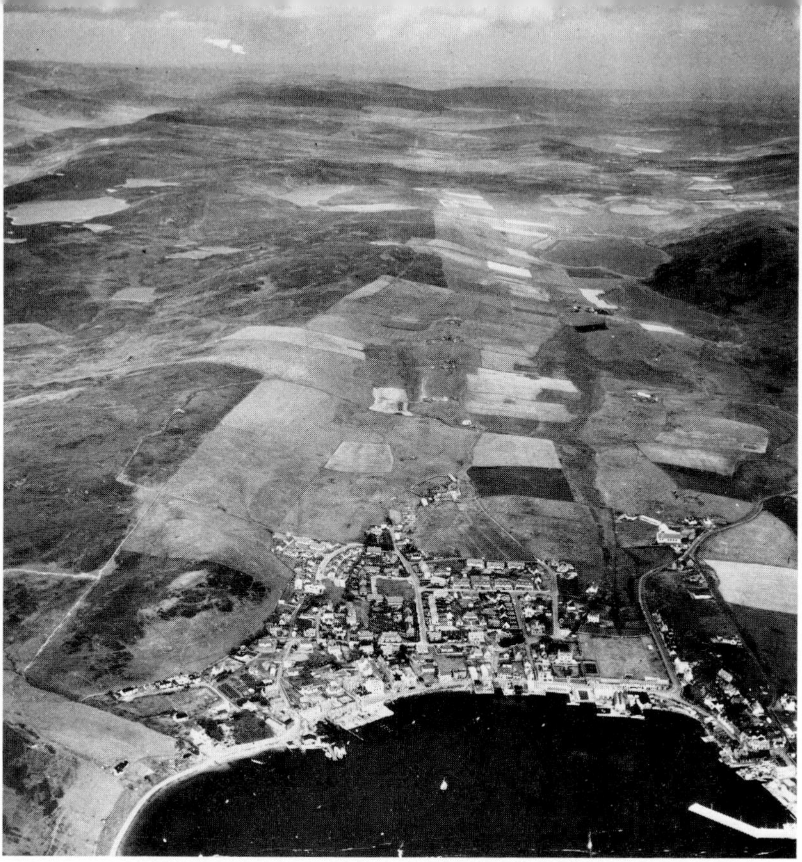

PLATE 10. *Studies on the Autumnal Rustic moth in Shetland (page 171)*

The Tingwall Valley, a fertile limestone band running south-eastwards across the Mainland of Shetland, and forming a barrier to migration for this heather-living moth.

View northwards across the Tingwall Valley, showing great differences in *f. edda* frequencies over a comparatively small area.

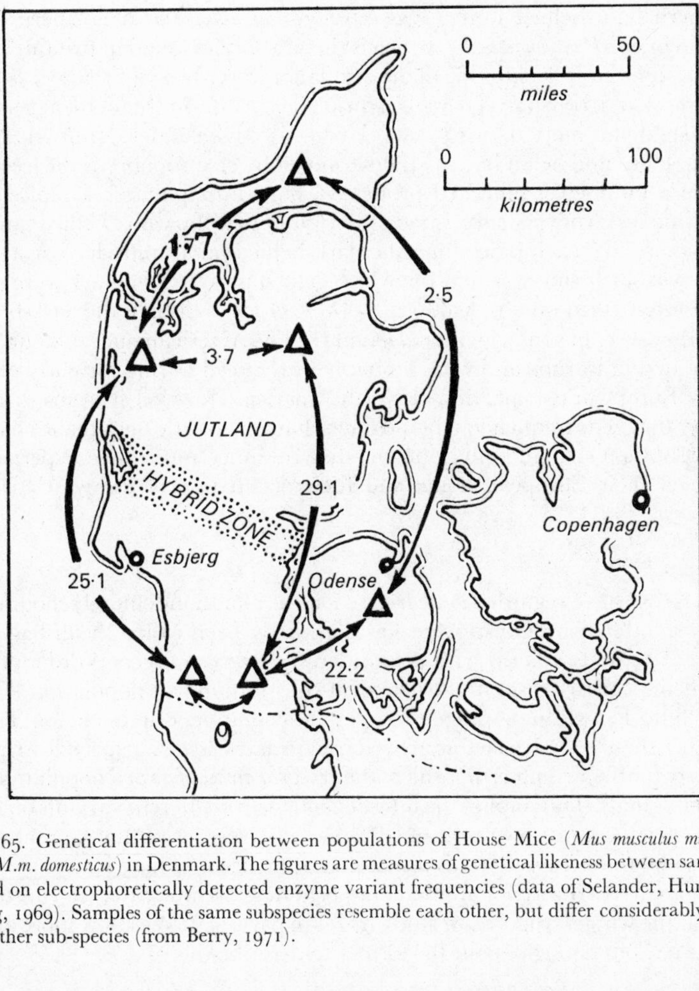

FIG. 65. Genetical differentiation between populations of House Mice (*Mus musculus musculus* and *M.m. domesticus*) in Denmark. The figures are measures of genetical likeness between samples, based on electrophoretically detected enzyme variant frequencies (data of Selander, Hunt and Yang, 1969). Samples of the same subspecies resemble each other, but differ considerably from the other sub-species (from Berry, 1971).

was to the south of the Scottish Highlands, but now it runs over the highest parts of the Cairngorm Plateau (Meise, 1928; Cook, 1975). Cook (1975) has shown that on Arran there are few signs of Carrion characters over 800 feet, but an increasing proportion of Carrions (including 'pure' Carrions) occurs at lower altitudes, only 10% of birds at sea level being 'pure' Hoodie. Cook believes that the change in distribution of the two forms is in part the result of a tendency towards milder, wetter weather in the past few decades.

Despite the change in position of the hybrid zone, the two crows remain completely distinct over most of their range. The same conservatism occurs in House Mice. There are two common sub-species of House Mice in Europe, a southern light-bellied form (*Mus musculus musculus*) and a northern dark-bellied one (*M. m. domesticus*), which is the sub-species found in Britain. These forms seem entirely inter-fertile in the laboratory, but right across Europe they meet and form only a narrow hybrid zone, which in Denmark at least has been stable for more than 40 years (Ursin, 1952). Selander, Hunt and Yang (1969; Hunt and Selander, 1973) determined by electrophoresis the frequencies of a number of inherited protein variants in population samples from Danish mice. Not very surprisingly they found that the white bellied samples were more like each other than the dark-bellied ones (and vice-versa), but what was surprising was that the white-bellied mice in Denmark were more like white-bellied mice in California (*M. m. brevirostris*) than the dark-bellied ones they were in contact with in Denmark. The American mice undoubtedly originated in Europe and were probably introduced during the early phases of the European colonization of North America. However it seems extraordinary that gene frequencies should have changed so little despite the changes in population size and environment that the mice must have experienced between their European home and their occupation of the west coast of America.

So far we have regarded gene frequencies as almost indefinitely modifiable given suitable conditions – what has slightingly been called 'bean-bag genetics'. We are now faced with a considerable degree of second-order effects producing conservatism in the genetical constitution of a population. These bring into focus a new range of factors affecting gene frequencies. Taken together these cohesive factors are referred to as *coadaptive*: a fairly descriptive analogy for their result is to speak of the *genetical architecture* of a population (or an individual). This implies the interdependence of different variants on each other, producing its own restraints in the same way as a human architect is limited by the materials at his disposal and the needs of his client.

There are two ways of approaching genetical architecture: we can try to look at the whole structure, or study in detail parts of it. We know more about the latter, but enough about the former to describe this first.

CORRELATION AND COADAPTATION

We have already noted examples of interaction between different gene-loci. Most of these examples arise for understandable biochemical or developmental reasons: the genes affecting the vertebrae in a mammal can only operate normally if those controlling the early growth of the nervous system have done their part; the control of growth depends on the sufficient materials being synthesized at the right time; and so on.

Sometimes a single gene may produce a host of symptoms. For example,

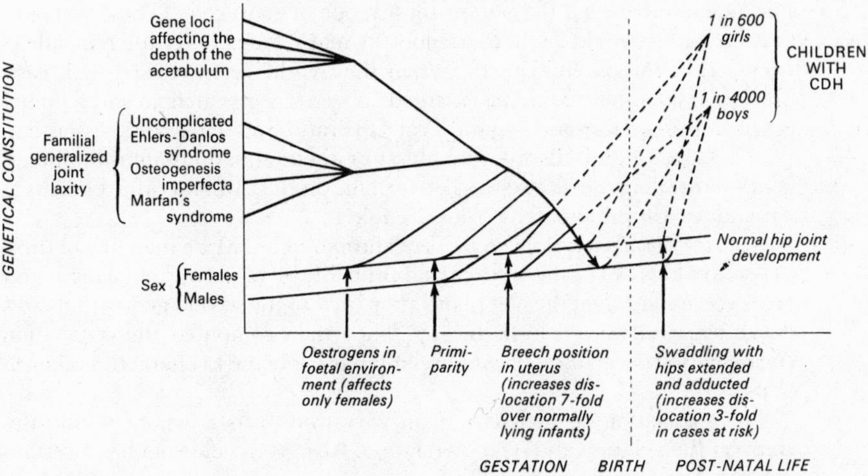

FIG. 66. Interactions between genetical and environmental factors in the causation of congenital dislocation of the hip (CDH) in man (based on Berry and Berry, 1971).

there is a dominantly inherited disease in man called osteogenesis imperfecta. The basic defect is in the collagen fibres, but this leads to a great variety of symptoms: brittle bones, blue tint to the whites of the eyes, deafness, and over-extensible joints. Here one gene affects many reactions. In contrast congenital dislocation of the hip (where the hip joint comes out of its socket in young babies) depends on genes affecting the depth of the hip joint, the tightness of ligaments, and even sex (since girl babies respond to oestrogens in their mother's circulation by relaxing their hip ligaments, while boys do not react in this way). Many genes interact to produce the possibility of abnormality, and by inference, the usual state of affairs is for alleles at the same loci to interact to produce normality.

If we take an overall view of the innumerable interactions that contribute to the growth and life of an animal and plant, we find that since particular inherited traits depend on each other, they will be correlated together. A picture of the associations between structures or behaviour in an organism will show us what the functional architecture of that organism is, and go some way towards describing its genetical architecture. One of the best-worked examples along these lines concerns *Potentilla glandulosa*, an American relative of Cinquefoil (Clausen and Hiesey, 1958).

In California there is a costal race of *Potentilla* which grows the whole year round, and an alpine race found high in the Sierra Nevada which is dormant in the winter, even when transplanted to a lowland garden. Clausen and Hiesey crossed the two races. The first generation hybrids were almost exactly intermediate between the parents in a range of characters. These first generation plants would be heterozygous at many loci where different alleles occurred in the parental races. When the hybrids were crossed with each other, considerable variation emerged as genes segregated to form homozygotes and heterozygotes again. From the ratios in their second generation plants, Clausen and Hiesey were able to estimate the minimum number of genes controlling the differences between the original two races for 14 traits as being due to approximately 51 loci (Table 17).

Now *P. glandulosa* has seven pairs of chromosomes. More than one of these 51 loci are likely to occur on each and probably all of the seven. Clausen and Hiesey scored each of the 992 plants they bred in the second generation cross for the 14 traits listed in Table 17. They then computed the correlation coefficients between the 91 possible combinations of the 14 characters taken in pairs.

The correlation coefficient (r) can vary from 0 to 1, where 1 indicates complete correlation and 0 no correlation. Any given value of r has a certain

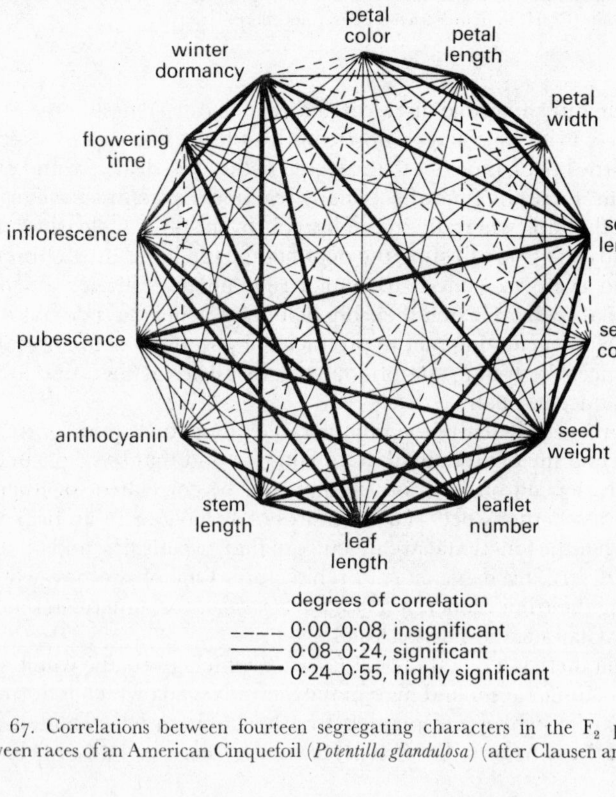

FIG. 67. Correlations between fourteen segregating characters in the F_2 progeny of cross between races of an American Cinquefoil (*Potentilla glandulosa*) (after Clausen and Hiesey, 1958).

TABLE 17. Estimated minimum number of loci controlling the differences between Coastal (Santa Barbara) and Alpine (Upper Monarch Lake) races of *Potentilla glandulosa* in fourteen characters (after Clausen and Hiesey, 1958)

Character	*Loci*	*Character*	*Loci*
Winter dormancy	3	Leaflet number	1
Flowering time	many	Seed weight	6
Density of inflorescence	1	Seed colour	4
Glandular pubescence	5	Sepal length	5
Anthocyanin	4	Petal width	2
Stem length	many	Petal length	4
Leaf length	many	Petal colour	1

probability of being due to chance alone, which decreases as the size of the sample increases. In the present case where each paired comparison is based on 992 individuals, weak correlations of less than $r = 0.08$ have a 1 in 100 chance of being due to chance alone, while values of r higher than 0.08 are statistically meaningful (or 'significant').

Sixty-seven of the 91 *Potentilla* comparisons were above $r = 0.08$. Most of them were between 0.08 and 0.40, representing incomplete correlation, although some were as high as 0.70.

Significant correlations may be due to pleiotropy (the same gene affecting different systems) or genetical linkage, or both. Pleiotropy would be expected to play a part, and probably a large part in the correlations between such characters as petal length and sepal length which are the end products of common or similar growth processes. Correlations resulting from pleiotropy are likely to be high; the lower correlations observed between developmentally unrelated characters are best explained as the result of weak linkage between the determining genes.

No character in the *Potentilla* experiment segregated independently of all the other characters that were studied. On the other hand, each character was uncorrelated with some of the other characters. The various characters overlap in their interrelations with each other: thus two characters (like sepal length and seed colour) which are not correlated together may both be linked independently of one another to a third character (pubescence), and this in turn may be linked to a fourth character (anthocyanin content) which segregates independently of the first (sepal length).

The picture of an interlocking system of weak linkages tells us something of the situation in the genotype. The genes (or some of them at least) responsible for each multifactorial character difference between the original races will lie on different chromosomes. Some of the loci controlling character A will be linked with some of those controlling character B, while others will segregate independently. So characters A and B will only be incompletely correlated in inheritance. Now any genes for character A that are not on the same chromosome as any of the genes for character B may occur on the same chromosome(s) as some of the genes controlling character C. Characters B and C may segregate independently of each other but they will be separately and incompletely linked with A.

Obviously both genetical and developmental considerations mean a high degree of interdependence in the genetical constitution of a population, and it is totally false to think of the genes involved behaving like beans in a bag. But when we think of the factors that affect the interaction of characters, we are forced to go even further, and envisage forces which will actively increase the complexity of the genetical architecture.

Consider two loci *A*, *B* affecting the same or associated characters, with each locus having two alleles (*A*,*a*; *B*,*b*), and assume that *A*, *B* influence the character(s) in question in one direction, while *a* , *b* influence it in the other. The multifactorial system determining height (page 44) would be an example of such a situation. The two loci may give rise to nine genotypes, and if we give *A*, *B* 'values' (*i.e.* effect on the phenotype) of 1 and *a*, *b* 'values' of 0, we have:

	Value
AABB	4
AABb	3
AAbb	2
AaBB	3
AaBb	2
Aabb	1
aaBB	2
aaBb	1
aabb	0

If the two loci are on different chromosomes, the chances of extreme phenotypes occurring will depend only on the frequencies of the segregating alleles; if the loci are linked, there will be an additional restriction provided by the amount of recombination between the two. If we make one further assumption that extreme expressions of the phenotype are disadvantageous to its possessor in a particular environment, important consequences follow.

The character we are concerned with produces the greatest fitness for its carriers when it has a value of 2. This can be produced by two homozygous and two doubly heterozygous genotypes:

A b		a B		A B		A b
	or		or		or	
A b		a B		a b		a B

The first two (homozygous) genotypes have no inherited variation, and might lead to extinction if the environment changed; the two (heterozygous) genotypes combine present fitness with phenotypic variation, and will be of greater long-term benefit to the population. However, the second heterozygous genotype (where the loci are said to be 'in repulsion') will be less likely to produce extremes than the other double heterozygote (which is 'in coupling'): the less the amount of crossing over between the two loci, the rarer will be the proportion of extreme variants.

So linkage in repulsion will combine both fitness and flexibility (Fisher, 1930). Any mechanisms (such as chromosomal inversions and translocations) which promote the accumulation of chromosomes heterozygous for balanced (for 'high' and 'low') factors will be favoured (Mather, 1943). A secondary consequence of this situation is that selection will tend to adjust such chromosomal complexes to yield their optimum effect when heterozygous, since they will only be deleterious when in the homozygous state, and that will occur relatively rarely (Parsons and Bodmer, 1961).

It is not easy to test whether these theoretical predictions are fulfilled in practice, but the experiments that have been done certainly support them (reviewed by Bodmer and Parsons, 1962; Lee and Parsons, 1968). For example, genes affecting the gain or loss of bristles in *Drosophila melanogaster* are scattered along all the chromosomes, and are interspersed with ones directly affecting viability (Breese and Mather, 1960; Thoday, 1961). However the implications of the existence of balanced chromosomes have been amply confirmed by inference from selection experiments. In particular, the repeated observation that selection for a particular trait leads to decreasing viability (as the chromosomes become 'unbalanced'), but that viability is restored if selection is relaxed for a few generations (giving the chromosomes time to recombine, and re-establish the balanced state), after which further selection may be possible and effective – indeed there may be accelerated responses (as particularly favourable recombinants are produced) (see also page 292). Inbreeding depression can be seen as the consequence of increasing homozygosity in a system which works best in the heterozygous state, while hybrid vigour is the result of a cross producing greater heterozygosity. The lessons and restrictions of genetical architecture are important for animal and plant breeding; what concerns us here are the wider effects in nature.

ARCHITECTURAL FEATURES

When we turn from the overall construction of the 'genetical building' to particular features, there are three bits of the structure which are of general interest.

1. Linkage disequilibrium

When two loci are linked, recombination between them should eventually lead to all four combinations (*i.e. AB*, *Ab*, *aB*, *ab*) appearing in the population

Parents		**Offspring**					
		From non-recombinant gametes		*With recombination in one parent*		*With recombination in both parents*	
Chromosomes	Phenotypic score	Chromosomes	Phenotypic score	Chromosomes	Phenotypic score	Chromosomes	Phenotypic score
Repulsion							
A b / a B	2	Ab/aB	2	AB/aB	3	AB/ab	2
		Ab/Ab	2	AB/Ab	3	AB/AB	4
A b / a B	2	aB/aB	2	ab/aB	1	ab/ab	0
		aB/Ab	2	ab/Ab	1	AB/ab	2
Attraction							
A B / a b	2	AB/ab	2	AB/Ab	3	Ab/Ab	2
		AB/AB	4	AB/aB	3	Ab/aB	2
A B / a b	2	ab/ab	0	ab/Ab	1	aB/Ab	2
		ab/AB	2	ab/aB	1	aB/aB	2

FIG. 68. Effect of linkage on the incidence of the extremes of manifestation of a continuously distributed character. If *A*, *B* produces a phenotype effect of 1 and *a*, *b* an effect of O, then matings between two repulsion heterozygotes are most likely to produce non-extreme phenotypes.

at frequencies related only to the frequencies of the alleles concerned. Often this does not happen: particular combinations of alleles are found 'too often' or 'too rarely'. The alleles concerned are then said to be in 'linkage disequilibrium' (Turner, 1968; Franklin and Lewontin, 1970). For example, the colour and banding loci are close together on the same chromosome in *Cepaea hortensis*, but the majority of snails tend to have either brown banded or yellow unbanded shells (Bantock and Noble, 1973); a similar imbalance occurs in another terrestrial snail, *Arianta arbustorum* in Derbyshire (Parkin, 1972). Linkage disequilibrium has also been found in cereal crops (Allard, Babbel, Clegg and Kahler, 1972).

We have described how inversions carrying favourable gene combinations will be selected to preserve the arrangements. Prakash and Lewontin (1968) found that certain alleles at some loci in *Drosophila pseudoobscura* were virtually restricted to particular inversions. This meant that the loci concerned were strongly out of linkage equilibrium with each other. The stability of this system was remarkable: the same inversions carried the same alleles no matter from what part of the species range they were drawn. Indeed the association between genes and inversion was found in both *Drosophila pseudoobscura* and the related species *D. persimilis*. These species separated from each other at least a million years ago.

Dobzhansky's work on the distribution and maintenance of inversions in *Drosophila* species has already been touched upon in the description of chromosomal variation (pages 86–87) (summarized Dobzhansky, 1970). Dobzhansky found that two inversions may be strongly heterotic (*i.e.* have a considerable degree of heterozygous advantage) in flies collected from the same area, but flies with the same inversions but from different areas are much less likely to show heterosis when crossed. This is a related point to Lewontin's findings, and implies that although an inversion may have a wide distribution, it will acquire extra properties by mutation which will be locally tested and proved for their contribution (or not) to local survival. Just as the roofs of houses in northern climates tend to be steep so that snow slides off them, so the genetical architecture of a species has to accommodate to varying conditions throughout its range. Linkage disequilibrium can be a sensitive indicator of the operation of natural selection.

2. Supergenes

If the order and nearness to each other of genes on chromosomes is an important method of maintaining an adaptive character, there should be many traits controlled by groups of closely linked loci. Obvious candidates for such an arrangement will be complicated patterns and behaviours which depend for their effect on their wholeness: any improvements may be incorporated, but disruptions will be disadvantageous and give lower fitness, resulting in their elimination. Complex gene arrangements of this nature often appear to behave as a single unit, and segregate like a single locus. For this reason such gene complexes are usually called 'super-genes'. They will frequently operate as major switches between different morphs, as in different mimetic forms in Batesian mimicry (Clarke and Sheppard, 1960; Clarke, Sheppard and Thornton, 1968).

The first example of a supergene to be analysed involved a very complicated system of colour forms in the American Grouse Locusts. In *Acridium arenosum* the colour patterns are controlled by thirteen genes which reassort easily despite being carried on the same chromosome. The same genes are recognizable in *Apotettix eurycephalus*, although here they form two closely linked groups of genes. However, in *Paratettix texanus* there has been complete suppression of crossing-over in 24 out of 25 of the colour-pattern 'genes' (*i.e.* chromosome segments) (Nabours, 1929; Nabours, Larson and Hartwig, 1933), so that each pattern is inherited as a unit. There is also yet another pattern 'gene' which is common and recessive, and which is heterotic in combination with the other (dominant) colour pattern gene aggregations. The dominants produce up to 40% disadvantage compared with the heterozygotes, when two are present in the same individual (Fisher, 1939).

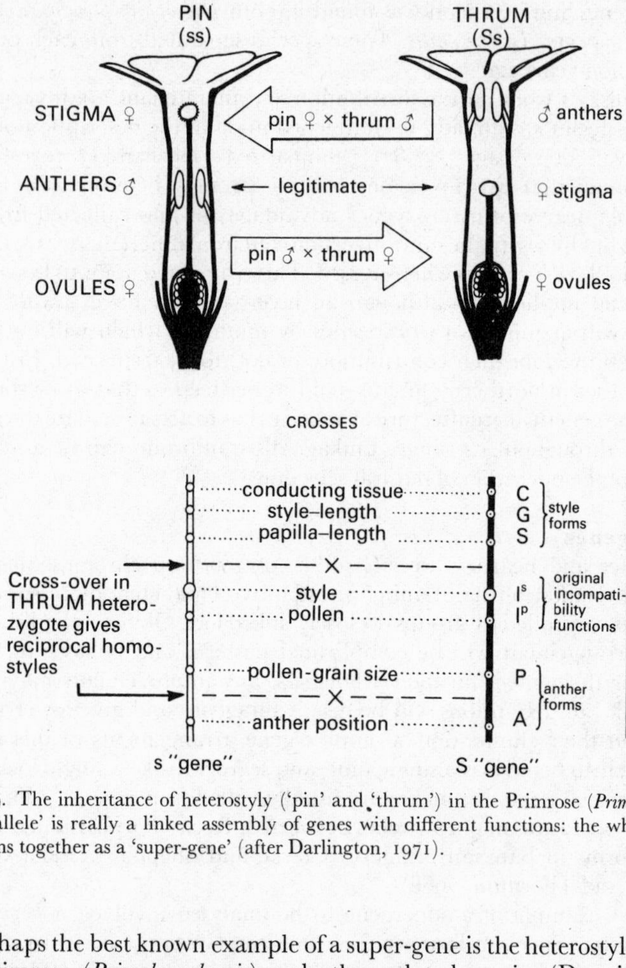

FIG. 69. The inheritance of heterostyly ('pin' and 'thrum') in the Primrose (*Primula vulgaris*). Each 'allele' is really a linked assembly of genes with different functions: the whole segment functions together as a 'super-gene' (after Darlington, 1971).

Perhaps the best known example of a super-gene is the heterostyle 'gene' in the primrose (*Primula vulgaris*) and other related species (Dowrick, 1956). There are at least seven loci involved in it: area of conducting tissue for pollen-tube growth; style length; papilla size; style and pollen incompatibilities; pollen size; anther height. The result of this complex acting together is to produce two 'super-alleles', *S* and *s*. Pin plants have long styles and low anthers and are genetically *ss*; thrum plants have short styles and high anthers and are *Ss*.

Normally thrums and pins must cross, and will produce equal numbers of each form – the mechanism is exactly parallel to the XX and XY sex determination in mammals. However, rare recombination within the super-gene may produce homostyle plants. Although these could be either long or short, in practice they are usually long. Either way, the result is the same: the stigma of the gynaecium is clasped by the anthers. Self-fertilization becomes almost obligatory, and if homostyles have as high a viability as heterostyles (and there is no obvious reason why they should not), they would be expected to spread because of their short-term advantages over heterostyles of invariance and certain fertilization (Crosby, 1949).

Two homostyle primrose populations have been studied in Britain. The main one was near Sparkford in Somerset, the other one 80 miles away in Buckinghamshire. Bodmer (1960) found that in one wood in the Sparkford area, the frequency of homostyle plants decreased from 82.4% to 73.2% between 1941 and 1948, and along a hedgerow not far away homostyles dropped from 34.5% to 20.1% between 1943 and 1944, although the frequencies were stable before and after this period.

Although the heterostyle-homostyle mechanism changes the mating system of primroses from a great deal of out-breeding to a relatively great deal of inbreeding, the evidence is homostyles do not spread through the population when they arise. The changes in agricultural practice during the last few centuries are probably sufficient in general terms to account for the existence of occasional areas where it pays primrose populations to be comparatively invariable – or inbred, but there is no known reason why homostyles survive in the Sparkford and Chiltern areas whilst remaining rare elsewhere. It is frustrating to have as much genetical information about a super-gene as we have and yet not know why it manages to persist in the face of forces breaking it up.

Another super-gene which has already been touched upon, is the system determining shell appearance in *Cepaea* species. This involves at least four loci: shell colour, number of bands, colour of bands, fusion (or 'spread') of bands. Genes for mid-banded (00300) and 00345 types are not included in it (Cain, King and Sheppard, 1960).

The super-gene inhibition of complete recombination saves *Cepaea* from undesirable individuals: if the unit involves two common chromosome arrangements, only two phenotypes will normally segregate and fewer unmatching phenotypes will need to be eliminated from the population.

The 'genes' for Rhesus blood groups and histoincompatibility in man are both complex. The Rhesus super-gene involves four loci at the end of a chromosome, and deletion may produce the loss of up to three of these in particular families. Any of the four loci *(C,D,E,F)* stimulate the production of antibodies, but the most important is *D*. This is the locus which is most commonly involved in Rhesus 'positive' and 'negative' reactions. However other combinations of *C-D-E-F* are found in varying frequencies, and are extremely valuable to physical anthropologists searching for population relationships.

Histoincompatibility genetics involves even more complex systems. The unravelling of these followed the stimulus of the problems encountered in tissue transplantation. It was known that grafts will take between identical twins but very rarely between random members of a population, and it became important to know the conditions favouring successful transplantation. It turned out that the rejection of grafts is mainly based on antibodies against antigens produced by a group of linked loci. The main loci involved are two *HLA* (human leucocyte antigen) ones, known as *LA* and *4*. This means that each of us has up to four different antigens. Close to one of the *HLA* loci (but not between the two) is an *MLC* (mixed lymphocyte culture) gene which also codes for individual antigenic differences, this time in one particular form of white cells. Finally there are a number of *Ir* (immune response) genes located between the two *HLA* loci controlling specific reactions to foreign proteins, including allergies. More than 30 *HLA* antigens have been described in man, and over 30 *Ir* loci have been mapped in mice (where the super-gene seems to be similar to that in man). The histoincompatibility segment of chromosome 6 is extremely important in medicine, both because of its direct relevance for transplantation, and also because of the number of diseases associated with particular alleles. (For example, both ankylosing spondylitis and Reiter's disease – a form of urethitis – are 13 times more common in carriers of one particular allele (W27) than in non-carriers.) There is no doubt that this segment is rapidly becoming one of the most-studied pieces of chromosome in any species.

3. Evolution of dominance

So far we have taken dominance or recessivity as somehow characteristic of a trait. At the biochemical level both alleles in a heterozygous genotype will be expressed, but some alleles may overcome or fail to influence a sister allele (*i.e.* they may be intrinsically dominant or recessive); while at the normal phenotypic level this genetical property may be obscured. Indeed, at one time it was believed that mutant alleles involved a loss of gene function since they were typically deleterious and recessive. During the 1920s and 1930s this produced a schism between geneticists and palaeontologists because the gross mutants studied by geneticists seemed to be so different to the gradual, progressive changes in the fossil record found by palaeontologists.

This apparent hiatus was resolved by R. A. Fisher's 'Theory of the Evolution of Dominance' (Fisher, 1928). Fisher argued that we ought to expect most mutations to be both recessive and deleterious if we think of the history of any species. His starting point was the fact that there is no intrinsic reason to expect a mutation occurring for the first time to be either dominant or recessive; the most probable thing is that it will be intermediate, with an effect somewhere between its expression in double dose (*i.e.* if homozygous), and the normal condition. However:

i. Mutations will repeatedly have occurred at virtually every locus in past times. The rare mutational events that we observe are recurrences of something that has happened thousands of times in the past.

ii. When a mutation occurs, it will almost always be present in the heterozygous condition: if an allele has become relatively common in a population so that a fresh mutation to it has a reasonable chance of occurring in an existing heterozygote, then other forces than mutation must be influencing its frequency.

iii. If a newly arisen allele has a beneficial effect on its carrier, combinations with it of other alleles that increase its effect will have a higher fitness than any which decrease it. This will be continually happening so that the architecture of the species will become modified to the extent that any new occurrences of that advantageous allele will always produce the maximum effect in its carrier, and this will be in the heterozygous condition. In other words beneficial characters will be selected for dominance – and will spread to replace the previous expression of the trait. Conversely alleles which are deleterious in the heterozygote are only likely to be transmitted in combinations where their effect is least, *i.e.* there will be selection for a small heterozygous effect, which means in the direction of recessivity.

Fisher put forward these ideas from first principles. They were received sceptically because of the difficulty in believing that selection pressures would be strong enough to allow genes which modified dominance to spread. In pre-1950 days primary selection coefficients were thought of as around 0.1 to 1.0%, and second order effects would be much less than that. Haldane suggested that dominance was more likely to be the effect of alleles with a margin of safety becoming the normal allele, since they could exercise an undiminished action when heterozygous, *i.e.* mutant alleles would be recessive and deleterious. However Fisher's theory has been proved right on a number of occasions and, although it may not apply for every allele at every locus, it has considerable historical significance in bringing together palaeontologists and geneticists, and has present relevance in highlighting an important factor influencing genetical architecture.

Fisher himself was the first to demonstrate the genetical modification of dominance, by crossing domestic poultry to wild jungle fowl for five generations. This changed the inheritance of certain characters so that a degree of heterozygous manifestation occurred where complete dominance normally prevailed (Fisher, 1935, 1938). Fisher reasoned that the dominance of the traits he studied had been attained during domestication as a result of selection for the more striking heterozygotes (see page 000).

A fuller demonstration of the influence of modifying genes on dominance was obtained by Ford (1940) by breeding from the greatest and least expressions of a variable yellow variant (*lutea*) of the Currant Moth (*Abraxas grossulariata*). Although the difference between *lutea* and *non-lutea* can be regarded as caused by a single allelic difference, after only three generations of selection Ford produced heterozygotes virtually indistinguishable from the

…gote in selection for the yellowest individuals, and ones like the …zygote in the white selection line. In other words he had changed …ote from a position of no (or intermediate) dominance to …inance or recessivity respectively. Ford then crossed his …ained heterozygotes back to unselected stock, and by the second generation (when the selected modifiers would have a chance of segregating independently) the original variable heterozygotes reappeared: he thus showed that it was the response of the organism rather than the gene itself that had changed.

Laboratory experiments of this nature have been carried out on a variety of organisms and a range of characters. One that has been particularly informative about genetical architecture in the wild was carried out by Kettlewell (1965) using British and Canadian Peppered Moths (These are named *Biston betularia* and *Amphydasis cognataria* respectively, but are fully interfertile.) Although there is a melanic form (*swettaria*) in the North American species, it is comparatively restricted in its distribution, and only pale (or typical) moths occur over vast tracts of Canada.

When a melanic heterozygote and a typical homozygote of British origin are mated, the offspring are clearly dark or light: the melanic character is a straightforward dominant. Even when the typical moths come from extreme south west England where melanics have never been reported, only slight loss of complete dominance occurs – and that only after several generations of crossing melanics back to Cornish stock (*i.e.* taking a melanic from a melanic × Cornish cross and mating this with a 'pure' Cornish parent). This modification consists of some white dots on the normally jet black wings of the heterozygote. Perhaps significantly, this slight heterozygous expression of the gene gives specimens similar to those caught in the early days of the spread of Peppered Moth melanism in the mid-nineteenth century and now prized collectors' specimens (page 129). This white speckling has long since disappeared in wild-caught British specimens, and modern heterozygotes are indistinguishable from melanic homozygotes. Melanism has become fully dominant over the past decades.

However there has been no opportunity for dominance to evolve in Canadian Peppered Moths. When British melanics are crossed to Canadian stock (from areas where *swettaria* does not occur) the first generation progeny segregate as dark or light in the same way as in a cross between British moths. In the first generation, the dominance modifiers in the British parent will be carried on the chromosomes in the same order as in British stock and produce dominance in the same way. In the next generation, the gametes contain chromosomes which have crossed-over between the British and Canadian grandparents. Consequently in the second generation, the 'switch' between pale and black forms does not operate so efficiently. Kettlewell crossed heterozygotes from a British × Canadian mating with Canadian moths, and repeated this back-crossing for four consecutive generations. At the end of this

FIG. 70. Effects of genes modifying dominance: a black (*carbonaria*) Peppered Moth from Birmingham (top row) crossed with:
Left, a typical Peppered Moth from Birmingham, produces a clear-cut distinction between typical (normal homozygote) and melanic heterozygote.
Right, typical Peppered Moth from Canada produced a range of forms of varying darkness. (from Berry, 1971, from specimens of H. B. D. Kettlewell, based on Kettlewell, 1965).

time, the heterozygotes ranged from black to pale – there was no sign at all of dominance. He then reversed the procedure, and mated his 'broken-down' melanics to British typicals. At once the dominance of the condition was re-established: the architecture of the British chromosomes shaped a clear segregation between a dark heterozygote and a pale homozygote.

We have already noted that virtually all industrial melanics in Britain are inherited as dominants. If this dominance has evolved rather than being some intrinsic property of melanism itself (which it is not in the Autumnal Rustic and Peppered Moth at least), this implies that melanism must have had an advantage at some stage in the history of the species. New mutations which occurred and spread in polluted areas as industrial melanics would be modified by the genetical background in the populations in which they occurred. Because they were inherited as dominants, their spread was much more rapid than it might otherwise have been. It was the search for this pre-industrial selection of melanics that led to the research on melanism in Shetland, and it has inspired other studies of geographical melanism.

For example, in the small areas of indigenous pine and birch forests remaining in Scotland there are a number of melanic forms. Kettlewell (1973) studied the Mottled Beauty (*Cleora* = *Boarmia repandata*) in the Black Wood of Rannoch. This has a dominantly inherited melanic (*nigricata*) similar and possibly identical with the industrial melanic (*nigra*) which occurs at about 30% in the London area but has apparently replaced the typical form completely in Bradford. The average frequency of *f. nigricata* is around 5% in the Black Wood, but rises during the flying season to about 10%. This could be due either to later hatching or higher survival than the typicals.

When sitting on tree-trunks, *f. nigricata* is undoubtedly more conspicuous than the typical to human eyes. However in flight the darker morph is much more difficult to see than the lighter one, and birds have been seen catching the moths in flight. The importance of the flight conspicuousness of the typical is that every day up to a half of the resting moths may be disturbed by ants or direct sunlight (which in summer strikes two-thirds of the tree circumference at the latitude of Rannoch), and take to the wing: the cryptic disadvantage of the black form whilst resting is balanced by its benefit whilst flying.

The Mottled Beauty situation may be taken as suggestive of the sort of circumstances which may have moulded the dominance characteristics of particular traits. Every trait will differ in its history and vicissitudes, but there is no justification for assuming that a mutation has always possessed the same fitness as it does when it occurs in a contemporary environment – blackness of moths in the Black Wood of Rannoch affects survival differently from blackness in Bradford.

Another property that should not be taken as necessarily characteristic of a trait is that its dominance will automatically be affected by the same modifiers in different parts of the species range. Differences in the control of dominance modification have been found in the inheritance of a dark form (*curtisii*) of the

PLATE 11. Pale *(above left)* and Dark *(right)* morphs of the Arctic Skua *(Stercorarius parasiticus)* in Shetland (page 142).

Common Guillemots *(Uria aalge)* on a cliff at Bullars of Buchan, Aberdeen. The recessively inherited bridled condition can be seen clearly on two of the birds (page 168).

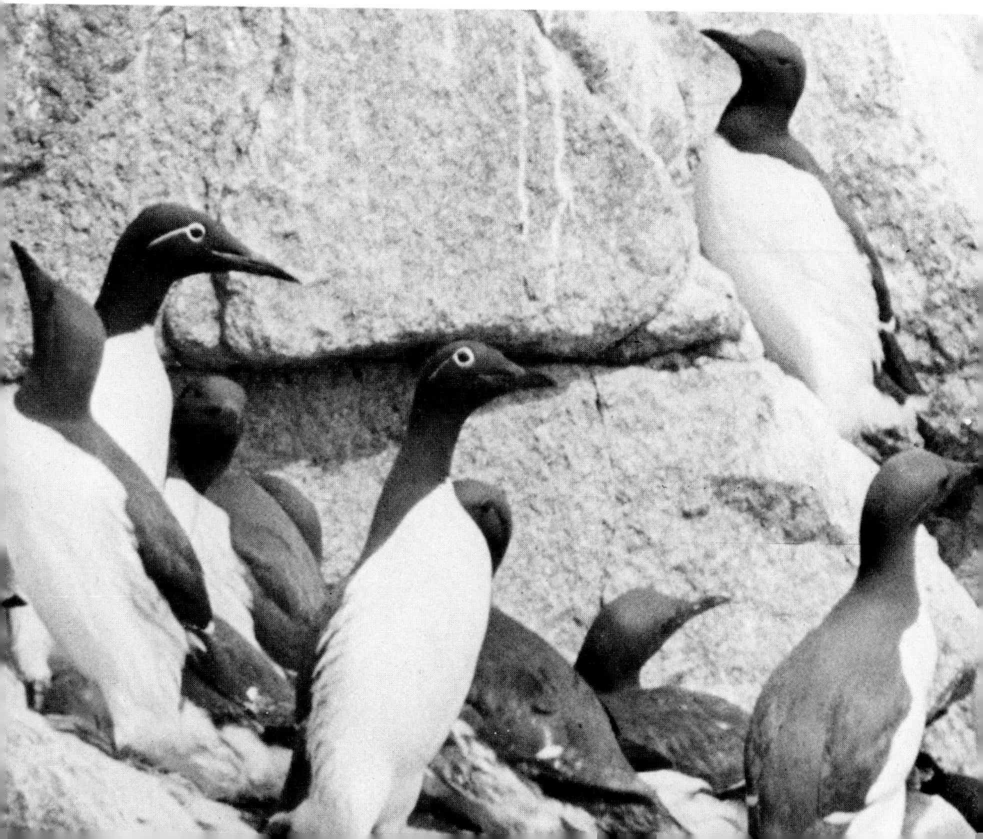

PLATE 12. Meadow Brown butterflies *(Maniola jurtina)* showing the spots on the underside of the hind-wings. The two left-hand columns show males with 0–5 spots, the two right-hand ones females with 0–4 spots (page 217).

Black and brown (agouti) rabbits on Skokholm. The black animals are much more conspicuous than browns, and are caught by predators (especially gulls) much more readily than are the wild-type brown.

Lesser Yellow Underwing Moth (*Triphaena comes*). Throughout most of Britain the species is monomorphic, but in central and northern Scotland *f. curtisii* occurs at up to 50% frequency. In the valiant early days of Mendelian genetics, Bacot (1905) showed that *curtisii* was produced by a single allele which was almost completely dominant: heterozygotes are nearly always distinguishable from the recessive homozygote (normal *comes*), but the appearance of the heterozygotes and the other homozygote overlaps. Ford (1955) crossed moths from Barra (at the southern end of the Outer Hebrides chain) with ones from the Orkneys (which lie to the east of the north coast of Scotland). Although the moths came from places only about 100 miles apart, Barra is 25 miles from the nearest point on the mainland and 15 miles from Skye (which is effectively a mainland peninsula), whilst Orkney lies six miles north of Caithness. The two populations were completely isolated from each other, although not phenotypically distinguishable. However the progeny of the crosses showed complete dominance breakdown, just as in the progeny of British and Canadian Peppered Moths. In other words, dominance of *curtisii* in Orkney and in Barra has been achieved by the selection of a different set of modifiers. The situation is parallel to the cline in the Autumnal Rustic, where each local population is a gene-pool unto itself.

MODELS OF ARCHITECTURE

Genetical architecture provides glorious scope for theoretical biologists (Wright, 1952, 1965*a*; Clarke, 1966 O'Donald, 1968*b*; Lewontin, 1974). In the simplest form, the three genotypes produced by two alleles at a locus can be considered to be modified by alleles at a second locus whose only effect is to change the fitnesses of the primary phenotypes. Using this simple basis, Clarke (1966), searching for a basis for area effects in *Cepaea*, found that the slope of a cline could in theory change steeply even though no environmental discontinuity occurred. The problem about such a single modifier model is that if both an allele and its modifier are rare, the rate of spread of their combination will be very slow at first – too slow to ever happen according to some critics. O'Donald (1967) applied some apparently realistic assumptions and got the same answer. He assumed that the expression of a character, and hence its selective value, will depend on the whole of the accompanying genes and not merely on a single locus. Sheppard (*q.v.* Sheppard and Ford, 1966) has pointed out that dominance evolution is a form of disruptive selection.

'Area effects' in *Cepaea* (page 152) have provided the grist for a great deal of speculation and a certain amount of work on population architecture. The original suggestion for an unvarying distribution of morphs over wide areas terminated by sharp frequency changes occurring over short distances apparently independently of habitat, was that the frequencies were controlled by climatic differences (Cain and Currey, 1963*a*). Goodhart (1963) put forward an alternative idea: he suggested that each 'area' was the result of an independent colonization by a small group of founders which built up its own architecture, and faced hybrid breakdown when populations expanded to meet. This would be a similar situation to the Hooded-Carrion Crow boundary in Scotland or the House Mouse subspecies in Denmark.

Cain and Currey (1963*b*) objected to this explanation on two grounds: firstly that the different gene complexes would not be equally successful, and one would eventually replace the others unless each was best suited to its own particular area; and secondly that different gene frequencies changed at different places, *i.e.* an 'area' refers to only a single locus. In other words, they appealed again to natural selection as opposed to the random events implicit in the founder effect.

However it is easy to overstate the importance of chance factors. Imagine a large region inhabited by a series of identical polymorphic populations. Local environments will differ from place to place, and some alleles will be favoured in one part of the region but not others. Migration will produce clines between the different parts. In areas where a particular allele is common, selection will favour genes that are compatible with it, *i.e.* modifier genes, as in the dominance evolution situation. In this way an originally uniform series of populations may become differentiated into a series of different and coadapted gene complexes (Mayr, 1954; Wright, 1965*b*; Clarke, 1966, 1968; Slatkin, 1973). Different clines may have few modifiers in common, so that area effects for different loci would occur independently of each other. An element of chance is involved in this selection process, since different modifiers may be available in different areas.

Support for this interpretation has come from electrophoretic studies of snails collected from adjacent 'areas' (Johnson, 1976). These have shown that alleles at other loci besides the ones concerned with colour and banding are associated in different areas. Moreover the availability of a large number of biochemically detectable genes show that there are definite patterns of allele associations in particular areas, reducing the force of the argument that area effects for different loci are not coordinated. In other words, the evidence suggests that the history of an area effect might be differentiation in relative isolation followed by secondary contact. The development of genetical architecture has proved an important conservative force restaining gene frequency change.

If coadapted genotypes are bound by strong stabilizing forces, it will be expected that they may maintain their gene frequencies – and even more, their allelic associations – even if the environment changes. Obviously there must be a limit to this, or adaptation will be impossible, but some evidence should be apparent. We have seen that North American House Mice are more like European ones than one might expect (page 178); parallel similarities occur in human traits, particularly in the HLA system.

The HLA 'gene' consists of two linked loci: LA and 4 (page 188). If the whole human species was in complete genetical equilibrium, chromosomes with the different combinations of LA and 4 alleles would occur in numbers directly proportional to the frequencies of each allele. However if individuals with particular associations have reproduced or migrated more commonly than ones with others, their HLA group will serve as a marker of their origins. Recombination between the two loci will produce linkage equilibrium in time, although particular associations may be selected against and remain permanently under-represented in the species. Generation time (and hence opportunity for recombination) is so long in man that most linkage disequilibria can be taken as indicators of migrations. For example, the population of western Europe can be divided into three HLA association groups diluted by two further chromosomal allele arrangements probably representing an Indo-European immigration from central Asia, and another (and later?) one from the north, presumably characteristic of the Huns and Goths. The three original associations seem to be the remains of an original Celtic differentiation (Degos and Dausset, 1974).

Such data present grievous temptations for theoreticians. The niceties of mathematical models are outside the scope of this book. Extended expositions appear in a number of works (Li, 1955; Crow and Kimura, 1970; Cavalli-Sforza and Bodmer, 1971; Mather and Jinks, 1971; Jacquard, 1974; Lewontin, 1974). Their cumulative result is to emphasize that the gene is not the unit of selection, and to point to the paradox between the rapid adjustment of gene frequencies in different environments and the innate conservatism of the integrated system of developmental processes that constitutes the living organism. So finely tuned is this system that two heterotic chromosomes may recombine to produce a chromosome lethal to its bearer (Dobzhansky, 1946; Batten and Thoday, 1969).* Waddington (1957) has written harshly but not unfairly about the genetics of natural populations, 'One can probably take the mathematical theory as being the earliest aspect of the truly modern outlook to be developed. Its origin may perhaps be seen in a series of papers by Haldane, begun in 1924, and it soon became enshrined in three major works, by Fisher 1930, Wright 1931, and Haldane, 1932. Examining it after this lapse of time one finds, unexpectedly, that it did not achieve either of the two results which one normally expects from a mathematical theory. It has not, in the first place, led to any noteworthy quantitative statements about evolution. The formulae involve parameters of selective advantage, effective population

**Footnote*. Rather ironically in the face of all that has been learned about genetics from *Drosophila* species, British *Drosophila* seem to be extremely uninteresting. *Drosophila subobscura* is widespread, and apparently breeds in woods (Shorrocks, 1975) – but no evidence at all has been shown for coadaptation in it, either from heterosis in crosses between races or from any increase in variance of morphological traits in F_2 hybrids (McFarquar and Robertson, 1963). However one of the earliest experiments in ecological genetics was carried out on *Drosophila* in Britain: Gordon (1935) released 36,000 *Drosophila melanogaster* with the recessive gene for ebony body colour at a frequency of 50% in the grounds of Dartington Hall near Totnes. Four months later the allele frequence of ebony had fallen to $11 \pm 3\%$.

size, migration and mutation rates, etc., most of which are still too inaccurately known to enable quantitative predictions to be made or verified. But even when this is not possible, a mathematical treatment may reveal new types of relation and of process, and thus provide a more flexible theory, capable of explaining phenomena which were previously obscure. It is doubtful how far the mathematical theory of evolution can be said to have done this. Very few qualitatively new ideas have emerged from it. The outcome of the mathematical theory was, in the main, to inspire confidence in the efficiency of the process of natural selection and in the justice of applying this type of argument also to the realm of continuous variation'. Or, as Mayr (1959) has written, 'The main importance of the mathematical theory was that it gave mathematical vigor to qualitative statements long previously made. It was important to realize and to demonstrate mathematically how slight a selective advantage could lead to the spread of a gene in a population. Perhaps the main service of the mathematical theory was that in a subtle way it changed the mode of thinking about genetic factors and genetic events. . . . The interpretation in Mendelian terms of the more sophisticated modern views on the interaction of genetic factors, on coadaptation, and on genetic homeostasis would have been impossible without the foundation laid by the mathematical theory, as over simplified as it may appear in retrospect'. We shall see in the last chapter how true this judgement has proved.

CHAPTER 6

ECOLOGICAL ISLANDS AND NICHES

Charles Darwin had an immediate success with the *Origin of Species* because of his simple mechanism for natural selection, rather than through the evidence he had collected for the fact that evolution had occurred. His book was based on *two* deductions from *three* facts:

Fact 1: All species have a great potential for increase.
Fact 2: The numbers of most species remain approximately constant.
DEDUCTION 1: There is a struggle for existence.
Fact 3: A considerable amount of inherited variation exists in all species.
DEDUCTION 2: (Made by putting together Deduction 1 and Fact 3) Natural selection.

We are beginning to be able to draw similar conclusions about the forces determining the distribution of variation in populations. In the last chapter we saw:

Fact 1: Population sub-division, especially in an environment which changes in time or space leads to a fine selective adjustment of gene frequencies, and
Fact 2: Coadaptation (or genetical architecture) imposes constraints on the amount and speed of genetical change.
DEDUCTION 1: Either way, natural selection will determine gene frequencies, and these will be unevenly distributed over the range of a species.

In this chapter we shall be largely concerned with:

Fact 3: The alleles present in a population derive from unpredictable processes of mutation and colonization, and these determine both the method of adaptation and the fate of individuals in that population.
DEDUCTION 2: Combines both directed and chance aspects – a local population may have unexpected characteristics due to its mode of origin, which will affect any future genetical changes that occur in it, and thus provide an essential part to the understanding of why an organism carries the genes it does.

ISLANDS

In the last chapter, we examined a number of examples of clines, where characters change in frequency with distance. Now a cline may result from the simple diffusion of genes, just as a leaking oil tanker is a focus of pollution which decreases with distance from the leak. Notwithstanding we have seen that a stable cline is more likely to be the sum of a whole series of partially connected islands. It follows that we may learn important lessons from real islands, where immigration is negligible and selection pressures often harsh.

The British Isles are peculiarly well suited to this type of analysis. The largest island (Great Britain) has been cut off from land contact with continental Europe for 8000 or so years, and many Continental species do not extend into Britain. The smaller islands that lie to the south, west, and north of Great Britain are even more sensitive indicators of genetical processes:

1. If a land species has to cross the sea to colonize an island, it is likely that fewer alleles will be represented on the island than on the mainland. There will be a founder effect (page 58) leading to a paucity of genetical variation.
2. Diseases and competitive interactions will be less common on islands than in a continental area.
3. The range of niches available for colonization is likely to be less than in the ancestral area.
4. Lack of gene-flow into island forms from the same species living under slightly different conditions on the mainland makes local adaptation easier and more precise. For example, organisms living on small islands have a perpetual risk of being blown into the sea, and both flightless insects and seeds with reduced dispersal potential are far commoner on oceanic islands than elsewhere (*e.g.* Carlquist, 1974).
5. The presence of few competing species provides increased opportunities for adaptation to a wide variety of environmental opportunities. The classical example of this is 'Darwin's finches' on the Galapagos islands (Lack, 1947), which provided a crucial stage in the development of Darwin's ideas on evolution.

These characteristics of island biology lead to two conclusions: that intensities of natural selection will probably be strong in isolated populations; and that these pressures will not necessarily lead to convergence of the island race with the ancestral form. However, the most important factor affecting the genetical constitution of island populations seem to be the founder effect. A moment's reflection leads to this being fairly obvious, since selection (or, for that matter, any of the other forces that change gene frequencies) can only operate on available variation. Adaptation cannot take place unless advantageous variants are available. In their absence, natural selection must lead to a less efficient form that in other parts of the species range; the rare favourable variety may produce a novel and exciting new possibility. It is not surprising, therefore, that island faunas and floras owe more to their relationships with their mainland relatives than to any common factors in the environment. For example, Ragge (1963) has described the British grasshopper fauna as migrating over the land-bridge with the Continent which existed once the

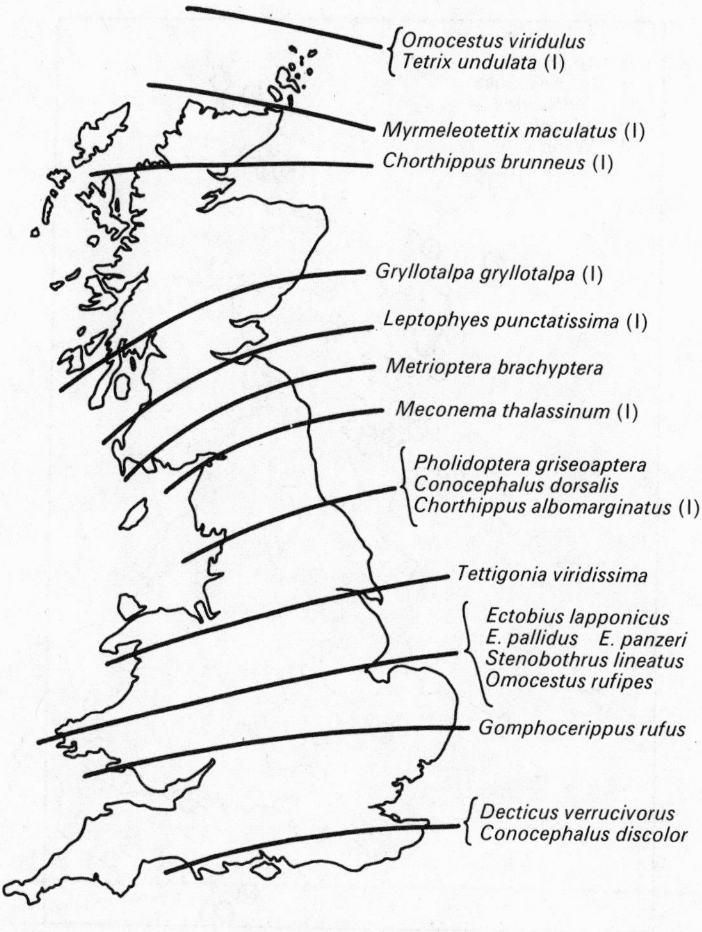

FIG. 71. The approximate northern limits of some of the British grass-hoppers. 'I' means that the species has also been recorded in Ireland: note that such species are the more northerly – presumably the hardier ones which entered Britain from Continental Europe first, and colonized Ireland before the land-bridge between Britain and Ireland was broken (after Ragge, 1963).

Pleistocene ice-sheet had retreated and permitted insects to survive again in northern Europe. Different species extended northwards to differing extents, presumably indicating their tolerance of adverse conditions and colonizing abilities. However, only the more northerly ones occur in Ireland although some of the others could presumably survive there. The Irish species must have emigrated via Galloway and Ulster before the land connection between Scotland and Ireland was broken.

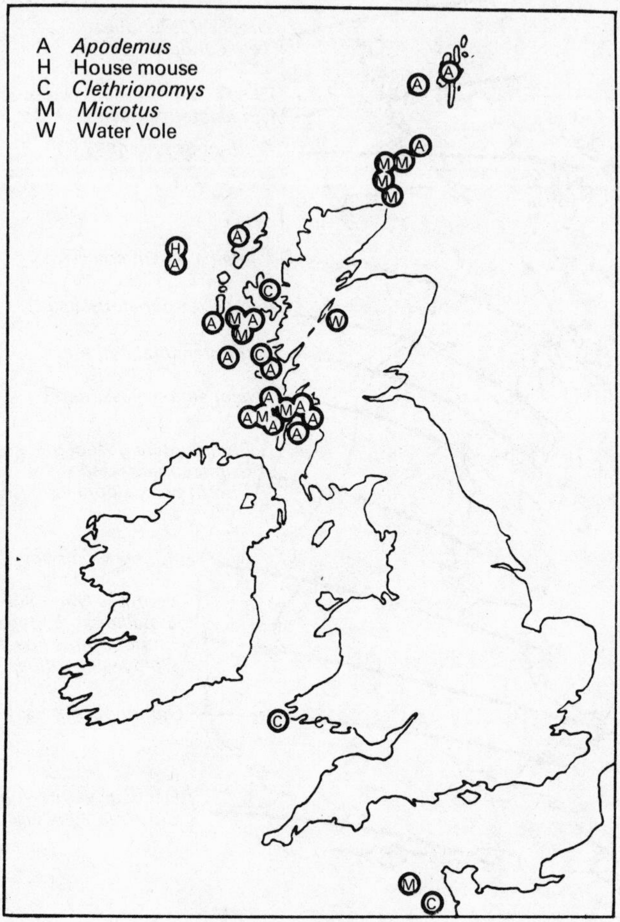

FIG. 72. Distribution of named races of small mammals in the British Isles: with one exception (the Scottish Highland Water Vole) all are on small islands.

A less simple but genetically more intriguing situation is provided by small mammals. Taxonomists have described British island races of the House and Field Mouse (*Mus musculus* and *Apodemus sylvaticus*), the Bank and Common Vole (*Clethrionomys glareolus*, *Microtus agrestis*, and *M. arvalis*), and the Water Vole (*Arvicola terrestris*) (Corbet, 1964). Most modern workers would regard these races as sub-species, although at times in the past some of the forms have been believed to be full species.

With the exception of the Water Vole of the Scottish Highlands, all the named races are confined to a single island (occasionally to neighbouring islands). The most polytypic species is the Field Mouse – rather oddly, since it is very constant over the greater part of its range from China westwards. In the British Isles, the species is represented by distinct forms on St Kilda, on eleven of the Hebrides, and on three of the Shetland group. It is instructive to consider this diversification in some detail, because it illustrates many of the factors leading to unevennesses in gene frequencies.

BRITISH RACES OF *Apodemus Sylvaticus*

In 1895 de Winton described a 'sharply differentiated local form' of the field mouse living in the Outer Hebrides, having a larger size but smaller ears than mainland mice, and being rather dark. He called this *Apodemus hebridensis*. Four years later Barrett-Hamilton described another species (*Apodemus hirtensis*) from Hirta of St Kilda. This was 'closely allied to *Apodemus hebridensis* from which it differs in its slightly larger size, and also in the greater amount of buff or yellowish-brown coloration on the underside'. A year later Barrett-Hamilton (1900) surveyed all the specimens of *Apodemus* in the British Museum (Natural History) and decided that *A. hebridensis* and *A. hirtensis* should be regarded only as sub-species, and they became known as *A. sylvaticus hebridensis* and *A. s. hirtensis*.

In 1906 Kinnear acquired six *Apodemus* specimens from Fair Isle. These had a 'longer and narrower brain-case than typical *A. sylvaticus*', and were put into a new species, *A. fridariensis*. By now the splitting and naming fever was in full flow. Hinton (1914, 1919) added another seven sub-species, Montagu (1922) four, and finally Warwick (1940) one, making a total of fifteen (Table 18). But as Harrison Matthews (1952) wrote 'If we are to be scientifically honest, we must acknowledge that the evidence upon which all this is raised is not good enough. The diagnoses of many of the island forms have been drawn up after the examination of much too few specimens, so that we do not know the range of individual variation that may occur in them. Even the limits of variation in the mainland form are not defined with accuracy, nor have the skull characters, which are much used in diagnosis, been studied in the light of the allometric (differential) growth principle'. So obviously the next step was to examine as many specimens as possible in order to determine what variation exists in each race, and to what extent this variation is common to a number of localities.

Delany (1961, 1964) began this task in the Hebrides and western Highlands. He trapped mice from a range of places and habitats, and made a number of measurements on each animal. Besides the traditional body weight and lengths of tail, head and body, hind foot, and various parts of the skull, he added the length of the dark spot lying between the forelegs, and two estimates of coat colour. Next he used the amount that the teeth were worn as

TABLE 18. Criteria used to distinguish the British and Icelandic subspecies of Field

Name and origin		*No. of specimens in original description*	*External measurements*			
			Head and body	*Tail*	*Hind-foot*	*Ea*
A. sylvaticus sylvaticus (British mainland)		—	*c.* 95	*c.* 90	*c.*22	15
A. s. butei (Bute)	Skull small in comparison with *A. s. sylvaticus*, but relatively broader and deeper	17	88.7	77.8	22	14
A. hebridensis hebridensis (Outer Hebrides)	Combination of large size, small ear, and dark tints sufficient to distinguish as 'a sharply defined local form'	*c.* 20	95.8	87.8	23.2	15
A. h. nesiticus (Mingulay)	Mesopterygoid fossa narrowed anteriorly	41				
A. h. hamiltoni (Rhum)	Skull distinctly larger and more massive than *A. h. hebridensis*, and also relatively longer and narrower	5	103.8	95.6	24.2	15
A. h. tirae (Tiree)	Size greater than the other rufous mice (*A. h. cumbrae* and *A. h. ghia*)	4	102.5	84.2	23.1	13.
A. h. maclean (Mull)	Distinguished from *A. h. cumbrae* by larger size, bigger feet, shorter tail and ear	5	97	87.4	23.2	14.
A. h. tural (Islay)		14	93.6	84.6	23.1	14.
A. h. larus (Jura)		5	92.6	83.7	22.8	14.
A. h. ghia (Gigha)		4	95.7	89.7	23.4	15
A. h. fiolagan (Arran)	Differs from *A. h. hebridensis* in lengthening of pterygoid region of skull	9	98.3	84.5	23.5	14.
A. h. cumbrae (Gt. Cumbrae)	Skull much like that of *A. h. hebridensis*, but smaller	6	93	90.3	22.8	15.
A. hirtensis hirtensis (St Kilda)	Similar to *A. hebridensis*, but skull larger, ears relatively longer, and colour darker	4	110.9	105.5	25.0	17.
A. fridariensis fridariensis (Fair Isle)	Brain-case longer and narrower than in *A. sylvaticus*. Coronoid processes of the mandibles noticeably less developed than in related species	6	108.5	102.4	23.6	16.
A. f. granti (Yell)	Distinguished from *A. f. fridariensis* by slightly smaller size and relatively shorter tail	5	100.9	88.2	23.6	15.
A. f. thuleo (Foula)	Differs from *A. f. fridariensis* in smaller size and larger hind feet	37	90	89.8	24.4	15.
A. s. grandiculus (Iceland)	Characterized by skull length, more powerful hind foot, and generally more pronounced colour of underparts	60	103	90.8	23.2	—

mus sylvaticus, A. hebridensis, A. hirtensis and *A. fridariensis*) (after Berry, 1969*a*)

-

ngth

Dorsal colour	*Ventral colour*	*Pectoral spot*
Wood brown, tinged with russet posteriorly; noticeable sprinkling of blackish hairs. Sides and flanks a russety yellowish wood-brown	Dull white, with throat and belly clouded by slate grey undercolour	Buff median area between the forelegs
Darker than *A. s. sylvaticus*		Faint trace
As in *A. s. sylvaticus*	Heavily washed with dull buffy brown	Usually rather longer than in *A. s. sylvaticus*
Slightly paler than *A. h. hebridensis*	Silvery, but dark underfur shows through	Prominent and elongated but not very bright
As in *A. h. hebridensis*	Irregularly darkened by slaty bases of the hairs	Evident, though not very bright
Rufous, back little darker than flanks	Silvery	Slight trace
Darker than *A. h. cumbrae*; flanks are light	Nearly clear silver	At most a feeble trace
Resembles *A. h. maclean* but less contrast than in that form	Silvery	Less distinct than *A. h. maclean*
Darkest form of *A. hebridensis*	Silvery	Faint
Rufous, but median back spangling	Silvery	Slight
Reddish	Silver, although darkened by slaty bases of hairs	No trace in 8 specimens; bright spot continued as yellow spot in 9th
As in *A. h. tirae*	Nearly clear silver	Hardly a trace
Peppery-reddish-brown, with dark dorsal line well-developed	White, although washed with brown along median line	Variable
As in darker, less russet specimens of *A. s. sylvaticus*	Peculiar dull bluish white; faint buffish tinge, sometimes present	Usually absent, but occasionally represented by tuft of brownish hairs
As in *A. f. fridariensis*	As in *A. f. fridariensis*	Small spot constantly present
As in *A. f. fridariensis*	As in *A. f. fridariensis*	Normally no trace
More vividly coloured than *A. s. sylvaticus*	Frequently washed with buff	May be present

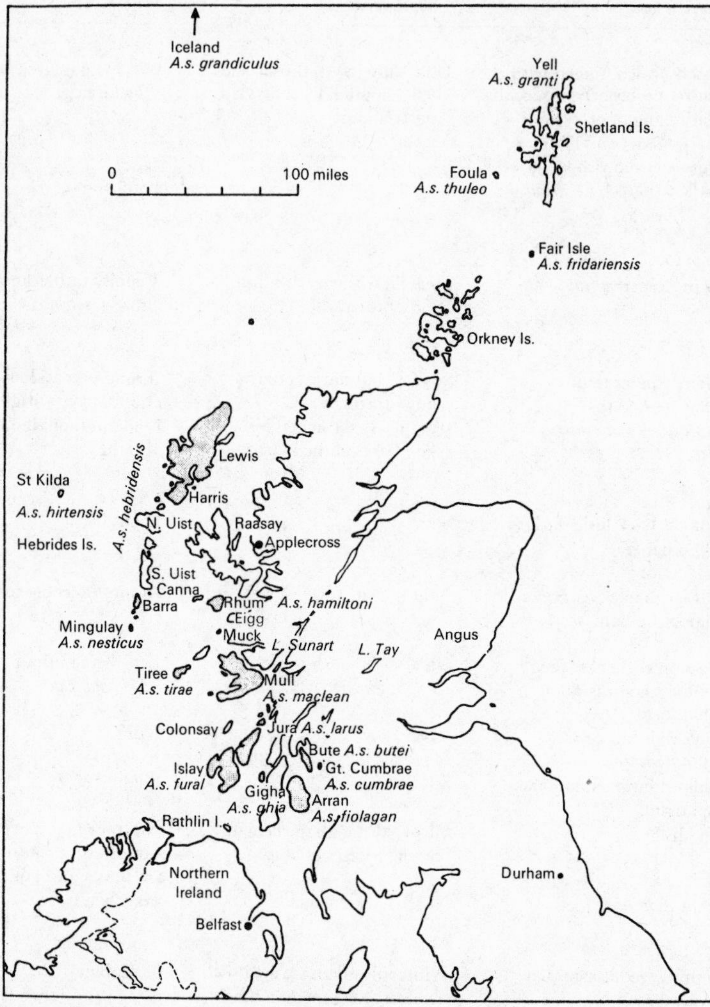

FIG. 73. Sub-species of Field Mice (*Apodemus sylvaticus*) which have been described in the northern part of the United Kingdom (from Berry, 1969*a*).

a method of ageing the animals in his samples (Delany and Davis, 1961). This was important. Previously workers had tended to take the largest animal in their collections as a 'typical' fully-grown one, ignoring the fact that there is a considerable amount of adult size variation in all rodent populations (just as in all mammal populations), or that a small sample may contain no old animals. Indeed at one time it was suggested that there was a distinct Irish

Field Mouse: but it turned out that the small, dark mouse that was the type of this was nothing more than a young adult which had not attained full size and typical coat colour.

To return to Delany's work: the mice he caught in any place differed in age and sex, quite apart from any inherited variation. Accordingly Delany 'adjusted' all his measurement with his tooth-wear criteria to take account of age differences, thus giving a revised set of measurements corresponding to animals of the same age. Next he further 'adjusted' the values in terms of skull length, so that the measurements of each animal were 'as if' it had a standard skull length. This meant he could detect differences in body proportion between samples. At last it was possible to compare accurately mice from different places.

A series of differences emerged from Delany's comparisons. For example, in the Hebrides 'the mice from Rhum are the largest with those from North Uist, Barra and Colonsay next in size. Tail length is variable, being longest in the mice from Rhum and shortest in those from Raasay. . . . The animals from the Outer Hebrides (except Lewis), Colonsay and Rhum are lighter coloured with those from North Uist particularly pale. The more richly coloured mice come from Raasay and Rhum and those with least colour from North and South Uist. . . . The Barra and Uist mice . . . are of a rather stockier build than the mice from Rhum – which appear of rather similar proportions as the mice from Applecross (on the Scottish mainland)' (Delany, 1964).

And so on. It is obviously possible to continue this sort of descriptive comparison almost indefinitely. Every population differs to a greater or lesser extent from every other one. Delany concluded that of all the island populations he examined, only the Rhum population was distinct enough to retain a subspecific name (*A. sylvaticus hamiltoni*). 'For the remainder, differences are so small between island and mainland populations that identification of individuals becomes impracticable'.

Is it possible to make any sense of the relationships between the island races? The classical theory of their origin assumes that they all came from the same stock (which was overcome and replaced on the mainland by a wave of immigrants after the various islands had become cut off from the main land mass). This would mean that there must have been a major population turnover on the mainland, and that the islands were separated before the 'improved' continental mice arrived, but after the original colonizers had penetrated to those mainland peninsulas which were going to become islands. There is no evidence for either of these inferences. Neither is there any support for the further implication that the island races are primitive relics, only able to survive when protected from their relatives*: many of the island voles and

**Footnote*. This argument is concerned with British mice only. There are, of course, plenty of examples of relics surviving on islands through their isolation from predators. The most spectacular is probably the South American marsupial radiation which was almost entirely eliminated when the Straits of Panama formed and the more efficient eutherian carnivores could cross from North America.

mice are less aggressive than their mainland relatives, but this is not necessarily a primitive feature.

The question of bridges between islands and mainland is one of the more confusing situations in the twilight where geologist and biologist meet. Virtually all land-bridges have been invented by biologists to explain the distribution of animals and plants. (Thus: 'How did mice get to St Kilda?'; 'As they cannot swim or fly, they must have walked'; 'Therefore there must have been a land-bridge'.) Geologists read the demands of the biologists and are impressed by them, so they state in their writings that land-bridges must have existed – and what biologist faced with even a guarded geologist's statement, can resist accepting a professional authority for a bridge?

The problem would be much simpler if no land-bridges had ever existed, but clearly they have – between Britain and Europe, between the Americas, and so on. However, it can be asserted fairly definitely that there has been no land suitable for mice to cross from mainland Britain to (at least) Shetland, the Outer Hebrides, and St Kilda since the Pleistocene ice-sheets retreated – and it is highly unlikely that any refuges persisted so far north to allow mice to survive the Pleistocene.

This means that mice (and, for that matter, the vast majority of the terrestrial and fresh-water animals and plants) must have colonized the outer islands by means other than walking. At this point we leave fact and can only speculate, because it is rare to know the history of a colonization. However there are ways to help us to make reasonably informed guesses. The most important piece of evidence we can adduce is the degree of relationship between the present day races – between the mice on different islands, and between islands and mainland: if the island races are relics, any differences will be small and mainly the result of adaptation after their isolation; if on the other hand, even if a particular difference is large, if it is significantly less than others involving the same population, this will be strong circumstantial evidence for common ancestry between the two. This in turn will suggest that the distinctiveness of the island race has arisen through colonization by a small number of founders from the other population.

Delany failed to find any common pattern of similarities between Hebridean mice. The reasons for this were that he only examined animals from comparatively few islands, and that his statistical methods were designed to describe variation within a population and only secondarily relationships between populations. This does not mean that there is no pattern of relationship. Using the frequencies of non-metrical skull variants for comparisons I obtained a clear-cut pattern, showing that the closest mainland relatives of Hebridean and Shetland Field Mice are in Norway, and that they seem to have been introduced in the first place to centres important for the Scandinavian voyagers. In population language the Scottish island mice are 'Viking mice' (Berry, 1970*b*). But before describing the Viking mice colonizations, it is necessary to digress briefly to explain the meaning and value of 'non-metrical' variation in the skeleton.

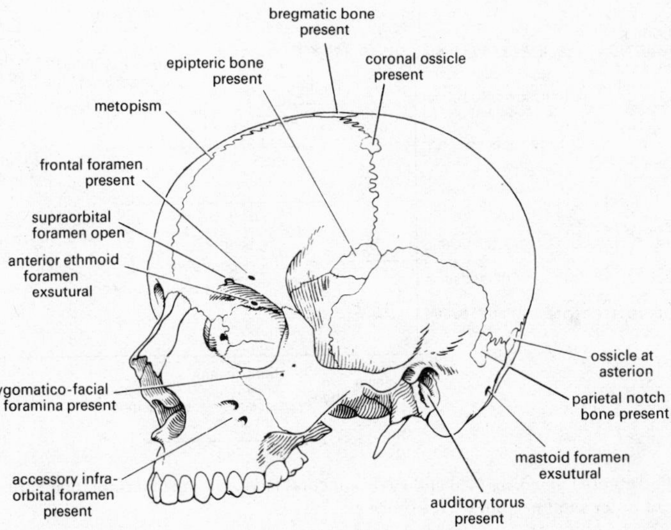

FIG. 74. Non-metrical variants in the human skull. All the traits labelled may be present or absent, and their frequency (of 'presence') is an inherited characteristic of the population (from Berry, 1968*a*).

Human anatomists have long recognized that minor variation occurs in the skeleton with characteristic incidences in different populations. One well-known example is metopism, that is the persistence of the mediofrontal suture of the skull in adult life rather than its obliteration when the bones become ossified. Grüneberg (reviewed 1963) pointed out that similar variants occur in laboratory mice, and that different inbred strains (*i.e.* mice brother-sister mated for many generations so that they can be assumed to be homozygous at effectively all their loci: they are sometimes called 'pure lines') also had characteristic frequencies. He crossed strains differing in the incidence of particular variants, and used a combination of genetical, embryological, and experimental techniques to work out how the variants were inherited. It turned out that they were 'threshold variants', dependent on the size of an embryological rudiment which might or might not reach a sufficient size to form a particular structure (such as a bony process, or the division of a single foramen into two). The size of developing organs is controlled by many interacting genes, and though environmental factors interact with genetical ones to determine how often the threshold is reached in any particular situation, the overriding determination seems to be genetical (Howe and Parsons, 1967).

One variant studied in detail in the mouse was the inheritance of 'missing third molar' (*i.e.* wisdom tooth loss). The actual 'decision' whether a tooth

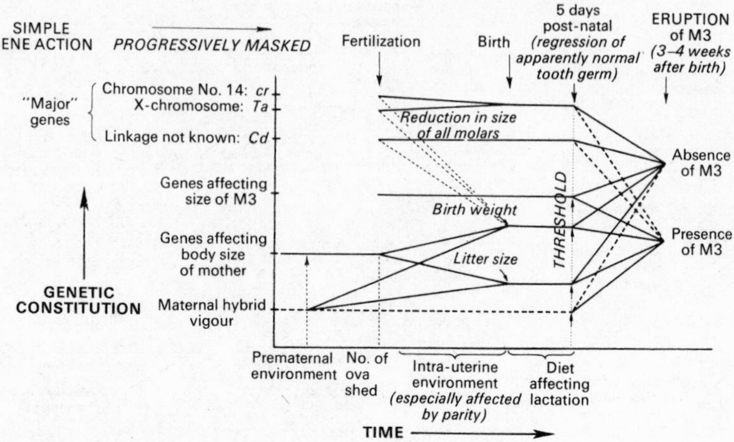

FIG. 75. Interaction of genetical and environmental factors in the determination of third molar (M3) loss in the mouse (from Berry, 1968*a*).

develops or not depends on the size of the tooth rudiment five days after birth. This is affected by genes controlling both tooth and overall size, as well as environmental factors (such as nutrition and maternal health) controlling growth before and after birth. The inheritance of the character is apparently the same in man (Berry, 1968*a*; Berry and Berry, 1967).

Grüneberg originally called non-metrical variants of this sort 'quasi-continuous' to draw attention to the distinct phenotypes produced despite the multifactorial determinants. Berry and Searle (1963) reviewed the spectrum of this sort of variation in a number of vertebrate species, and followed Waddington (1953*b*) in calling it 'epigenetic polymorphism' to emphasize the interaction between genetical and developmental (*i.e. epi*-genetic) events in individuals with different genotypes. Whatever the semantic niceties, non-metrical variants provide a useful way of comparing gene frequencies in different populations (since each variant is determined by a number of loci, and the occurrence of any one variant tends to be poorly correlated with the occurrence of any other). It is worth noting that many of the common congenital diseases in man (such as spina bifida, anencephaly, congenital heart disease, etc.) are examples of epigenetic polymorphism, so the method of inheritance has more than purely academic interest. Congenital dislocation of the hip (page 179) is a particularly well-worked out example.

To return to the British island races of field mice: I classified cleaned skulls for 20 non-metrical characters (Berry, Evans and Sennitt, 1967; Berry, 1969*a*, 1973), using the material collected by Delany, large series in museums, and some samples trapped by myself. This gave 20 frequencies for each population sample. Combining the differences between these frequencies for every pair of populations, *i.e.* comparing each sample with every other one, it was possible to calculate a single 'Measure of Divergence' for every comparison. Thus it was easy to determine the close and distant similarities for each population.

Delany had found that the Rhum mice were the most distinctive he looked at. This proved to be the same on the non-metrical skeletal statistic, both when the Rhum sample was compared with the Scottish mainland and with samples from the Outer Hebrides. When compared with the mice on all the main neighbouring islands, the Rhum population turned out to be much

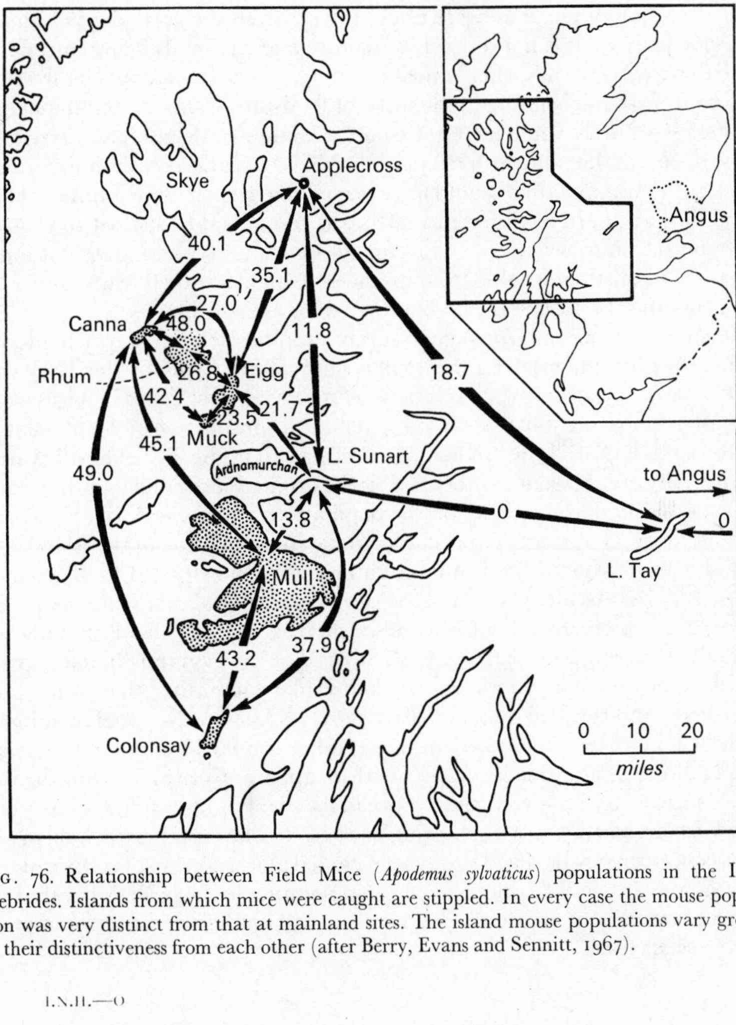

FIG. 76. Relationship between Field Mice (*Apodemus sylvaticus*) populations in the Inner Hebrides. Islands from which mice were caught are stippled. In every case the mouse population was very distinct from that at mainland sites. The island mouse populations vary greatly in their distinctiveness from each other (after Berry, Evans and Sennitt, 1967).

more like those of Eigg than anywhere else. Furthermore the other two islands in the group (which geologically is the remains of an ancient volcano south of Skye) were also more like Eigg than other populations.

We can now turn to recorded history: Eigg was a political centre from the early days of human organization in northern Britain. In the eighth century the Vikings raided there; by the thirteenth century the island was a seat of the Lord of the Isles: the Vikings must have had considerable dealings with Eigg in their exploratory journeys before they turned to conquest. In contrast Rhum never had a permanent settlement before the seventeenth century and it seems likely that the smaller islands of Canna and Muck were only occasionally visited from Eigg.

If we assume that there were mice on Eigg from the early days of human settlement there, it is not difficult to picture a few animals being introduced onto the nearby islands when someone landed – perhaps taking over livestock in a boat, with mice hiding in the straw or foodstuff. If this is what happened, only a few animals would have got ashore, and they could not have carried the same alleles in the same proportions as in their parental population. Straight away the new population would have been different at a large number of loci. It might have been so different that it was recognizably distinct to classical taxonomists, and we will have instant sub-speciation; alternatively it might have been indistinguishable from its ancestors in traditional traits, at least in the genes that affect colour, body proportions and the like.

If colonizing animals manage to survive and breed in an empty habitat in which they find themselves, they will occupy all the available territory with descendants resembling themselves: any further mice that land in future years are unlikely to contribute much to the genetical constitution of the population established, if indeed they manage to breed at all in the face of hostility from territorially organized incumbents. It is the original successful founding event that stamps its characteristics on the population: it is ironical that the Rhum population is so distinct yet probably young in occupation of the island.

The Rhum-Muck-Eigg-Canna argument can be extended to all the northern islands. As already noted, all the island populations are more like Norwegian mice than Scottish ones, whereas the Highland samples are virtually indistinguishable from each other. The mice on the Shetland group are distinct on each island, but are more like each other than mice from elsewhere; and the Yell ones are fairly closely related to Norwegian animals. Both St Kildan and Icelandic mice are rather similar to mainland Norway. The Hebridean situation is complex with an apparently primary introduction from Norway, and then secondary spread locally, but fits into the same story. The Uists are particularly interesting because the channels between them and Benbecula are only believed to have developed about 5000 BC, yet their mouse populations are very distinct – once again pointing to the probability that the present populations are descended from a small founding group in comparatively recent times.

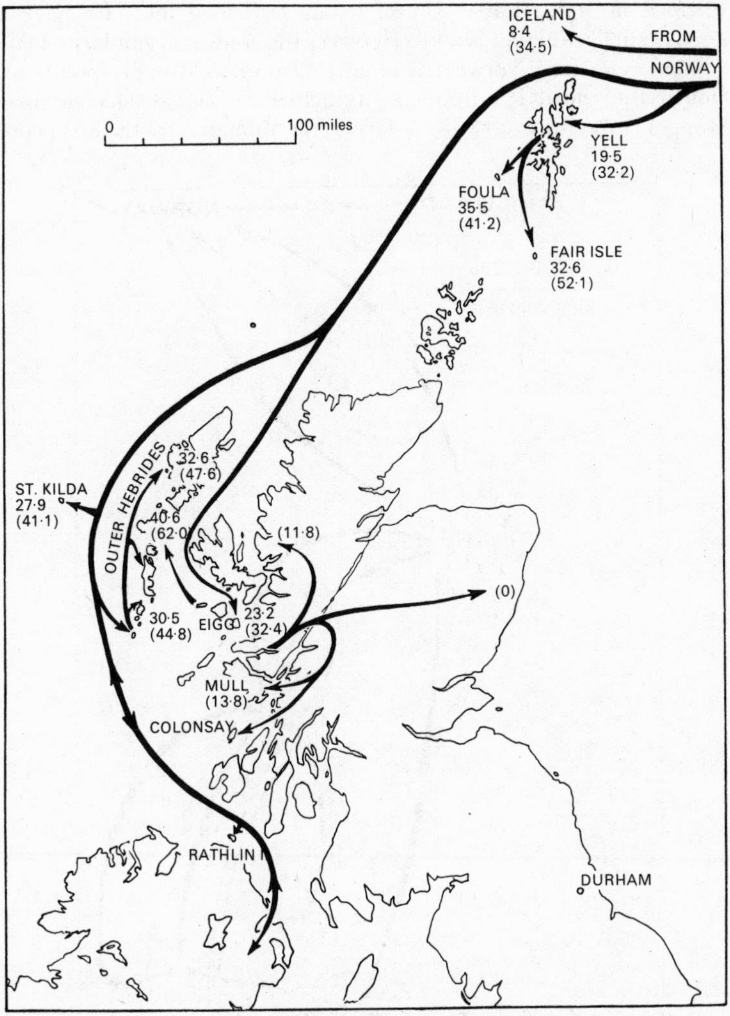

FIG. 77. Map to show the suggested routes by which the colonization of Shetland and the Hebrides by Field Mice (*Apodemus sylvaticus*) took place. The figures are estimates of divergence × 100 (the higher the value, the more distant the relationship), those without brackets being the 'distance' from the closest related population, and those in brackets the 'distance' from the Loch Sunart population (taken as typifying Scottish Highland Field mice) (from Berry, 1969*a*).

There is no indisputable way of telling how long the mice have been isolated on their different islands. However the consistent similarity between the island mice and Norwegian mainland animals strongly points to the Vikings as the culprits in introducing the field mice. This idea has circumstantial support from the history of sea-faring: the Vikings were the first people to

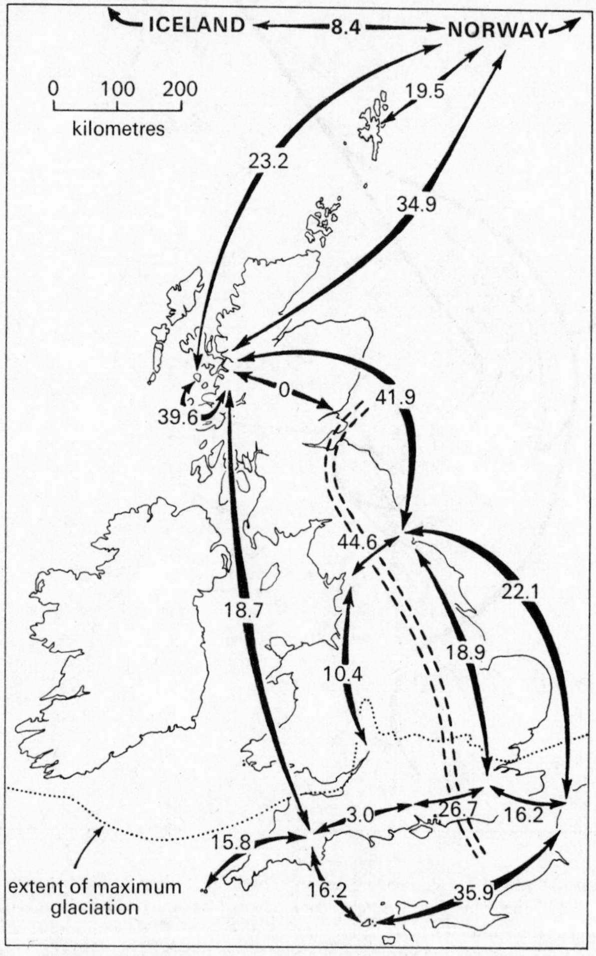

FIG. 78. Genetical distances between Field Mouse (*Apodemus sylvaticus*) population samples collected on mainland Britain and neighbouring islands and countries. Eastern and western British populations apparently form distinct groups; populations on northern islands are descended from a few founder individuals in historical times, but those on the larger southern islands seem to be relics of animals which survived the Ice Ages.

build boats fit for routine voyages in northern seas. Although itinerant farmers and monks colonized the off-shore islands from early times, it was only when efficient boats were made that it was possible to actively colonize and trade. The early Vikings must have often carried mice inadvertently, and some of these would be the ancestors of the present day populations.

The islands to the south of Britain (the Scillies in the west, and the Channel Isles to the east) are a test of the interpretation of the mouse relationships on the northern islands. The farthest extension of the ice in the Pleistocene was approximately from the Severn to the Thames; it is possible that truly relic populations may have survived on the southern islands.

When the inter-island relationships are determined (using the same characters and statistics as in the north), it turns out that the larger islands (Jersey, Guernsey, and St Mary's in the Scillies) have field mice not greatly different from those in Cornwall and west Britain generally. The small islands (Alderney, Sark, Herm, and Tresco) have much more distinct populations. The simplest explanation of this is that the population on a small island is more liable to extinction than the larger population of a big island, and that the small islands are now occupied by mice introduced from elsewhere and not the original pre-glacial survivors. Support for this comes from the close similarity of the Alderney, Herm and Sark populations to that on their nearest big island (Guernsey), and of Tresco to St Mary's. The situation in the southern islands is not fundamentally different from that in the northern ones.

Comparison of the island mice with British mainland samples revealed an unexpected set of relationships: all the island races are closer to mice from the western side of Great Britain itself, and when 'eastern' and 'western' British mice are compared with each other, they are found to be surprisingly distinct. In fact there seem to be two partially isolated forms of the common Field Mouse in mainland Britain, freely breeding among themselves so that samples within each 'race' are similar to each other, but have only restricted (or perhaps no) breeding success when they meet. No genes affecting the external appearance of the animals are involved in the difference between the forms. The line of separation between the two runs approximately from Edinburgh along the line of the Pennines to the south coast. The eastern race is much closer to mice on Continental Europe than is the western one.

The obvious hypothesis to explain the existence of the mainland races is that they are descended from groups which diverged in Pleistocene 'refuges' (or rather the western form would have done; the eastern form may have retained breeding links with the main part of the species in unglaciated Eurasia). Indeed Beirne (1952) believed that all British Field Mice were confined to a 'Celtic Land' refuge during the height of the Ice Age, in the area of the Atlantic coast of Britain and including a large area of land now under the sea between the present south coast of Ireland and the north-west coast of France. The separated races would have evolved differently, and developed their own genetical architecture. This would automatically reduce their inter-fertility. The irony of following Beirne in this guesswork, is that he was an arch-advocate of successive colonizations and multiple relics to account for differentiation between British races of animals. He wrote 'The Long-Tailed Field Mice of the genus *Apodemus* apparently invaded (the British Isles) on three occasions. The first invasion is represented by *Apodemus fridariensis*, of the Shetlands; *Apodemus hirtensis* of St Kilda; and *Apodemus hebridensis* of the Hebrides. They represent a population that inhabited the northern part of the Celtic Land and which was later divided into a number of smaller populations that in turn developed independently of each other. . . . A later invasion is represented by *Apodemus flavicollis* (the yellow-necked mouse) and a post-glacial invasion by *Apodemus sylvaticus*'. Now Beirne was wrong with his idea of simple splitting and adaptation, and it is *a priori* probable that the eastern British race represents not so much an invasion as a survival from the time when Britain was physically connected to the Continent. Nevertheless the way the present-day situation has arisen can be seen as a development from his over-simple 'land-bridge and relic' evolution.

ORKNEY VOLES AND MEADOW BROWN BUTTERFLIES

A parallel to the Field Mouse story, is the one concerning the Orkney Vole. This was originally described by J. G. Millais (son of the painter) who 'returning one evening from fishing (in Orkney Mainland) . . . noticed what looked like a Water-Vole running swiftly along the sheep track' (Millais, 1904). He subsequently obtained some dead specimens and described a new species *Microtus orcadensis,* nearly twice the size of the common British Vole (*Microtus agrestis*). It was later shown that this 'species' was a member of the Continental *Microtus arvalis,* and not of the species found in Britain, *Microtus agrestis.* Elsewhere in the British Isles *M. arvalis* only occurs in Guernsey, where it was also believed to be a distinct species (*M. sarnius*). However, Orkney, Guernsey and European Voles readily mate and produce fertile offspring (Zimmerman, 1959), and comparison of the different races using non-metrical skull variants (as with the Field Mouse races) showed that the Guernsey and Continental forms were virtually indistinguishable, although the Orkney Voles were more distinct and were more like Yugoslav voles than any others (Berry and Rose, 1975). Vole remains have been found in the earliest levels of the Orkney Stone Age village of Skara Brae, and it seems likely that they were imported by the earliest human inhabitants of the islands who were megalith builders from the Eastern Mediterranean. The Guernsey Vole is probably a true relic from the time when the Channel Isles were part of continental Europe.

Reconstruction of evolutionary sequences is a fascinating and futile activity, in that it can suggest mechanisms of genetical change but never prove them. However the *Apodemus* situation is important because of the implications that have been read from it into the distribution and differentiation of

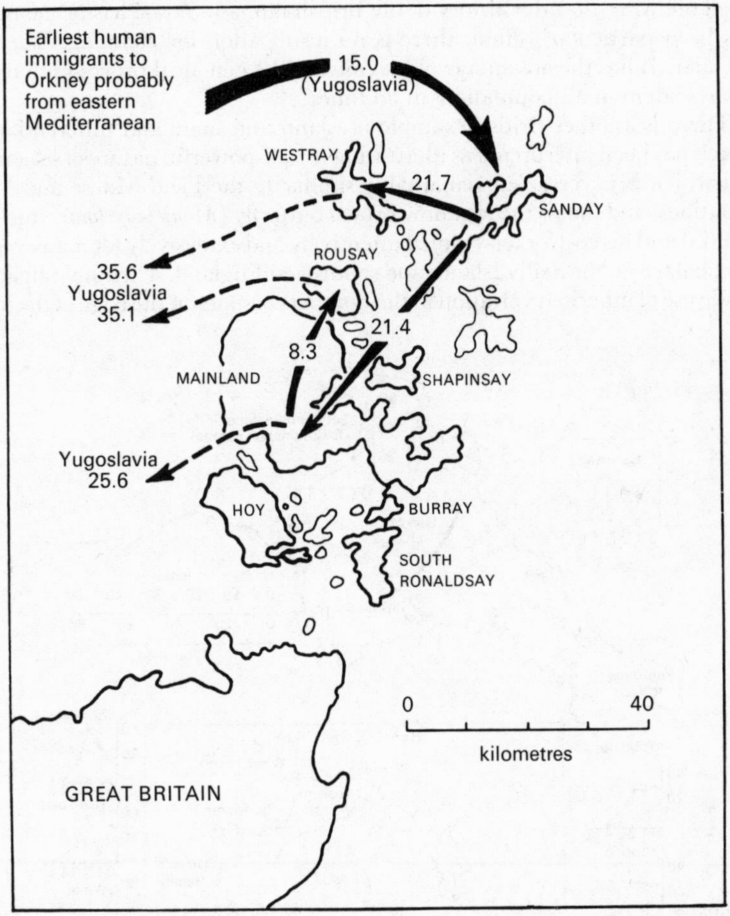

FIG. 79. The nearest relatives of the Orkney Vole (*Microtus arvalis orcadensis*) are apprently the voles of the eastern Mediterranean. Presumably the earliest human inhabitants of Orkney inadvertently introduced voles to the islands (probably to Sanday), and the voles spread to Westray and Mainland, and from Mainland to Rousay (from Berry and Rose, 1975).

the island races. For example both E. B. Ford and Charles Elton have attempted to describe and date the entire fauna of the Shetland Islands from an assumption that *Apodemus* 'walked' to the islands immediately after the retreat of the main ice sheet, while Kurten (1959) calculated the rate of possible evolution from the size difference between Norwegian and Icelandic mice. The truth is that populations are nothing like so ecologically and genetically constant over long periods as might appear from fossil lineages.

Even allowing that the history of the British fauna in post-Pleistocene times has been particularly fluid, there is no justification for claiming that it is peculiar. It has the advantage of magnifying the genetical processes that are always latent in all populations at all times.

There is another British example of island and mainland differentiation which has been interpreted as illustrating the all-powerful nature of selection, but which is nevertheless remarkably similar to the Field Mouse and Vole situations, and that is the Meadow Brown butterfly (*Maniola jurtina*) which E. B. Ford and his co-workers studied intensively and extensively for many years, particularly in the Scilly Islands and south-west England. Their investigation made use of inherited variation in the numbers of spots on the edge of the hind

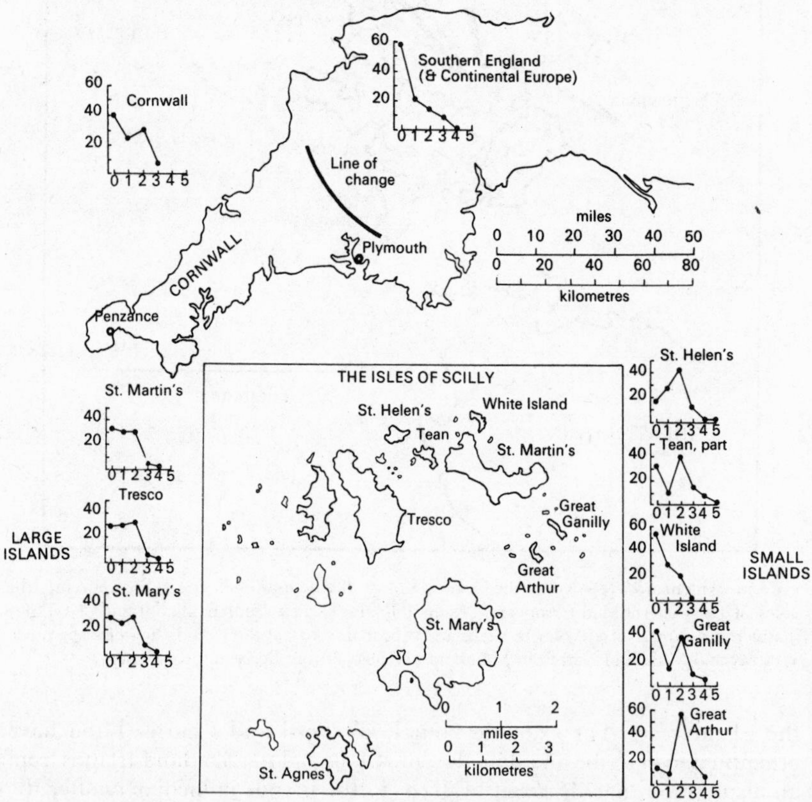

FIG. 80. Distribution of spots on the hind wing of female Meadow Brown Butterflies (*Maniola jurtina*) in Devon, Cornwall, and different Scilly Isles. The vertical axis of the graphs represent percentage of moths with spot numbers 0–5 (horizontal axis) (from data of Ford, 1955, 1964; Creed, Dowdeswell, Ford and McWhirter, 1970. Figure from Berry, 1971).

wing. The number of spots in males is variable but unimodal, most individuals having 2. Spot number in females differs in different parts of the range: from Devon eastwards across England and in continental Europe, most females have 0 spots; whilst in east Cornwall, the females have a bimodal distribution, one in five having 1 spot but more having 0 or 2 spots. In west Cornwall and the Scillies other expressions of the spotting character are present. The main 'southern English' pattern has been almost stable for over 20 years. Largely from this geographical evidence, Ford concluded that the genes affecting spotting were controlled by natural selection. Dowdeswell (1961) found a more direct connection between phenotype and survival: caterpillars collected near Winchester soon after hibernation gave rise to butterflies with a spot pattern very similar to that in butterflies raised from eggs in the laboratory, but very different to that found in the field. On the other hand, caterpillars collected late in the season were heavily parasitized with the hymenopteran *Apanteles tetricus* and suffered a high mortality rate; the few butterflies that hatched were identical to those flying in nature. Selection acts *via* differential elimination of different spotting phenotypes by the parasite, and can be calculated to have a strength of about 70%. Other populations of the species are subjected to even higher selection pressures as a result of bacterially determined mortality (McWhirter and Scali, 1965; McWhirter, 1969).

On different islands in the Scillies, the female spot patterns are different from those on the mainland. Those on the three large islands are similar to each other, while on five small islands, the distributions are each distinct). Following the interpretation argued for Field Mice, this would suggest that

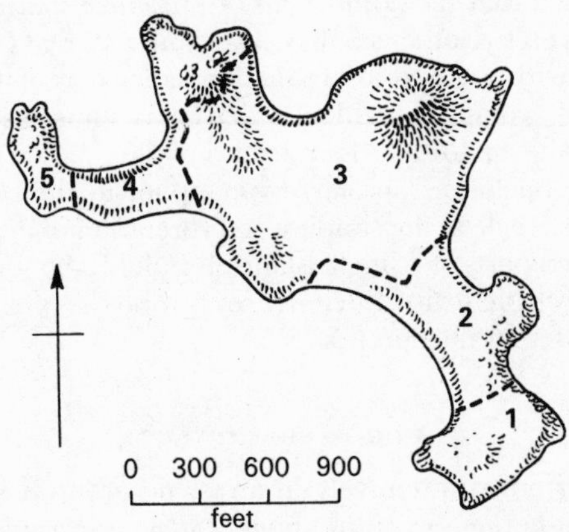

FIG. 81. The island of Tean in the Isles of Scilly showing the areas which at different times supported discrete populations of the Meadow Brown Butterfly (*Maniola jurtina*) (after Dowdeswell and Ford, 1953).

the original colonization by Meadow Browns led to a spot distribution similar to that found on the larger islands, while on the smaller islands, the chance genetical constitution of the founders has led to differing populations. Ford (reviewed 1971) has argued strongly against this interpretation, basing himself most importantly on the Meadow Brown situation on Tean.

Tean is one of the larger of the small Scilly islands, being about 1000 yards long with an area of 35 acres. Before 1953, there were two distinct habitats on the island: land suitable for *M. jurtina* where gorse, bracken, bramble, and long grass predominated; and two windswept necks of lawn-like turf respectively 200 and 150 yards long where butterflies were never found (although they were often seen setting out to fly over the 'lawns', they always turned back after about 10 yards). Ford's estimates of the butterfly populations on the favourable habitats of Tean were about 3000 imagines (area 1), 15,000 (area 3), and 500 (area 5). From 1946 to 1950 the female spot-frequencies were similar in all three areas each year – being bimodal with a lesser mode at 0 and a greater at 2. In 1951 the middle region changed to give a greater mode at 0 and a lesser at 2, the other two areas remaining constant.

Now about the time of the change (autumn 1950), a small herd of cattle which had been maintained on Tean, was removed. By 1953 the grass on the two 'lawns' was long, and *M. jurtina* was found flying freely over both. However the centre of the island had also become an impenetrable jungle of gorse and bracken through the removal of the cows, and the island butterfly populations were still not continuous, although there were now two sub-populations in place of the previous three: one group was unchanged in its spotting pattern from that previously existing, the other had become unimodal at 2 spots (Ford, 1964). This situation then persisted.

Ford has argued that the adjustment of *M. jurtina* spotting on Tean (and similar examples on two other small Scilly islands – Great Ganilly and White Island, together with a temporary reduction in one part of the larger island of Tresco) provides 'complete evidence that it is unnecessary to appeal to "intermittent drift" or to the "founder principle" . . .' (Ford, 1964). He is justified in this conclusion, insofar as it applies to the *changes* in spotting frequencies. Nevertheless, the simplest interpretation of differences between islands (or isolated parts of a single island) is indubitably that different areas were colonized by genetically different groups, and that selective adjustments in each led to different phenotypes.

CUCKOO SPIT INSECTS

Another species studied intensively in an island group has been one of the polymorphic frog-hoppers or spittle-bugs, *Philaenus spumarius*, in Finland. In pre-adult life *Philaenus spumarius* is protected by 'cuckoo-spit', produced by forcing air into a fluid exuded from the anus. Adults are highly polymorphic: about 80% of the adult females and 95% of the males have the typical

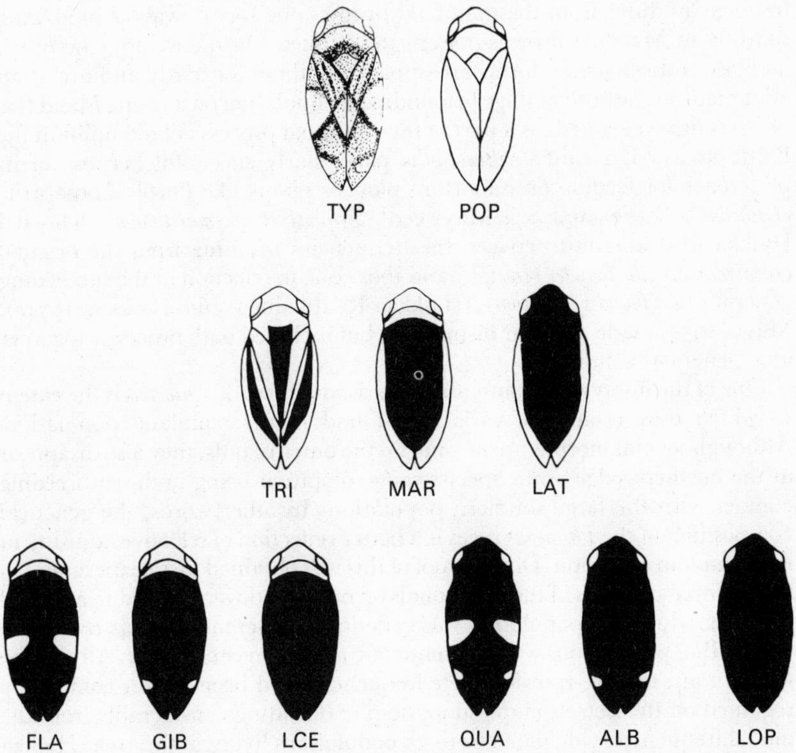

FIG. 82. Colour morphs in the spittle-bug, *Philaenus spumarius*. The common form is *typicus* (TYP); the other forms are *populi* (POP), *trilineatus* (TRI), *marginellus* (MAR), *lateralis* (LAT), *flavicollis* (FLA), *gibbus* (GIB), *leucocephalus* (LCE), *quadrimaculatus* (QUA), *albomaculatus* (ALB), and *leucophthalmus* (LOP) (from Halkka *et al.*, 1973, who describe also the inheritance of the different phenotypes).

greenish brown marking of the species, but there are at least 10 distinct morphs differing from *typica* in colour and marking. Some of the morphs are probably maintained by frequency dependent selection by bird predators (Thompson, 1973). The morphs are determined by a series of alleles, with *typica* being recessive to other morphs (Halkka, Halkka, Raatikainen and Hovinen, 1973). Throughout southern Finland, the frequencies of the alleles in females deviate little from a situation where p^t occurs at 91%, p^T at 1.5%, p^L 3%, p^M 2%, p^c 1%, p^o 1%, and p^F 0.5%, but p^t (determining the *typical* morph) increases at the expense of the others to the north, while p^c and p^o increases at the expense of p^t in more humid areas (Halkka, Raatikainen and Halkka, 1974*a*). In the Baltic islands to the south of Finland, the allele frequencies differ from the mainland in the same sort of way as hind-wing spotting in Meadow Browns does in the Scillies. On the islands, the *typica*, *lateralis*, and *marginellus* forms are apparently the most hardy and are at an advantage in the critical stage of founding a population on a young island (*i.e.* one recently vegetated, as a part of the continual process of land uplift in the Baltic area). Thereafter *trilineatus* is particularly successful because of its preference for feeding on important pioneer plants like Purple Loose-strife (*Lythrum salicaria*) and Sea Mayweed (*Triplospermum maritimum*). This led Halkka to distinguish between the frequencies resulting from the original colonization (*the founder principle*) and those due to selection in the succeeding generations (*founder selection*) (Halkka, Raatikainen and Halkka, 1974*b*). Mayr (1954) made a similar distinction, but included both processes together as a 'genetic revolution'.

One of the interesting points about the island races of *Philaenus* is the extent to which they retain the variability found in the mainland population. Although several morphs are missing on the outer islands, they also disappear at the northern edge of the species range despite it being in direct breeding contact with the large southern population. In other words, the genetical composition of the *Philaenus* races is a better reflection of selective adjustment than random accretion. Direct proof of this was obtained in an experiment in which three-quarters of the individuals on one island were moved to a second island in which the population was genetically different, and *vice versa*. The Spittle Bug populations were estimated at 11,000 on each island. After three generations, the pre-transfer allele frequencies had been almost completely regained on the islands in question, despite deviating considerable from the modal frequencies calculated from 35 populations living in the area (Halkka and Halkka, 1974).

DOG WHELKS

A species which shows apparently 'arbitrary' variation around the British Isles is the Dog-whelk, *Nucella lapillus*. The strength of selection acting on variations of whelk shapes has already been described (page 138). However whelks also manifest visible variation in shell colour and banding. Moore (1936) claimed that the common white or yellow shells could be changed to purple by a diet of mussels. Whilst this is certainly an over-simplification (for example, white individuals are frequently found in a mass of coloured whelks feeding on mussel beds), there may be some truth in it: coloured shells could be the result of a mutation which prevents dietary pigments being digested. A parallel for this would be the allele in rabbits which prevents its possessors removing the colour from xanthophyll, or caterpillars which store poisons from their food plants, but are harmless to predators if reared on innocuous plants such as lettuce or carrots (Rothschild, 1972). On the other hand, shell banding in Dog-whelks seems to be a truly inherited trait, although the details are difficult to sort out because whelks breed in clusters and females are alleged to submit to multiple mating. However, all who have reared whelks in a laboratory agree that the frequencies of banded forms in young whelks are similar to those in the parental generation (Largen, 1966; Coombs, 1973).

Most Dog-whelks around Britain are non-banded, but there are certain areas where banded forms reach high frequencies. The chance of finding banded whelks is greatest in areas exposed to heavy wave action, but the correlation between banding and exposure is only 33%, and there are so many exceptions to it (*e.g.* highly exposed populations which are entirely unbanded), that it is unlikely to be causal.

The simplest form of banding in the Dog-whelk is an even pattern of thin pigmented lines running round the shell. This may be modified by the suppression of particular bands, and by the partial or complete fusion of bands to produce one or more broad bands (usually about three). The intriguing observation is that almost every locality that has banded whelks,

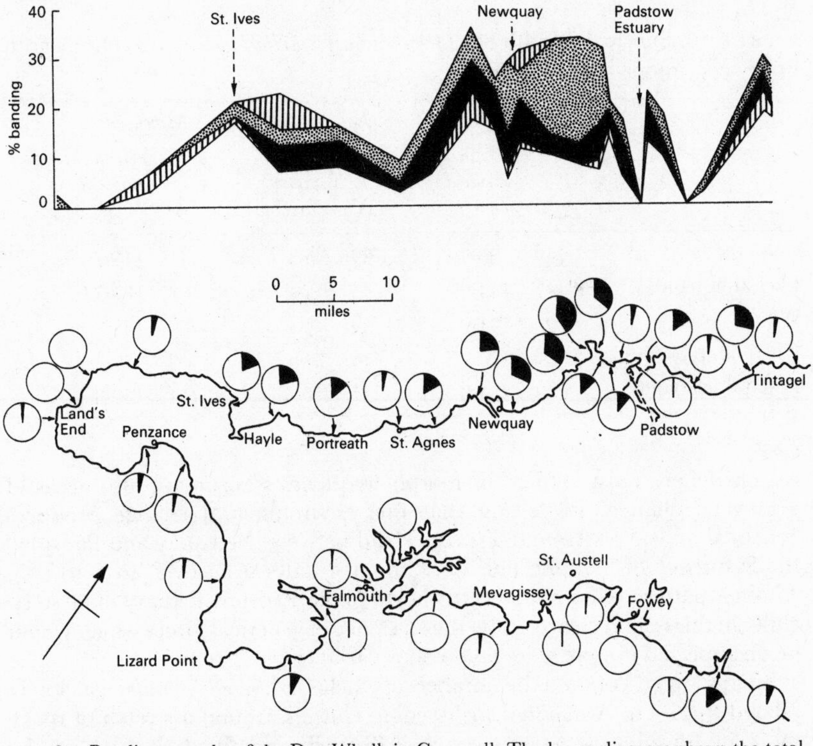

FIG. 83. Banding morphs of the Dog Whelk in Cornwall. The lower diagram shows the total frequency of banded shells; the upper one the frequency of different banding morphs along the north coast of Cornwall (for details see Berry and Crothers, 1974).

has a unique pattern of fusion which is found nowhere else. Moreover in a number of cases the same unique form occurred in the same place 50 or more years ago as revealed by old collections in the British Museum.

The most spectacular example of the persistence of local forms is found around Newquay in north Cornwall, where 15 distinct morphs can be recognized (Berry and Crothers, 1974). One of these (a black form with thin white bands) was illustrated by Cooke in the *Mollusca* volume of the Cambridge Natural History (published in 1895) and is still commonly found around Newquay, but nowhere else in Britain. Cooke believed that the variation in the Newquay region might be related to crypsis on the underlying rocks. This is unlikely, both because mass predators of Dog-whelks do not exist, and because the south coast of Cornwall has very similar geology to that of the north but few banded whelks. The most likely explanation is that the banding morphs are merely visible expressions of allelic variants concerned with survival traits. In the Newquay area there are clear differences between morphs in their resistance to desiccation in the laboratory (Table 19), and

TABLE 19. Survival of different Dog-whelk (*Nucella lapillus*) morphs at 50°F in dry conditions

	Newquay I		*Newquay II*		*Minehead*	
	N	*Mean survival (hrs.)*	*N*	*Mean survival (hrs.)*	*N*	*Mean survival (hrs.)*
Coloured	36	99.7	30	65.3	41	117.7
Even thin bands	27	114.0	35	65.4	32	128.7
White	61	121.0	36	62.2	50	144.3
Unequal thick bands	38	117.2	45	70.9	0	—
Equal thick bands	20	154.9	27	80.0	14	136.3

regions where most change in morph frequencies occur are also areas of geological change, suggesting that fine environmental change produces genetical pressures. Around Trevose Head between Newquay and Padstow, the frequency of banding falls over about 3 miles from over 40% to 0%. Another pattern of banding occurs north of the Padstow Estuary. The steep clines in this region clearly show the existence of genetical forces which would be unsuspected if there were no obvious variants.

Spight (1973) counted the numbers of whelks (of *Nucella lamellosa*, a closely related species to *N. lapillus*) in breeding clusters around a stretch of rocky shore in Washington State, western USA. He identified sites where the clusters occurred in successive years, and by marking individuals, concluded that the whelks found apparently continuously along the shore were divided

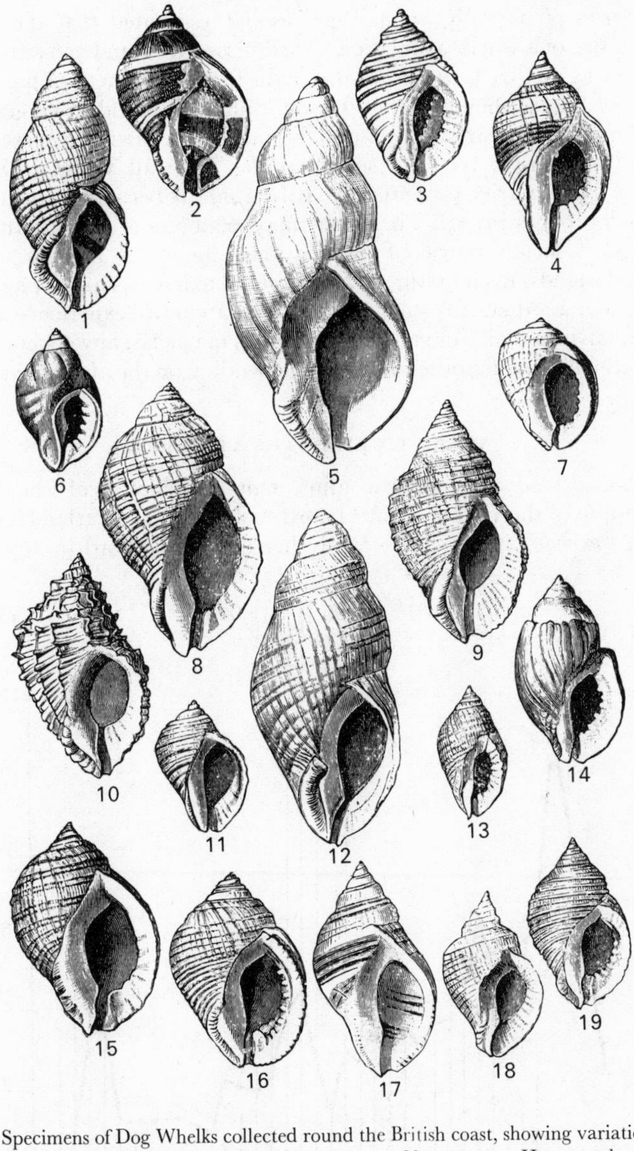

FIG. 84. Specimens of Dog Whelks collected round the British coast, showing variation in size, shape, and colour: 1, Felixstowe, sheltered coast; 2, 3, Newquay; 4, Herm, rather exposed; 5, Solent, very sheltered; 6, Land's End, exposed rocks; 7, Scilly, exposed rocks; 8, St Leonards, flat mussel beds; 9, Robin Hood's Bay, sheltered under boulders; 10, Rhoscollyn, oysterbed 4–7 fathoms; 11, Guernsey, rather exposed rocks; 12, Conway estuary, very sheltered; 13, 14, Robin Hood's Bay, very exposed rocks; 15–17, Morthoe, rather exposed rocks; 18, St. Bride's Bay; 19, Loch Swilly, sheltered (from Cooke, 1895).

into a series of partially isolated groups. He estimated that the effective breeding size of a whelk population varied between 10 and 700 animals.

This brings us back to one of the main facts about real populations of animals or plants: they need not live on a physical island to be genetically isolated from their nearest population. Distance or ecological preference may be equally effective. We have seen that the Tingwall Valley in Shetland acts as a flight barrier to Autumnal Rustic moths because of firm habitat preferences (pages 171–3). There are many examples of animals and plants having such locally-restricted distributions – living, as it were, on their biological islands. Even marine organisms tend to live in shoals or aggregates of one sort or another. Any such isolated population will experience a slightly different environment from any other, and, as in whelks, any genetical forces generated will produce different results depending on the available genetical variation.

SCARLET TIGER MOTHS AT COTHILL

The classical study of genetical adjustment in a biological island, is the investigation of the Cothill (near Oxford) colony of the Scarlet Tiger Moth (*Panaxia dominula*) begun by R. A. Fisher and E. B. Ford in 1939. Most

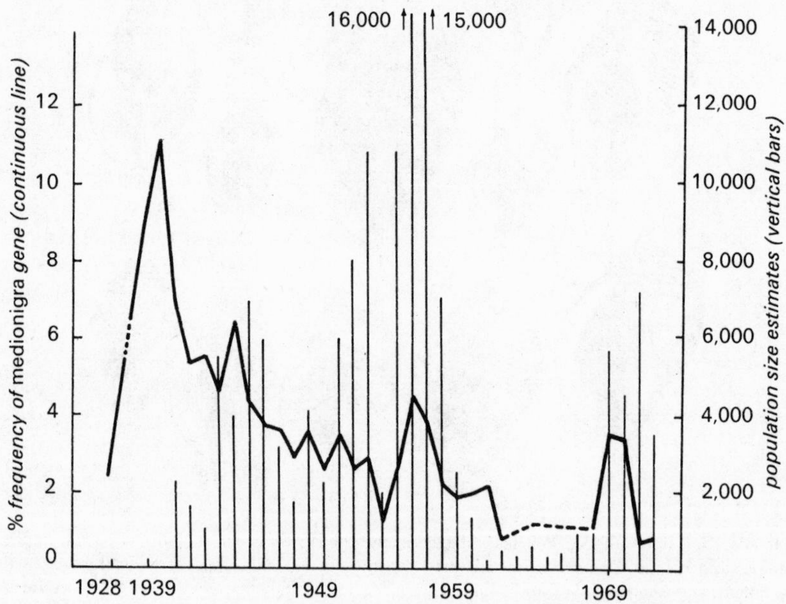

FIG. 85. Variation in the size of a colony of the Scarlet Tiger Moth (*Panaxia dominula*) at Cothill, Berkshire (vertical bars), and the frequency of the *medionigra* allele in the colony (continuous line) (after Berry, 1971, based on Fisher and Ford, 1947; Ford, 1964; Ford and Sheppard, 1969).

colonies of this species are relatively invariable for wing pattern and colouring, but the Cothill population has a variant *medionigra* unknown elsewhere (Fisher and Ford, 1947).

During the period when the colony was actively studied, the adult population of the moth in different years fluctuated from an estimated 15,000 to less than 1,000, while the frequency of the *medionigra* allele ranged from 11% to 1%. The important finding was that there was no simple relationship between colony size and *either* gene frequency *or* change in gene frequency. In fact the observed change in *medionigra* frequencies is consistent with the phenotype being at an advantage until 1939, and a net disadvantage thereafter until about 1962 except for two years (1955–57) when it again conferred an advantage on its possessors. One of the factors keeping the gene in the population is mating preference producing an excess of crosses between dissimilar phenotypes (page 65). The other known pressure on *medionigra* all tend to act against it: selective predation by birds (although there could be frequency-dependent selection favouring the rare phenotypes); a 50% mortality of eggs and larvae in *medionigra* when compared with the typical form; and reduced fertility in *medionigra* males. However, the actual selective pressures do not matter in the present discussion; what does matter is that they override any possible effects of drift producing random fluctuations in frequency.

EDGE EFFECTS AND COLONIZERS

Particular genetical processes act at the edges of species and population ranges and complicate the picture of simple islands, but do not change it. The position is that as a species reaches the limits of the conditions to which it is adapted, it is increasingly likely to meet circumstances with which it cannot cope. Examples of this are small warblers faced with low temperatures in high northern latitudes, marine organisms subjected to fresh-water estuaries, or cryptic coloured organisms on the edge of a polluted or geologically different area. In such circumstances, a high degree of variability would probably be advantageous, because individuals able to tolerate particular adverse conditions would be more likely to occur. This is rather nicely illustrated by some species of *Drosophila* which have a lower number of inversions at the edge of their range than at the centre (Carson, 1955, 1959). Since the effect of inversions is to reduce crossing-over and maintain particular gene sequences, the absence of inversions results in freer recombination and a greater release of inherited variation. In the metaphor of genetical architecture, the rigidity of an 'approved design' is relaxed in favour of more flexible structures.

Individuals in species nearing the edges of their ranges are difficult subjects for research – it is often difficult to distinguish between active members capable of breeding with the main body of the species, and erratic vagrants like American waders in Britain or exotic plants escaped from gardens; or even to identify adaptations to extreme conditions. For example, *Cepaea nemoralis* reaches its northern limit in Britain between Montrose and Stonehaven on the east coast of Scotland (Jones, 1973*b*). No brown shells occur in the area, although that morph is at a high frequency in frost hollows in the Marlborough Downs, and has been suggested to be particularly cold resistant. No clear association occurs between the frequency of yellow or banded shells and height, aspect, or slope of the sites sampled.

In other cases only the widest generalizations can be made. For example, many colonizing (or weed) plant species forswear sexual reproduction as a means of extending and consolidating their ranges; aphids have one of the most effective mechanisms with alternating winged and wingless generations. A frequent tactic is for a species to change from a dispersal to a residential phase at different stages of its life-cycle. An alternative manifestation which is particularly relevant for peripheral members of a species, are those forms in which different individuals have different capabilities or inclinations for colonizing or dispersing ability.

There is an intriguing situation in the large colony (*c.* 1,000 breeding pairs) of Eider Duck (*Somateria mollissima*) on the Forvie reserve north of Aberdeen (Milne and Robertson, 1965). About two-thirds of the population migrates southwards after the breeding season and overwinters largely on the Firth of Tay; non-migrants remain at Forvie throughout the year. Pair formation takes place in both groups before the migrants return so that the migrant and sedentary birds are reproductively isolated. The migrant birds have an allele frequency of 27% of an electrophoretically detected variant of the egg albumen; the frequency in the sedentary group is 14%. Both groups have the number of heterozygotes expected on the Hardy-Weinberg distribution. The frustrating thing is not knowing whether the genetical difference determines the migratory difference or is merely a secondary result of reproductive isolation.

However there is growing evidence of genetical involvement in the population cycles of *Microtus* voles, where the increase phase is marked by a restlessness and increased movement of animals and the decrease phase by retrenchment into the most favourable habitats. Lemming 'migrations' are probably no more than an extreme example of the former effect.

Vole cycles in Britain have been studied particularly by Dennis Chitty (1952). He systematically eliminated all relevant environment variables as possible causes of population fluctuation (*e.g.* weather, food availability, disease, predation), and was forced to postulate the existence of 'intrinsic' factors in the voles themselves as the causes of the 3–4 year cycles. The hypothesis that best fitted the facts involved natural selection acting in opposite directions during the increase and decrease phases of a single population cycle: as numbers grow and contacts between individuals increase, aggressive animals will have a disrupting effect and a low fitness; in contrast, docile animals will be at a disadvantage when numbers are low, resulting in an increased fitness for wide-ranging, aggressive phenotypes. Assuming that the relevant behavioural patterns are inherited, this will produce a complex interaction of behaviour, reproduction, and genetics that might produce wide and regular fluctuations in number (Chitty, 1967).

Chitty's ideas have been explored by C. J. Krebs (Krebs, 1964; Krebs, Gains, Keller, Myers and Tamarin, 1973). He found that fencing a group of voles in a field so that they could not escape (although predators – foxes, cats, weasels, snakes, owls and hawks – and all other extrinsic forces were acting on the voles) resulted in a continuous increase in numbers until all food had been eaten, and the voles starved – a situation which does not ordinarily occur under natural conditions. In other words, dispersal and 'spacing behaviour' is part of the normal method of population control. This was borne out when Krebs studied unconfined populations: almost two-thirds of the animals were lost through dispersal during the time of population increase, but only one in seven during the decline. During the cycle, changes in the frequencies of two electrophoretically detected alleles also took place, and there is no doubt that dispersing and resident animals were genetically different from each other on the average.

Krebs worked in North America; Semeonoff and Robertson (1968) found a parallel situation in Voles (*Microtus agrestis*) in the Carron Valley, Stirlingshire. They studied an esterase polymorphism in blood taken from voles living along the valley. One allele was consistently commoner in part of their area. The interesting point is that this part was periodically flooded and had to be recolonized – and the colonizers were consistently different from the main population.

The weak point for the acceptance of the Chitty-Krebs hypothesis is our ignorance about the genetical control of aggressiveness in voles. Krebs has shown that aggressiveness of individuals is different at different stages of the population cycle, but has so far been unable to link observed genetical differences with the putative driving behavioural changes. Indeed Lidicker (1973) has argued from a 13-year population study of Voles (*Microtus californicus*) on a small island in San Francisco Bay that the factors controlling population size are more complex than Krebs wants to believe. Lidicker maintained that the simple annual cycle of small mammals (in which there is a limited breeding season when the population increases, and a non-breeding season during which the population declines) is produced by largely density independent factors (weather and food); that this is lengthened into a two-year cycle by density-dependent factors in which the population reaches high levels only in alternate years; and that the two-year cycle may be extended into a three or four year one through predation pressure. As we shall see in the next chapter, marked genetical changes may take place during a simple annual cycle in House Mice, and this also complicates the interpretation of Krebs' results. Notwithstanding, there are clearly important genetical influences acting within vole populations, and their elucidation should provide a fascinating insight into the control of population processes.

Many (if not all) widely distributed species are genetically heterogeneous and their heterogeneities are often magnified under island or colonizing situations where selection may be particularly intense (Anderson, 1970). So far we have been concerned mainly with spatial discontinuities, although in the last chapter we saw a number of examples of 'genetical islands' – local adaptation of the Autumnal Rustic Moth in Shetland and area effects in *Cepaea nemoralis* were two examples described in detail. Another example which nicely links an odd genetical situation with territorial limitation is the introduction of male House Mice carrying a *t*-allele onto Great Gull Island in Long Island Sound. Recessive alleles at the *t*-locus have the peculiar property that although the mice are entirely normal in appearance, male heterozygotes transmit the *t*-allele to about 95% of their offspring instead of the expected 50%. Transmission by females is normal. Homozygotes die before birth due to a defect in neural tube (nervous system) development, but the allele is bound to spread in a population because of its transmission rate by males.* Most American populations of mice contain *t*-alleles, but none were present on Great Gull before six males were released at one end of the small (18 acre) island (Anderson, Dunn and Beasley, 1964). Although both the resident and introduced animals roamed over all the island, they were apparently much more restricted in their breeding. The introduced allele became established in the immediate area of release of its carriers, but spread very slowly – much more slowly than would have occurred if the Great Gull population was breeding as a single breeding unit (Bennett, Bruck, Dunn, Klyde, Shutsky and Smith, 1967).

Reference has already been made to the highly polymorphic state of littoral winkles and whelks. Many other shore species are similarly polymorphic – but only those with non-planktonic larvae. The Edible and Small Periwinkles (*Littorina littorea* and *L. neritoides*) both spend their young stages floating in the sea, and show very little colour or banding variation. When the extent of local restriction in movement is recognized, it becomes apparent that 'pools' of genes are far more common than 'oceans' (Ehrlich and Raven, 1969). As John and Hewitt (1969) say in discussing the implications of a finding they had made of an unequal distribution of extra

**Footnote*. When the properties of *t*-alleles were first discovered, their describer L. C. Dunn wrote 'here were animals which refused to distribute their genes in the way called for by the binomial calculus on which all gene frequency and equilibrium estimates in populations had been based. It was a horrid thought that if other animals or plants had learned to do this trick, but had skilfully concealed it from the investigators, then the results of hundreds of experiments in population cages or breeding plots might have a meaning other than that attributed to them. Of course no one had very serious doubts about this: Mendel, Hardy, Weinberg, Wright and Dobzhansky probably could not all be wrong at the same time' (Dunn, 1960). A number of other examples of similarly distorted segregations (through *meiotic drive*) have been described, but the phenomenon does not seem to be particularly common (Peacock and Miklos, 1973).

chromosome segments within a population of the Meadow Grasshopper (*Chorthippus parallelus*) in the New Forest: 'Even large populations are frequently broken up into breeding units of small size. Under such circumstances the actual amount of inbreeding that results is largely a function of the relative mobility of the organism (animals) or else of its gametes (plants). The mating possibilities are thus always greatest for individuals in the immediate vicinity of one another and tend to decrease with separation'. In its most extreme form this fragmentation of large groups is exemplified by the very rapid change in inherited characters found by Snaydon and Bradshaw at the boundaries between different soils (*q.v.* Bradshaw, 1972). It is shown also by neighbouring populations of Tormentil (*Potentilla erecta*) adapted to living in communities dominated by either Purple Moor-grass (*Molinia caerulea*) or by the Fescues (*Festuca* spp.); these are as genetically distinct as if they were separated by 200 miles (Watson, 1970; Watson and Fyfe, 1975). It has even been demonstrated in lichens (Culberson and Culberson, 1967).

Population sub-division seems to be almost as common as individual variability. One major consequence is that local groups become relatively inbred (Wright, 1965*a*). Most attention has been paid to this as a result of its implications for plant and animal breeding, and here a complex quantitative theory has been erected (Fisher, 1949). Notwithstanding, *any* group of animals or plants consists of a number of families or lineages sharing the same alleles because they share a common ancestor.

The correlation (or similarity) between the genetical make-up of individuals in the same area, is a measure of their common ancestry or inbreeding (in contrast to their *out-breeding* with unrelated members of the species). Consequently, it is possible to use measures of inbreeding in a population to identify the amount of subdivision and local differentiation (Wright, 1965*a*). For example, commercial breeds of cattle tend to be inbred because of the relatively small number of founders that have contributed to them, and the tendency to use proven bulls to inseminate a large number of cows. British Shorthorns twice went through a genetical bottleneck through the influence of the bulls Favourite (born in 1793), and Champion of England (born in 1859). By 1920, the inbreeding coefficient (*q.v.* page 000) had reached 45% (McPhee and Wright, 1926).

KIN SELECTION AND GROUP SELECTION

Now individuals in an animal population are related behaviourally as well as genetically and spatially. From Darwin onwards, attempts have been made to explain the evolution of behaviour beneficial to the group rather than the individual. Repeatedly geneticists have had to emphasize that it is individuals and not groups who survive and breed, and that there is no such thing as 'group selection', *i.e.* selection for traits beneficial to the group at the expense of the individuals (*e.g.* Williams, 1966; Lack, 1968). The classical expression of this intellectual frustration was Haldane (1955), who pointed out that if he dived into a stream to save a drowning child (which would be a *good* action from the community's point of view), he would be reducing his own chance of passing on his altruistic genes because he might not survive the rescue attempt, while the selfish character who watched the child drown without risking his life would have a higher fitness.

However, there is a way 'altruistic' behaviour can evolve, and that is if the drowning child and his rescuer are related, so that the risk to the genes carried by the rescuer is more than outweighed by the genes 'saved'. This was pointed out by both Fisher and Haldane; Maynard Smith (1964) has called the process 'kin selection', defining it as 'the evolution of characteristics such as self-sacrificing behaviour (like injury-feigning in birds or sterility in social insects) which favour the survival of close relatives of the affected individual'. Hamilton (1964, 1972, 1975) has formalized kin selection by defining a quantity called 'inclusive fitness', which adds the investment an individual has in his own genes to those he shares with his living relatives. Hamilton has shown mathematically that natural selection will tend to maximise this inclusive quantity in the same way as it maximizes individual fitness; selection will favour the activities of an individual which benefits his relatives and repress those which harm them.

In a book published some years ago, Wynne-Edwards (1962) argued persuasively for the existence of behavioural mechanisms for regulating the density of animal populations, but he was unable to suggest a plausible way for them to have evolved (Wynne-Edwards, 1963). His book was described by a scientist who declined to review it as 'either the most important biological work since the *Origin of Species*, or utter rubbish'. Since its publication, evidence has accumulated in a number of species for the reality of the effects he described. The most elegant work has been on the Red Grouse (*Lagopus scoticus*) in the Scottish Highlands, where density is regulated by a flexible territorial system, from which 'subordinate' males are excluded in proportion to the amount and quality of available food (Watson, 1970). Nevertheless, the problem remained that there was no known mechanism for making group selection work. It was only when the fact dawned that virtually all natural populations are heterogeneous compounds of families just like a country village, that the problem was resolved. Group selection involves much more than kin selection (Brown, 1966), but the understanding of population structure that has come from experimental work has shown that kin selection provides an acceptable explanation of the evidence about group behaviour collected by Wynne-Edwards. He himself accepts this synthesis, and it has encouraged him to draw a moral for our own species, and for all 'those working on the biology of dominance and aggression . . . on the process by which individuals are denied the rights of existence or procreation by their fellows. In real life animals do not get to the top just on the strength of a pugnacious temperament' (Wynne-Edwards, 1974).

CHAPTER 7

GENES, SURVIVAL, AND ADJUSTMENT IN AN ISLAND POPULATION OF THE HOUSE MOUSE

We have now reached the point of finishing the description of the main processes and problems relating to gene action and inherited pressures in natural populations. We have seen how genes segregate in families and determine the characteristics of populations, and discussed the operation of factors controlling stability and change of allele frequencies. The exposition of these topics has been illustrated by a range of examples from a variety of animal and plant groups. This method is all very well for those familiar with the species concerned, but it may have been confusing for the many whose acquaintance with natural history is with one group only, or largely *via* television. The purpose of this chapter is to cover many of the ideas set out in earlier chapters, but to draw them together by summarizing work on a single population of a species known to everyone, the House Mouse (*Mus musculus*).

In the 1950s, Grüneberg and his collaborators at University College London showed that many minor skeletal variants were inherited in laboratory mice (*q.v.* page 206f). Weber (1950) found the same characters in small samples of mice trapped in London buildings. It seemed possible that the same variants might be usable as markers to investigate genetical relationships between wild-living mice, and in 1960 I began collecting mice from a range of British localities for this purpose, aiming to get 50 mice in each sample so that accurate frequencies of the variants could be determined. The easiest way to do this was to hand-catch animals fleeing from corn-ricks when they were broken down for threshing in early spring. Field-living mice take refuge in the warm, dry habitat of ricks when they are built in the autumn, and live in a virtually ideal habitat through the winter. The colonizers of each rick can be regarded as a random sample of the local population, and the possibility of catching every animal means that no additional sampling errors are introduced. An incidental result of the change of farming practice from threshing stacked corn to combine-harvesting has been to reduce considerably the national population of House Mice.

The most informative results from the first studies came from a single farm at Odiham in Hampshire. The distribution and composition (*i.e.* wheat, oats or barley) of different ricks made it possible to analyse the effect of both biological and microgeographical factors on the occurrence of 36 variants (Berry, 1963). No obvious environmental factors affected the incidence of any

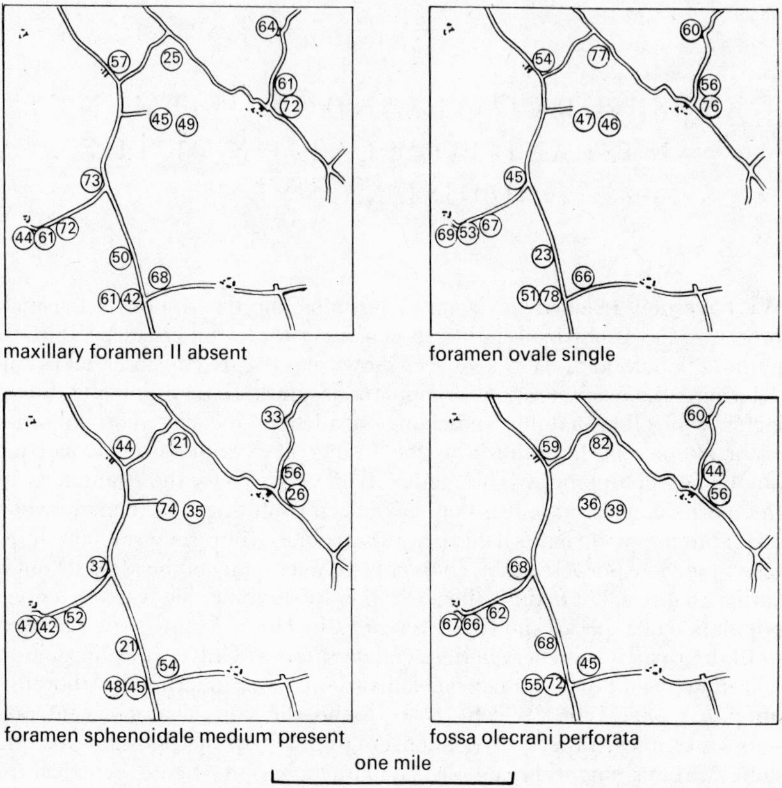

FIG. 86. Percentage frequencies of four non-metrical variants in 15 isolates of House Mouse (*Mus musculus*) populations in corn ricks – all derived from a single breeding population (after Berry and Berry, 1972, from data given by Berry, 1963).

variant, but it was disconcerting to find that many variant frequencies differed greatly, even between adjacent ricks. Subsequent experimental work on laboratory mice in Australia has shown that maternal health can influence the occurrence of particular variants, but overall the spectrum of non-metrical variants possessed by a mouse is an apparently valid indicator of a substantial portion of its genome.*

The problem was to interpret these observed inter-population differences. When the 20 Odiham samples were combined, the pooled sample fitted well into the general southern English picture, suggesting that it was the process of sub-division that had produced the inter-rick differences through the founder

*Genome is a useful word to describe the whole genetical composition of an organism.

effect (pages 58, 210–14), but there was no way of proving this. Pest control workers had described the basic ecology of mouse rick populations, but they left many genetically important imponderables: the degree of genetical heterogeneity in agricultural House Mice; amount of individual movement (and gene-flow); genetical distinctiveness between colonizers and refuge-seekers; the numbers of mice that invade different ricks; intraspecific selection in the closed community of a rick; and similar tantalizers. The answer to the problem set by the Odiham ricks had to come from a genetically closed population where environmental, demographic, and genetical pressures could be measured. This pointed to a long-term (*i.e.* several generation) investigation of an island population.

SKOKHOLM AND ITS MICE

At this point, I was introduced to Skokholm, which had a mouse population living apparently independent of man.

Skokholm is a remnant of the coastal plateau of the Welsh county of Pembrokeshire (now part of Dyfed) cut off from the mainland by marine erosion. It is 2½ miles from the mainland, and is 1¼ miles long by ½ mile wide, having an area of 242 acres. The surface of the island is relatively flat (apart from a number of rocky outcrops), sloping from about 150 feet above sea level at the Lighthouse (south west) end of the island, to less than 50 feet at the Neck end. The coastline is composed of steep rugged cliffs, at the foot of which are

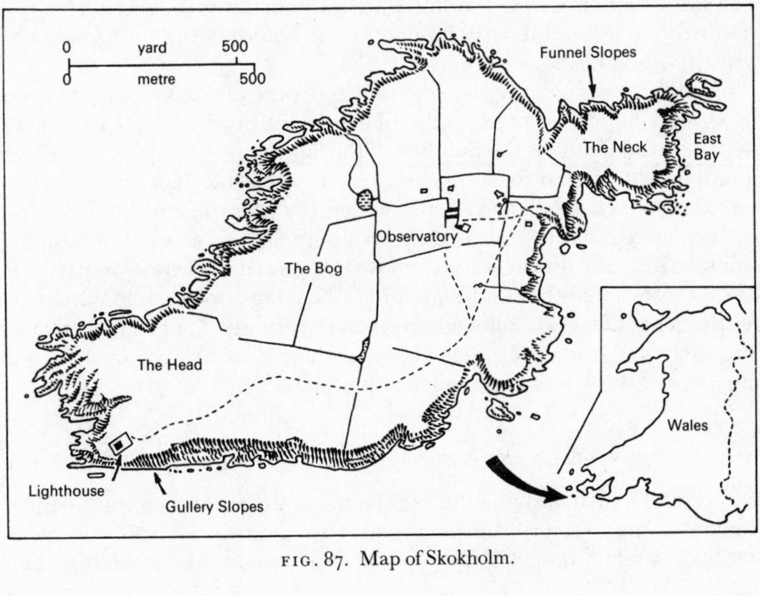

FIG. 87. Map of Skokholm.

immense boulders. The less exposed cliffs have growths of grass and Sea Pinks (*Armeria maritima*); while the top of the island is covered with maritime grass heath (Goodman and Gillham, 1954). During the winter, surface water collects in the Bog, and slowly drains away during spring. The only permanently inhabited human dwelling is a lighthouse, but an old farmhouse is used as a bird observatory during the summer months.

One of the advantages of the Skokholm population was that the origins of the mice were apparently known. There were no mice there in 1881 when the island was last farmed intensively. The naturalist R. M. Lockley lived on Skokholm during the 1930s and recorded an anecdote that mice were so common when the lighthouse was being built in 1913–15 that special precautions had to be taken to render the building mouse-proof. He also tells of the introduction of mice, 'The story goes that the house mouse was accidentally brought to the island in a boat which had been filled with straw on the mainland beach of Martinshaven (3 miles to the north) in readiness for loading a young horse required for use on the island farm. An eye witness has since told me that, when the next day they sailed across and unloaded the boat in South Haven (on Skokholm), the colt in getting up, kicked the straw about until several mice darted out and were lost immediately in the crannies of the island cliffs. A few years later they were abundant and caused an influx of owls in the winter' (Lockley, 1943, 1947).

No one lived on Skokholm from 1881 to 1907. In 1907 the farmer who imported the mice with his horse, one Jack Edwards, took possession of the island and lived there for some years, retaining the lease until 1927 when it was made over to Lockley. The latter's account of the introduction of the mice presumably refers to Edwards moving to the island; one of his first wants for the farm would have been a horse.

Now if there were really no mice until 1907, the time is very short between the arrival of a few animals on the island and the great number there must have been at the time of the building of the lighthouse (it is even shorter if the influx of owls referred to by Lockley occurred earlier than 1913). Furthermore there was a succession of cold winters in 1907–09 and the island mice are very susceptible to cold (see below). The failure of mice successfully to colonize Skokholm during the years it was farmed (for at least 150 years before 1881), suggests that the conditions for establishing a mouse population there are stringent. The years following 1907 were certainly not propitious for this. Thus there are grounds for believing that there were already mice on Skokholm when Edwards took his horse to the island.

CHARACTERISTICS AND RELATIONSHIPS OF THE SKOKHOLM MICE

The original aim of studying the Skokholm mouse population was to discover whether changes occurred in non-metrical variant incidences from year to year and to relate these to environmental factors. However the first sample trapped on Skokholm in May 1960 proved to have such a markedly different array of frequencies to those in southern British mice that the distinctiveness of the island population seemed worth investigating in its own right.

The obvious hypothesis to explain the differences between the island and mainland phenotypes was that they represented adaptation to two environments. The island habitat was clearly different from the mainland. The variant frequencies were broadly similar over areas of the mainland too large for much gene flow, which implied that some measure of stabilizing selection was at work.

Now if the divergence between Skokholm and its mainland neighbours (and closest relatives) had been produced by adaptation to feral island life,

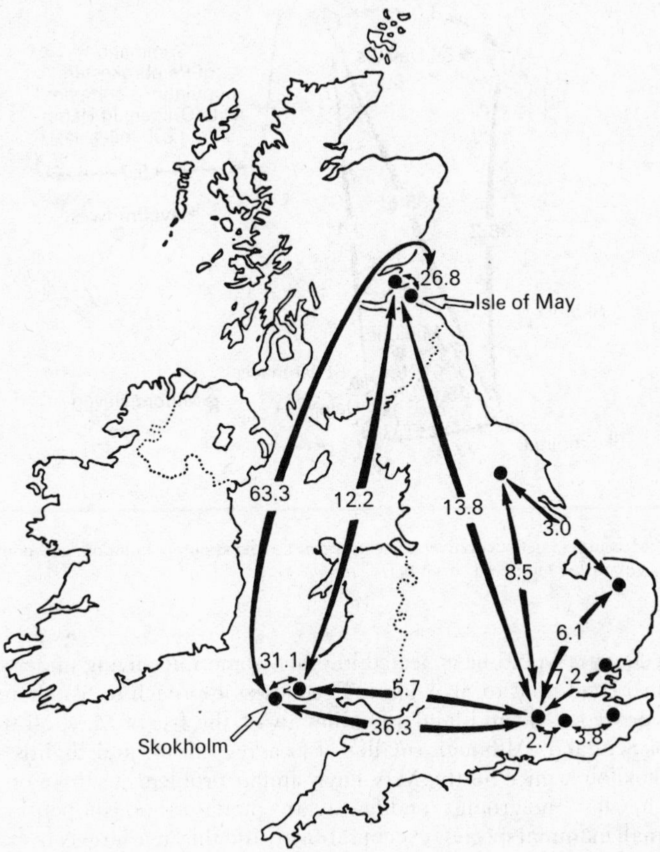

FIG. 88. Measures of genetical divergence between the Skokholm mouse population and other British populations (from Berry, 1967).

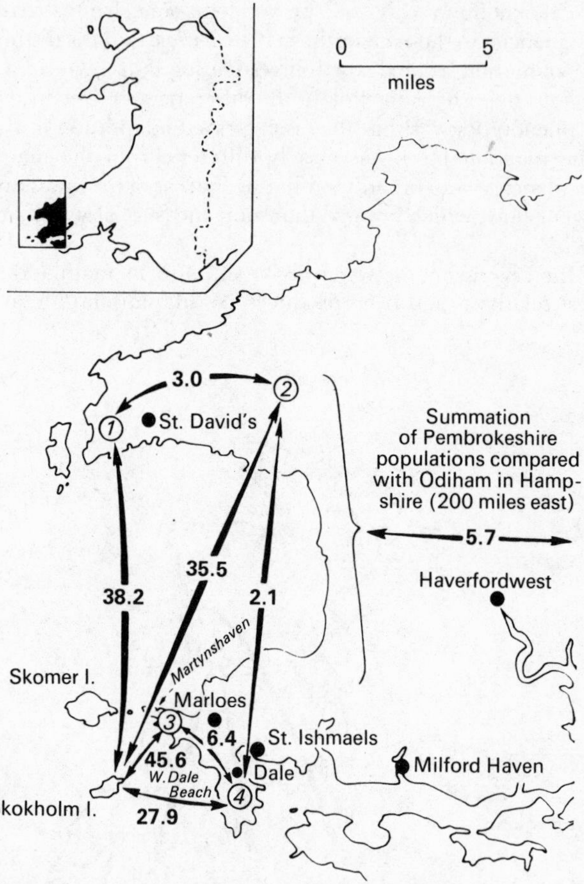

FIG. 89. Measures of genetical divergence between the Skokholm population, and populations on the neighbouring mainland (from Berry, 1965*b*).

similar changes would be expected in other island mice living under similar ecological conditions to Skokholm. The closest approach to Skokholm conditions proved to be an island 400 miles away, the Isle of May, off the east coast of Scotland. Although smaller (150 acres), drier, and slightly cooler than Skokholm, mice on the May have similar problems to those on Skokholm: they have no ground (and few avian) predators; no competition from other small mammal species (except rabbits); and they live largely in cracks in the cliffs and in dry stone walls on both islands.* It is not known when mice got to the May. The first mention of them is in a letter written by a lighthouse-keeper in 1868: 'We have no rats, but legions of mice, and most impudent mice they are, for they sit and look in your face, and even gnaw at the legs of your trousers with a composure that is quite amusing' (Muir, 1885). On both Skokholm and the May, mice grow to a length and weight about 15% greater than the mainland norm for free-living mice – an increase in size which has happened in all island races of small mammals in Britain. It was therefore surprising that a comparison between the two island populations in terms of their non-metrical variant frequencies showed that they were almost as different from each other as possible. There was no sign of convergence between the two, although both islands had the same degree of genetical distinctiveness from their nearest mainland (Berry, 1964). Apparently adaptation had taken place in the two populations, but it had led to two very different genomes. The reason for this became clear when the details of the relationships of Skokholm with its nearby mainland were considered.

The historical mainland link of Skokholm was usually *via* Martinshaven, the best local landing place. According to Lockley, the original Skokholm mice came from there. However modern mice from the nearest farm to Martinshaven are significantly more *unlike* the island population than samples from elsewhere in the neighbourhood – even from ones caught some distance away in the main farming area of north Pembrokeshire. In contrast mice from near Dale are much more like the Skokholm population than others. All the mainland samples are similar to one another and to mice from other parts of southern Britain.

Now for a time in the 1890s Skokholm was rented by a farmer from St Ishmaels, who kept sheep there and cropped the rabbits, until he left to fight in the Boer War. This man often kept his boat near West Dale. Hence it seemed conceivable that mice may have originally got to Skokholm during his tenancy, besides which:

a. A successful introduction at this time would have allowed the mice to be well established before the plague(s) of 1913–15.
b. There was a period from 1895–1900 of very mild winters which would have allowed immigrants time to adjust to the demands of island life, and in particular to survive the cold conditions of the 1900s.

It was encouraging therefore that another story of mouse introduction to Skokholm came to light after the mouse skeleton work was complete: Howells (1968) tells that 'when the island was held by Gilbert Warren Davis of St Ishmaels ... mice were found in some sacks which had been brought across by rabbit trappers for making knee-pads and as protection over their shoulders when it was raining, as well as for carrying rabbits. Normally they could get

**Footnote*. The value of comparing fine genetical differences in two populations as far apart as those on Skokholm and the May is debatable. It was for this reason that the Field Mouse (*Apodemus sylvaticus*) study was undertaken, because in that species mice living on geographically close islands could be compared (page 201).

such sacks from the farm where they were catching but, as no general farming activities were taking place on the island at the time, they obviously brought some with them from a farm on the mainland'.

Skokholm mice differ from British wild mice mainly in high incidences of three skeletal variants: presence of interfrontal bone, dyssymphyses of thoracic vertebral arches, and foramina transversaria imperfecta of the sixth cervical vertebra. Interfrontal presence is a very rare variant in Britain, but it (and the other variants) occur in Dale mice. Presumably these traits have become common in the Skokholm population through the original colonizers happening to carry them, so that subsequent selection has confirmed them in the genome (page 210). Mice taken from Skokholm and sib-mated for a number of generations in the laboratory maintain these characteristics (Berry and Jakobson, 1975*a*). Unusual alleles are common in the genetically closed populations because the presumed selection acting after isolation has had to use the available variation. In other words, the Skokholm (and, by inference, the Isle of May) mice are examples of the operation of the founder effect.

Whether the Skokholm and May populations would converge if both were isolated for much longer periods is unknown: all one can say in the short-term is that they have achieved sufficient efficiency for survival – and that is the acid test of evolution.

INITIAL AND SUBSEQUENT DIVERGENCE

There can be little argument that the bulk of the Skokholm mouse genes derive from the animals which ran from the rabbit-catchers' sacks in the mid-1890s. A more important problem is to determine the extent of change (and its causes) since the original isolating event.

The Skokholm mice were monitored fairly intensively for a 10-year period (1960–69), and showed changes during that time which can be related to the genetical influences acting on an isolated and established population. A sample trapped in 1962 was skeletally (and therefore genetically) different from ones trapped at the same time of year in 1960 and 1961. So the question arose: what had caused these changes? Negligible breeding occurs on the island during the winter months (October to March), and the number of survivors to spring each year are sometimes few. This implied that random drift might occur in any year. Indeed, Elton (1930) went so far as to suggest that drift during times of minimum population size might be a valuable opportunity to try out new genetical combinations. Consequently it became important to discover the true population sizes on Skokholm – and if possible to make estimates from them of the effective breeding size of the population. This seemed worth the effort since some authors have suggested that the breeding unit in House Mice is very small indeed –perhaps as few as four.

Mark-release-recapture experiments were begun in 1964, and over 3000 mice were marked and released during the next 5 years. Combining Lincoln

Index (*i.e.* proportion of marked to unmarked individuals in any sample) and 'trap-out' estimates (*q.v.* Hayne, 1949; Berry, 1968*b*, 1970*c*), it was possible to

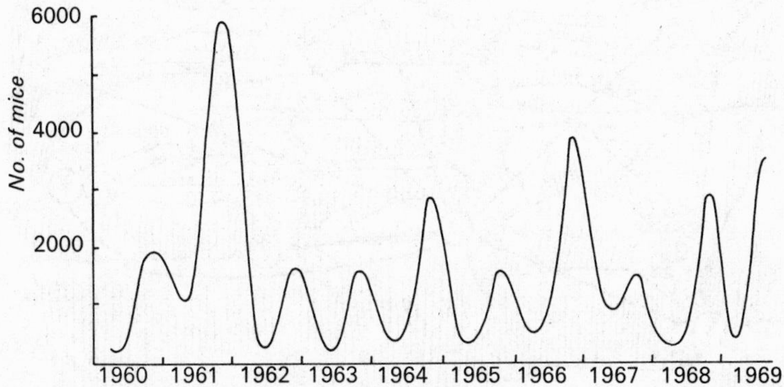

FIG. 90. Numbers of mice on Skokholm 1960–69 (from Berry and Jakobson, 1971; methods of estimation of population size given by Berry, 1968*b*).

deduce the number of individuals on the island at all times of the year. In general, population size increased eight to ten fold in most years. The lowest number observed was about 75 pairs in Spring 1963. Since virtually all the females roamed only within the territory of individual males and about two-thirds of female winter survivors in any year became pregnant, this meant that the smallest breeding size of the population between 1960 and 1970 was about 100 individuals.

Breeding mice on Skokholm are highly territorial, and it could be that the population is divided up into a large number of discrete mini-populations. If so, this would mean that much of the overall genetical composition of the population could be ephemeral chance. The key question is whether territorial groupings persist from one generation to another, *i.e.* whether or not they are 'inherited'. The marking of individual mice undertaken as a preliminary to release and recapture for estimating total population sizes made it possible to determine how often mice moved their residence, and if there was any sub-division of the island population. Since virtually all animals in a particular area could be caught (as judged by varying the density of trappings; by 'trapping-out' an area and monitoring the movement of mice into it from elsewhere; and by recaptures on subsequent visits to the island, particularly in winter when food is presumed scarce and mice are quickly caught), it was possible to get apparently good estimates of population mixing.

It turned out that more than a quarter of the mice on Skokholm breed at a site other than the one at which they were born ('site' being defined as a

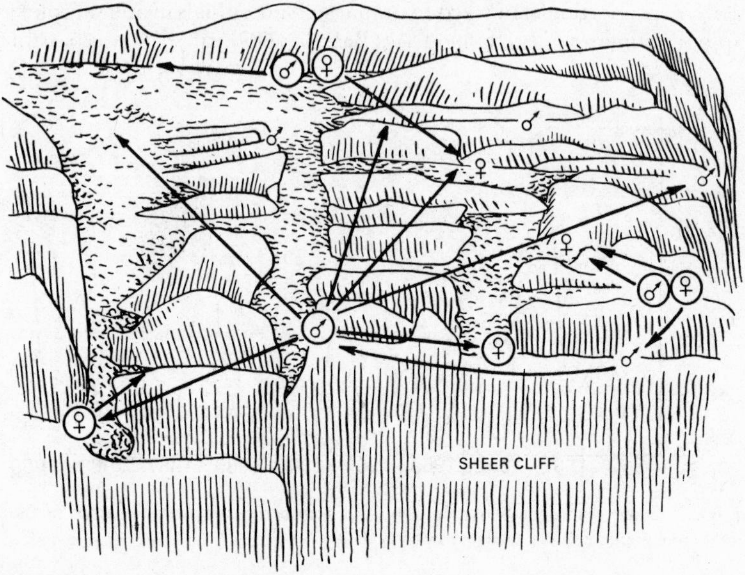

FIG. 91. Results of grid trapping with 40 traps for 11 nights on an area of 50 sq. yds. on Skokholm in July. If a mouse was caught in the same trap for three or more nights (indicated by ♂ or ♀, this was regarded as the 'home' trap; movement from this point (in the sense of being caught in another trap) is shown as a line radiating from the 'home' trap. Animals trapped only once or twice are shown unringed (from Berry, 1970*c*).

trapping area, usually at least 50–100 yards from another) (Table 20). Confirmatory evidence for this conclusion comes from a study of the spread of rare alleles detected electrophoretically (Berry and Jakobson, 1974). There can be no doubt that individual territories form only temporary barriers to gene flow, and for practical purposes the whole of the island mice can be regarded as a freely interbreeding population. Nevertheless there is enough restriction of movement for some local adaptation to be possible.

TABLE 20. Change in residence of marked mice which survive the winter (% of total population)

	Between autumn and the following spring	*Between spring and the following autumn*
Males	33·6	25·7
Females	29·7	23·6

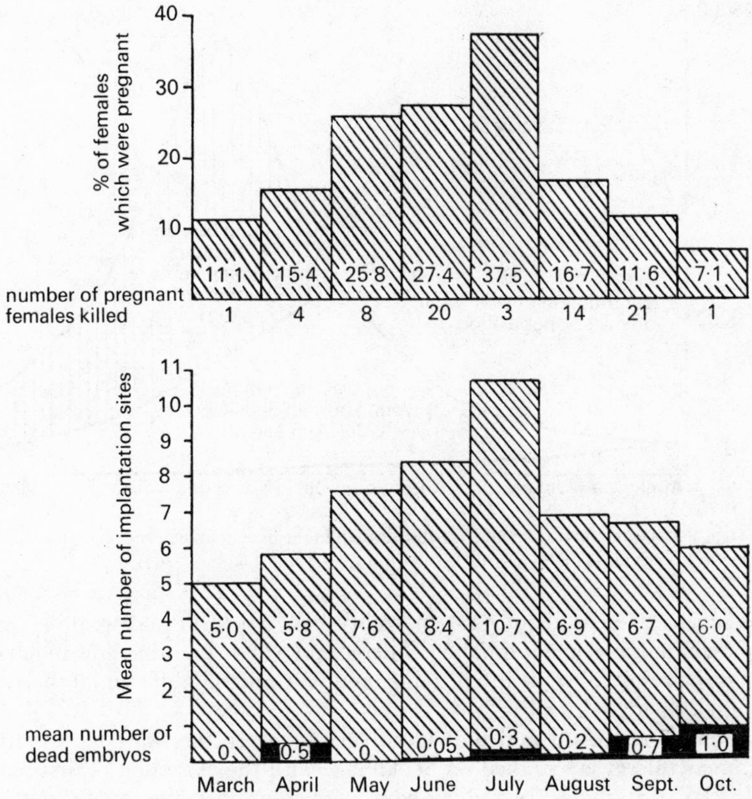

FIG. 92. Litter size and breeding intensity on Skokholm (from Berry, 1970*c*; modified from Batten and Berry, 1967).

POPULATION INCREASE AND DECREASE

Population size estimates were originally undertaken to test whether genetic drift was likely to be important in causing genetical change in the island population. They also permitted the measurement of individual survival, the establishment of the distribution of animals on the island, and the construction of a life-table like those used by actuaries for life insurance calculations (Berry and Jakobson, 1971).

Reproduction on Skokholm begins in mid-March and has usually ended by the end of September. A few pregnant females have been caught during the winter months, but no winter-born young has ever been found surviving in the spring. In most years the numbers of mice on Skokholm increase ten-fold during the breeding season. Since litter size and pregnancy rates are known,

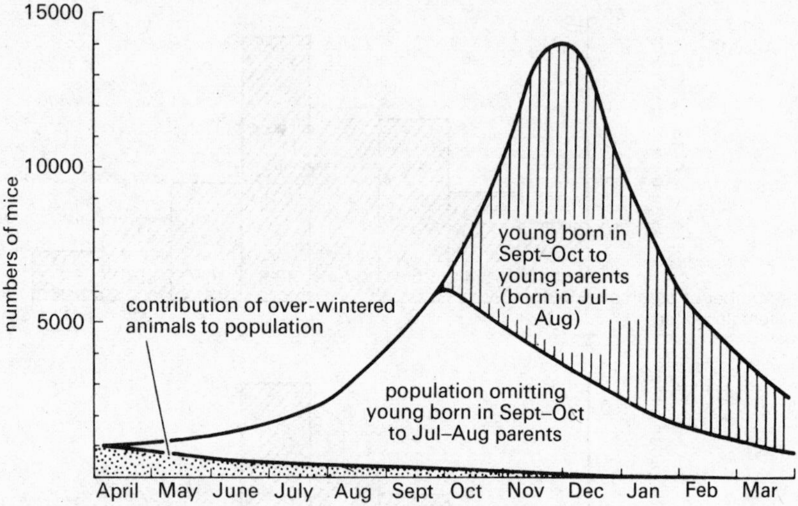

FIG. 93. Population fluctuation and age-distribution in the Skokholm mouse population, based on average survival estimates over five years (from Berry and Jakobson, 1971).

this means there is about 50% mortality between birth and weaning (*i.e.* when animals become trappable). There is no way of knowing how much of this is the result of whole litters dying, and how much is differential survival within individual litters.

Population growth and individual survival were the same for the five summers these were studied on Skokholm: with the exception of first-born (March–April) males, all animals have an average 60% chance of surviving two months. In contrast the mortality rates in different winters are highly heterogeneous, the proportion of deaths being much higher in a cold winter (Berry, 1968*b*). The maximum population size in the autumn of any year is directly related to the number of mice surviving the previous winter and high or 'plague' numbers have always followed a mild preceding winter (Table 21).

In the years we worked on Skokholm the proportion of the autumn population surviving to the following spring varied from 6% to 20%. During an 'average' winter, the average 2-monthly death rate of mice is 55% for first year animals, but higher for second year ones. No mouse has ever been found to survive two winters on Skokholm, although in equable commensal situations more than one in five live over two years, and pampered laboratory mice often survive into a fourth year (Varshavskii, 1949; Russell, 1966). The highest winter survival rate on Skokholm occurs on the cliffs where the territorial organization is strongest. During the winter littoral animals form an important part of the diet (particularly the amphipod *Talitrus saltator*).

TABLE 21. Climate and fluctuations in mouse population number on Skokholm

A. Records of mice from Skokholm archives

		Difference from mean temperature (°C)*		
		January	*February*	*March*
1895	'The mice were originally shipped	−2.7	−5.6	−0.9
1896	to Skokholm . . . in the mid	+0.3	+0.7	+1.3
1897	to late 1890s'	−2.1	+1.3	+0.3
1898		+2.2	+0.8	−0.8
1899				
1907	'Mr Jack Edwards rented the island	−0.3	−1.1	+0.5
1908	in 1907. . . . A few years later	−1.3	+1.3	−0.9
1909	(the mice) were abundant and caused an	+0.1	−0.5	−1.7
1910	influx of owls in the winter'	−0.3	+0.7	−0.5
1913	'The lighthouse was rendered proof	+0.9	+0.2	+0.4
1914	against the entry of mice; whilst it was	−0.9	+1.6	+0.3
1915	being built a plague of these mice	−0.2	−0.6	+0.8
	caused special precautions to be taken'.			
1936	In 1938. . . . 'Continues to increase,	−0.3	−0.8	+1.1
1937	apparently reaching a high peak of	+0.9	+1.4	−1.6
1938	population after its low numerical	+1.3	+0.3	+1.9
	status 3 years ago'			
1947	'After human reoccupation, they were	−1.6	−6.6	−1.3
	not observed until late in August'			
1948	'Very common throughout the year'	+1.5	+0.2	+4.2

B. Direct population size determinations

	Spring	*Autumn*			
1960	150–400		+0.2	−0.1	+0.2
1961	500–1000		+0.2	+2.6	+1.4
1962	150–400	1500–3000	+0.5	+0.4	−2.5
1963	100–300		−6.5	−4.2	−1.1
1964	200–500	2000–4000	−0.2	+0.4	−1.3
1965	150–400	1000–3000	−0.2	−1.1	−0.7
1966		2500–5000	−1.2	+0.9	+0.7
1967	300–600	1000–2000	+0.7	+1.1	+0.9
1968	250–400	2000–3000	+1.1	−0.9	+0.2
1969	200–300	2000–4000	+1.3	−1.9	−1.2

*10 year means: January 5.9°C
February 5.2°C
March 7.0°C

The interior of the island becomes very wet in the winter and virtually all the mice there die. The area is recolonized by young animals every spring. Only a very few animals make a reverse migration from the centre to the periphery when conditions become harsh in the winter.

The main causes of death seem to be 'physiological', that is to say there is no interspecific competition, no sign of obvious disease, and little predation. Thus we can summarize the life-cycle as two distinct phases (an increase and a decrease phase), with temperature-related factors playing an important part in mortality during the winter decrease.

GENETICAL CHANGES

The genetical situation in the Skokholm mice became clearer when electrophoretic techniques made it practicable to detect in blood a number of segregating genes (Berry and Murphy, 1970). The animals from which blood was taken suffered no harm, as shown by their subsequent survival.

We classified the genotypes of individual mice at six loci: haemoglobin-β (*Hbb*), serum transferrin (*Trf*), two peptidases (*Trip-1* and *Dip-1*), and two esterases (*Es-2* and *Es-5*). Fourteen alleles were segregating at these loci. This was surprising, since only one further allele is commonly found in British mainland populations, and a population founded by only a few individuals would be expected to have an acute poverty of inherited variation. Wheeler and Selander (1972) have commented that this implies 'the original propagule (if, indeed, only a single introduction occurred) was large enough to carry most of the variability present in the parental mainland population'. Unfortunately, it is impossible to know whether they are right, or if the extremes of environmental change which a Skokholm mouse undergoes during its life and the genetical responses this engenders (see below) have led to a high rate of incorporation of alleles arising since isolation.

There were also changes in alleles and genotype frequencies between the beginning and end of single breeding seasons, changes which were repeated in two consecutive years. Only rare variants occurred at 3 loci (*Trf, Trip-1, Es-5*), but the other loci all showed remarkable volatility (Table 22).

The most regular changes took place at the *Hbb* locus: frequencies of heterozygotes increased on the cliffs during the breeding season and decreased during the winter. Genotypic fluctuations in the island interior were irregular, and to some extent mirrored the excess of young produced by the coastal breeders. In the autumn of any year, the old mice in the population will be the survivors of the previous winter or their first born young of the spring, and the proportion of heterozygotes among them would be expected to be less than that in younger mice. This expectation is fulfilled.

Previous workers have reported a *deficiency* of heterozygotes in populations segregating for *Hbb*, but some selection is suggested by the widespread occurrence of a polymorphism at the locus throughout North America and

TABLE 22. Biochemical loci showing selective change

		Hbb			*Es-2*				*Dip-1*
			% excess of heterozygotes from expectn.			*% excess of heterozygotes from expectn.*			
	N	*Freq. Hbb*[s]	*Cliffs*	*Interior*	*Freq. Es-2*[b]	♂♂	♀♀	*Cliffs*	*Freq. Dip-1*[b]
Spring 1968	47	0.343	2.4	37.1	0.409	50.4	23.0	28.0	0.925
Autumn 1968	76	0.438	43.3	16.5	0.242	34.5	30.8	27.6	0.815
Spring 1969	32	0.386	27.0	13.0	0.312	56.5	26.5	30.8	0.807
Autumn 1969	67	0.464	72.8	45.4	0.296	42.9	17.6	30.0	0.663
1971: Older mice	56	0.260	−29.8	−10.0	0.216	33.3	20.7	30.0	0.989
Younger mice	61	0.281	9.2	14.6	0.271	35.8	42.9	47.5	0.981
1972: Older mice	27	0.593	−11.0	28.6	0.289	62.8	33.3	62.8	0.981
Younger mice	49	0.510	31.0	−43.4	0.135	12.4	7.1	16.3	0.980

Europe. Differences in reproduction and survival between *Hbb* genotypes have been found in field-living mice in Californa (Myers, 1974), and Berry and Peters (1975) recorded a significant differences in allele frequency between young and old mice on the sub-Antarctic Macquarie Island where the animals breed the whole year round.

Nothing is known of the *Hbb* alleles which helps us to understand any effect on fitness (Russell and Bernstein, 1966). One allele (Hbb^d) is a duplication, with the two polypeptide chains differing from each other at six positions; the other allele (Hbb^s) has three out of 196 amino acids different from that in the duplicated chain (Popp and Bailiff, 1973). A histocompatibility locus (*H-1*) is closely linked to *Hbb* (Russell and McFarland, 1974). If alleles at this locus affect survival in the same way as HLA alleles in man (page 188), it is conceivable that the haemoglobin polymorphism is the result of 'hitch-hiking' on an adaptive chromosome segment which includes the *H-1* gene.

In the years for which biochemical data are available for Skokholm the frequency of Hbb^s (which is the commoner allele in most mainland populations) ranged from 26% to 59%. Changes at other loci were almost as marked, but completely uncorrelated: the commonest *Dip-1* allele decreased from 92% to 66% during two breeding seasons, but then increased again to

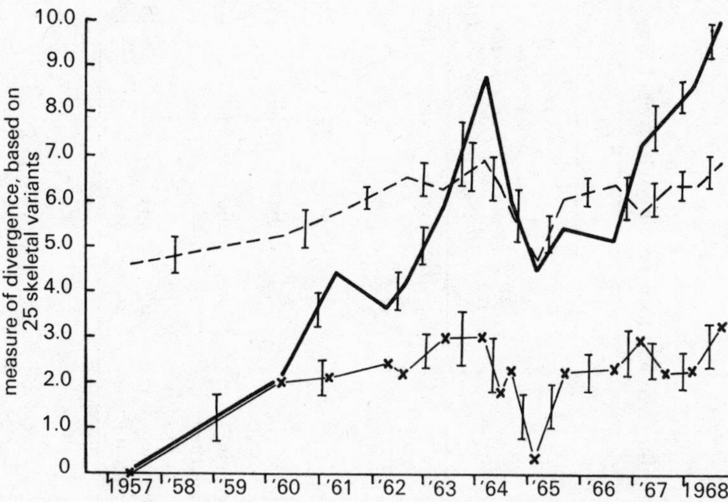

FIG. 94. Genetical adjustment in the Skokholm mouse population from 1957–1968 as measured by changes in the frequencies of non-metrical skeletal variations. The continuous line shows the Measures of Divergence, where each step is the mean change from the previous sample; -x-x-x- plots the Measures of Divergence from the 1957 sample; the broken line plots successive Skokholm sample differences from the neighbouring mainland (from Berry and Jakobson, 1975*a*).

98%. At the *Es-2* locus, an excess of heterozygotes in males increased in the 1968–69 winter, decreased in the summers in 1968 and 1969; in 1971 there was an increase in female heterozygous frequency most marked on the cliffs; and in 1972 there was a decrease in heterozygotes during the summer in both males and females, particularly on the cliffs.

The cyclical selection shown by biochemical markers are also detectable in skeletal variant frequencies (Berry and Jakobson, 1975*a*). During the period of study, samples collected in the autumn were more like each other than those from the spring – indicating that stabilizing selection acts during the summer months (as shown also by the *Hbb* locus changes), but that the heterogeneity revealed by direct measures of individual survival in different winters also led to genetical differences. One of the peculiarities of the summer selection is that the midline dorsal weakness which characterizes Skokholm mice when compared with their mainland relatives (gaps between the frontal bones of the skull and a failure of the neural arches of some vertebrae to fuse) actually increases. The incidence of the variants concerned then decreases again during the winter. This emphasizes the old and rarely regarded truth that selection is concerned with survival not perfection: the peculiar genome of the Skokholm population leads to odd correlated responses when particular traits are selected.

The genetical difference between Skokholm and Isle of May mice has already been noted. In contrast to Skokholm, the Isle of May mice are monomorphic for Hbb^{s} – indeed they showed no isozymic variation at 17 loci tested. This highlights the difficulties of relating genetical changes at a locus with the action *per se* of the locus: the May mice have similar ecological problems to the Skokholm ones, yet their genetical diversity is very low. Herein lies an advantage of morphological investigation: inherited traits are affected by many loci, and multivariate comparisons between samples are likely to be more sensitive detectors of genetical differences and pressures than ones based on a limited number of biochemically scored loci.

The pattern of skeletal variants on Skokholm changed progressively during the 1960s. This change was neither large nor continuous, but was enough to show that the genome was subject to adjustment over a number of years. It is impossible to predict whether an isolate like Skokholm will ever come into a state of balance with its environment that such genetical change will stop. Clearly both cyclical and long-term genetical adjustment form part of the present survival strategy of the population.

SURVIVAL AND SELECTION

Mortality in Skokholm mice is not random with respect to genetical constitution, so an obvious practical question is 'what causes the mice to die'? Since the death rate in the winter is related to external temperature, this has been investigated by measuring the responses of individuals to cold in temperature-controlled metabolism chambers, and then following their subsequent survival when released back into the population.

At first it seemed that particular traits (such as high basal metabolic rate) were correlated with winter survival. This proved too simple an explanation. Different physiological measures were more important in different years, although older animals were always less able to cope with environmental pressures than younger ones. The conclusion from direct physiological testing of individuals is in fact the same as that from statistical comparison of survival in different winters, that the characteristics which lead to a mouse surviving or dying in a particular set of circumstances will vary according to those circumstances.

It is possible to generalize more confidently about the effect of age. Second year animals invariably succumb to winter conditions (see above), even though their survival in the less harsh summer weather does not differ from that of younger mice. Consequently any measurement of the reaction of an individual of any genotype to its environment has to take age into account. We have been able to show that a whole range of phenotypic measures (organ weights, ionic contents of bones, haematological values, overall body sizes) are significantly correlated with age, even when the data are 'corrected' to eliminate the effects of size (by adjusting them to a standard body size). In a statistical analysis based on all these measures it was possible to separate young (under 3 months), adult, and over-wintered mice from each other with only 15% error. Since many animals must die from genuine accidents, this amount of imprecision is quite acceptable. The different age classes in this analysis were genetically alike. However it was also possible to produce a similar degree of separation between biochemical phenotypes *using only the same quantitative data* (Table 23). In other words, there are differences between

TABLE 23. Amount of mis-classification between biochemical phenotypes shown by a discriminant analysis using 17 quantitative measures (after Bellamy *et al.*, 1973)

	Males		*Females*	
Locus	*Number*	*Percentage incorrect*	*Number*	*Percentage incorrect*
Hbb	40	17.5	38	21.0
Es-2	40	12.5	37	10.8

phenotypes which are subject to different selection pressures, and which include characters which change with age – and age affects survival.

This led to a model of survival to emphasize the dynamic nature of the problems an animal faces at different stages through life. The aim of the

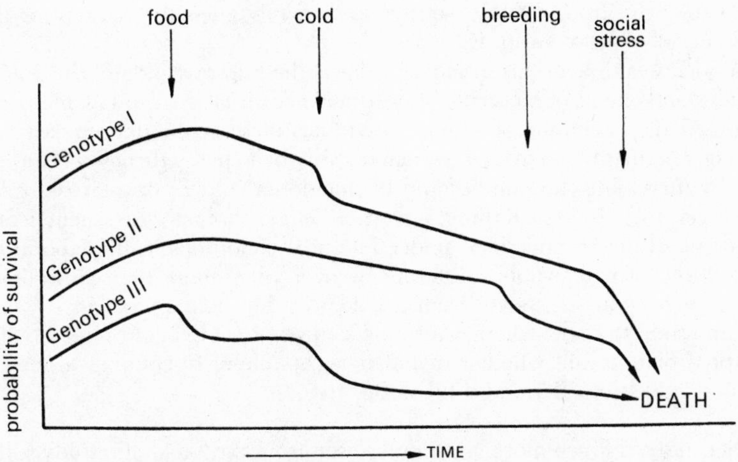

FIG. 95. A model to describe the interplay of individual history, environment, and genotype: different phenotypes (*i.e.* the resultant of genotype plus age) react differently to environmental factors (from Berry, Jakobson and Triggs, 1973).

model was primarily conceptual rather than predictive; *i.e.* to suggest problems for investigation rather than give a basis for a mouse's life insurance. It was important to proceed this way, because it is only too easy to characterize a population sample in terms of its genes, physiology, or morphology, and then invent an explanation that fits all the characteristics. This may work sometimes, but it will tend to naivety and may obscure important happenings. For example, the *Hbb* genotype frequencies in the September 1971 sample from Skokholm closely fitted the binomial (Hardy-Weinberg) expectation. Only because previous samples collected at different times of the year had shown that seasonal selection occurs at this locus were the data divided by age and habitat. This showed a change from a 30% deficiency of heterozygotes in older mice on the cliffs to a 9% excess in younger ones.

CONCLUSIONS

The Skokholm mouse study began as an apparently simple genetical investigation, and ramified into morphology, physiology, ecology, behaviour, reproductive, and ageing studies. It highlights two points:

1. Firstly, a continuous study of a population yields disproportionately more information than a series of 'snap-shots' of different populations. On Skokholm it was possible to follow the fate of individual mice from almost cradle to grave, and discover the advantages or disadvantages of particular genotypes at different stages of the life history. Far too often biologists have to deduce

what they can from a single sample, and merely guess the reasons for the existence of a particular trait.

A good example of this in House Mice is the interpretation of the potentialities and extent of metabolic adjustment (or *acclimation*) to low temperatures. At the beginning of winter individuals increase the oxygen carrying capacity of their blood to enable them to raise their energy turnover. They do this by increasing the number and haemoglobin content of their red cells. However, there is no such thing as 'normal' blood characteristics: genetically different strains of mice kept under identical conditions in the laboratory have levels of haemoglobin differing by over 10%, more than an individual changes when it moves from a cold to a hot climate, or *vice versa*. It is impossible to know whether a single sample of wild-caught mice is cold-adapted or not, and whether or not it is capable of responding to an environmental stress (Berry and Jakobson, 1975*b*).

2. Secondly, a much more important lesson from the Skokholm study is the futility of divorcing one part of an animal's biology from the whole. For example, a straightforward ecological investigation of the mice would have shown the environmental stresses that individuals have to face (cold, social, age), but not revealed anything about the differential responses of genotypes to these stresses, nor the amount of genetical adjustment that the population undergoes throughout the course of a year. Conversely a purely genetical study would have yielded information about allele frequencies and association between genotypes, but told nothing about the ability of genotypes (typified by the *Hbb* locus) to respond to the social problems associated with territory holding. There may come a time in the future when we can identify primary gene products easily and confidently, but this expertise will only explain the success or failure of an individual in the simplest situations, such as where an allele produces a defective or poisonous substance. The brash claims of molecular reductionists to give the complete answer to all biological problems in chemical terms is mere arrogant ignorance.

CHAPTER 8

HUMAN INFLUENCES

THERE is no dispute nowadays that man has a major and frequently devastating effect on 'this fragile space-ship, Earth' (Mellanby, 1970; Passmore, 1974). Words spoken and written about environmental damage wrought by man have become a major pollutant in themselves. In the face of these outpourings, it is surprising how little attention is paid to the factors which have made men important agents of genetical change. Whereas activities which affect the lives of individual animals or plants usually last only for the life-times of the organisms concerned, changes in the genetical composition or structure of populations may persist for tens or hundreds of generations. The thalidomide affair was a tragic episode, but its consequences will have disappeared when the last deformed person dies; if a chemical that causes mutation (*q.v.* pages 60, 256–7) is used as commonly as thalidomide was, its effects may not begin to appear for two or more generations, by which time considerable potential suffering may be stored for the future. Governments are aware of this danger, and it is now obligatory in many countries to test a new chemical (drug, food additive, or cosmetic) for its action on genetical material, as well as the more established checks for cancer-causing (carcinogenic) properties.

Historically and emotionally the most publicized genetical hazards associated with human activity have centred round ionizing radiation, particularly that produced by atom bombs and industrial uses of atomic energy. However the consequences of environmental change are genetically much more important to both man and other creatures.

MUTATION

We have seen that mutation is one of the factors that may change the frequency of an allele, but that the *rate* of mutation is normally so low that mutation can be neglected as a controller of gene frequencies (page 60). Notwithstanding, all new variation arises ultimately from mutational events (either through a change in the DNA 'code' or *via* a chromosomal rearrangement), and it is possible to imagine situations where the rate of production of new mutations is so high that few offspring are adapted to their environment. After all, although the spontaneous mutation rate is only about one in 50,000, we have about 50,000 genes, so one of them is likely to be newly mutated in each of us. Even worse, all the 50,000 pairs of genes in the single zygote nucleus are represented by several million copies before death, and mutated alleles may accumulate during life. Only genes present in the germ cells which give rise to the gametes will be transmitted to future generations, but somatic mutations (*i.e.* mutations in non-reproductive cells) are a frequent cause of cancer and ageing, since random changes are likely to decrease body efficiency, and their accumulation through life will lead to progressive inefficiency in many tissues (Curtis, 1971).

X-rays were discovered by Roentgen in 1895. Soon afterwards, it was realized that there exists a certain amount of natural radioactivity. Compounds of some heavy elements in the earth's crust, such as uranium and radium, spontaneously emit rays which have similar properties to X-rays, although they differ in their powers of penetration into matter. Some years after the original discovery, rays were identified which reached the earth from outer space. These were named 'cosmic rays'.

In 1921 Mavor found that X-rays could produce genetical changes both by increasing the amount of crossing-over and the incidence of chromosomal non-disjunction at cell division (page 77); by 1927 H. J. Muller had developed a technique for objectively estimating the number of sex-linked mutations in *Drosophila*, and discovered that the mutation rate itself is greatly increased by X-irradiation (page 102).

Muller's mutation-detecting technique was of great importance. An event like mutation which occurs only once at any locus in several thousand individuals is difficult to record accurately. A plant-breeder or animal fancier on the look out for a particular variant may be able to identify mutant individuals, but he may miss small or lethal gene expressions. Muller's trick was to use upsets in the sex ratio as a measure of mutations occurring in the sex chromosomes. He realized that any alleles in the X-chromosome which lower viability will manifest immediately in the hemizygous sex (*i.e.* the one with different or XY sex chromosomes; these are the males in insects and mammals, females in birds), but will only affect the homozygous (XX) sex if the alleles are present on both X chromosomes. The way Muller employed this property was to treat male *Drosophila* with a potential mutating influence (a *mutagen*), mate them to normal females, and then cross the progeny with each other. All the female offspring would carry a 'treated' X chromosome from their father, and half *their* offspring would inherit this chromosome. Any deleterious recessive mutations induced in the original male's X chromosome would affect up to half the male offspring of a particular female. Since equal numbers of males and females are expected, a deficiency of males indicates that mutations have been induced somewhere in the X-chromosome of the treated fly.

A similar argument can be used for sex ratio in the children of humans who have been exposed to a potentially mutagenic agent:

all the sons of a treated woman will inherit an X chromosome from their mother, so the genetical effect of the agent will be measured by a deficiency of sons;

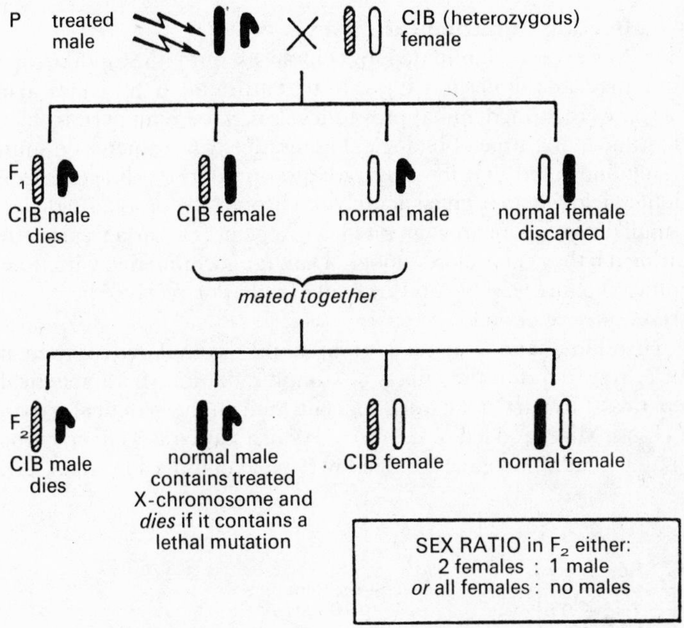

FIG. 96. The CIB method for detecting lethal mutations in the *X* chromosome of *Drosophila melanogaster*. In many cases an induced detrimental mutation may only kill a proportion of its carriers: this will reduce the proportion of males in the F_2 but not eliminate them entirely (from Berry, 1965*a*).

conversely, male children of an exposed man inherit less genetical material from their father than do his daughters (22 autosomes plus a largely inert Y chromosome to sons, as opposed to 22 plus an active X chromosome to daughters). A deficiency of *daughters* will measure the induction of deleterious dominant mutations in the sex chromosomes.

Muller improved the *Drosophila* testing system by assembling an X-chromosome carrying the gene Bar eye (*B*), which served as a marker for his untreated chromosome; a closely linked recessive lethal (*l*) which killed all females homozygous for it; and an inversion (*C*) preventing crossing-over between *B* and *l*, meaning that a *B* fly can be assumed to be carrying the *l* allele. The resulting *ClB* chromosome has been used in a vast number of tests to measure the effects of different agents. Similar tests exist for detecting mutation in other chromosomes of *Drosophila*, in mice, in fungi and in micro-organisms (Auerbach, 1962). Our concern here is with the deduction drawn from them about the nature and quantitative significance of mutations.

Factors affecting mutation rate

Mutation theory was dominated by physicists until the mid-1940s. It is important to recall this because it means that attitudes to the genetical effects of radiation were formed and approved levels of maximum permissible doses were laid down at a time of biological ignorance. Experiments on mutation before 1939 indicated that ionizing radiation produced a linear response of mutational yield whether given acutely or chronically, or as a single large or many small doses. There grew up an idea of a gene as a 'target' susceptible to rays hitting it if they came close enough. Damage accumulated with time, and the significant figure was the total amount of radiation received before the end of the reproductive period.

The Hiroshima and Nagasaki atom bombs stimulated a tremendous amount of research into the effects of atomic radiation (both genetical and non-genetical). In particular work concentrated on the genetical responsiveness of House Mice, and led to the discovery of a 'dose-rate' effect: chronic or fractionated irradiation gave a much lower yield of mutations than the same

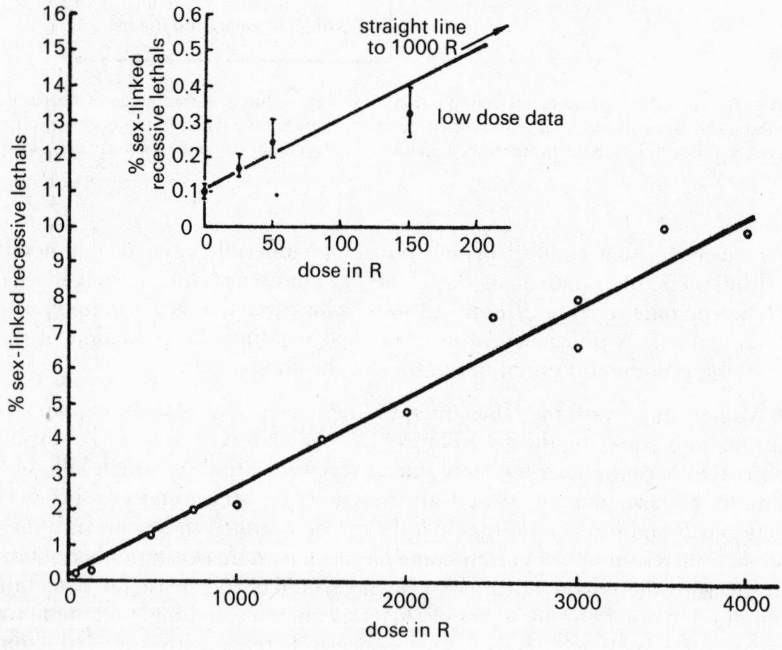

FIG. 97. The relationship between radiation dose and the frequency of induced sex-linked recessive genes in *Drosophila melanogaster* spermatozoa. R = Rads (from Berry, 1965*a*).

dose given in a shorter space of time. This suggested that the genetical material has a limited ability to repair damage caused in it, and that the repair system itself can be damaged by acute radiations. With hindsight it became apparent that virtually all the earlier experiments had involved the treatment of male flies carrying mature sperm – and a mature sperm is metabolically sluggish, so that repair processes do not work.

The discovery of the dose-rate effect meant a rethinking of mutation as a biological rather than a physical phenomenon. This was helped by the finding in 1940 by Charlotte Auerbach in Edinburgh that some chemicals may cause mutation, and by the elucidation of the structure of DNA in 1953, together with the breaking of the 'genetic code' that followed it (page 33f). In general, the amount of chromosomal breakage produced by a physical or chemical agent is directly proportional to the frequency of induced mutation (Brewen and Preston, 1974).

Spontaneous mutation rates vary between loci – some genes mutate 100 or 1000 times more frequently than others. Overall rates may be increased by various treatments (such as warming cold-blooded animals), and most mutagens seem to produce the same assortment of changes as arise spontaneously. Some chemicals affect particular sites on the chromosomes more than others, but compounds which cause mutation of specific genes are not known, nor are they likely to be found.

If the graph of increase of new mutations produced by (say) radiation is extrapolated backwards to no radiation, a sizeable number of mutations

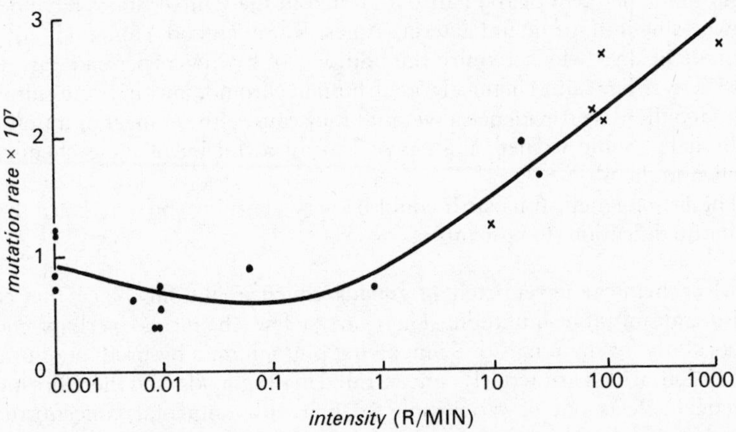

FIG. 98. The relationship between radiation intensity and the yield of induced mutations when spermatogonial stages (*i.e.* premeiotic germ cells) are irradiated in the mouse: × = X-rays, • = γ-rays, R/MIN = Rads/minute. Both the total response and the rate of increase of induced mutations are much less than when mature spermatozoa are exposed to radiation (after Searle, 1974).

would still apparently occur. In other words, no external agent seems to be responsible for all observed mutation – or even for most of it. The mutagenic agents we know merely increase the 'mistakes' which produce mutation in the DNA or chromosomes. It is here that the environmental problem occurs: many chemicals in the atmosphere (as well as both ionizing and ultraviolet radiation) are known to cause mutation. The most important at the moment seem to be the nitrosamines (formed from the reaction of nitrites and secondary amines both of which occur commonly in biological systems) and methyl-mercury (organically bound mercury formed by the metabolism of mercury used in seed dressings and industrial processes). Fishing has had to be banned in parts of Sweden because of the high concentration of mercury in lakes (Ramel, 1969; Ackefors, 1971). Other specific compounds are known – Captan, a commonly used fungicide is powerfully mutagenic in micro-organisms, although it is rapidly broken down by mammals (Bridges, 1975); cyclamates, frequently added to food, may cause chromosome breakage, as do many drugs used in the treatment of cancer; hycanthone, an antibiotic of choice in the treatment of the parasitic disease schistosomiasis is mutagenic in many organisms; there is a substance in Bracken (*Pteridium vulgare*) which if eaten causes cancer and probably mutations (this is almost certainly the cause of the high incidence of intestinal cancer in both cows and humans in north Wales, where milk cattle are often pastured on marginal hill pastures); and so on. Almost any chemical that is at all out of the ordinary may come under suspicion.

For example hair dyes have been used by women since the days of woad. Eighty-nine per cent of 169 hair dyes tested in the United States have been shown to be mutagenic in bacteria (Ames, Kammen and Yamasaki, 1975). Permanent dyes which require the addition of hydrogen peroxide are the most active. The same chemicals break human chromosomes in tissue culture, although there is no evidence as yet that they cause either cancer or mutation in humans. Some women must have had an awful lot of these chemicals applied to them!

The list of genetical hazards could be very alarming, but there are three major qualifications to remember:

1. Most chemicals never reach the gonads, which are the only places they can cause transmittable mutation. There are a few chemicals (perhaps most importantly for mammals those involving plutonium, a by-product of uranium fission) which are actively concentrated in the gonads, and therefore have genetical effects out of proportion to their environmental concentration (Beechey, Green, Humphreys and Searle, 1975).

2. A compound may be mutagenic in one organism, but not in another. For example, hydrogen peroxide causes mutations in the bread mould *Neurospora* but has no evident effect on plant chromosomes; phenols produce mutations

in *Drosophila* and chromosome breaks in plants, but no mutations in *Neurospora*. A demonstration that a chemical may cause mutations in a fungus or a plant does not mean that it will be dangerous to mammals. The converse difficulty is that a chemical which is shown to be harmless in a lower organism (or even in human cell culture) may have mutagenic activity in some higher organisms.

One of the more frightening stories of this sort is a food preservative AF-2, once widely used in Japan because of its antibacterial properties. This was cleared for use in human food as a result of tests on strains of *Salmonella*. It subsequently turned out that the chemical is highly mutagenic, and may have caused a large number of mutations before it was withdrawn (De Serres, 1974; Tazima, Kada and Murakami, 1975).

3. Most important (and more reassuringly), almost all genetical damage induced by the fairly low concentration of chemicals in most environments will be repaired before it becomes overt mutation.

Nevertheless it would be irresponsible to neglect the importance of environmental agents either as potential hazards or as the cause of new inherited variation in natural populations, although all studies of environmental mutagenesis have shown negligible effects (Berry, 1972).

The earliest search for genetical danger to humans was on the children of American radiologists. The argument was that in the early days of X-ray practice, doctors were ignorant or careless of radiation dangers, and a significant excess of congenitally deformed children was indeed found in the offspring of those using X-rays when compared with the families of other doctors. However, the investigation was carried out by a postal survey with a significant proportion of those exposed not replying. Since any of those approached who had affected children were more likely to volunteer information about their families than those with normal children, a result like this is suspect. Indeed a later study of essentially the same groups of doctors showed no difference between their families. Other workers have claimed that an unexpectedly large number of congenitally deformed children have been born to people living in parts of the United States where background radiation from the underlying rocks was particularly high, or to those from heights in the mountains where they were exposed to more cosmic rays than at lower levels, but other explanations are more likely for the observed unevennesses of distribution. For example, the reduced oxygen pressure at high altitudes is likely to have a greater effect in producing congenital malformation than are cosmic rays.

It has been argued that changes in the sex-ratio provide a more objective and sensitive indicator of genetical damage than congenital malformations, since these have a complex determination. Unfortunately the sex ratio (*i.e.* the proportion of male births) decreases with the number of children the mother has had previously, the age of the father, and the social status of the parents; while it increases in time of war (presumably due to changes in reproductive behaviour and maternal physiology) (Neel, 1963). (Some of the results from human groups where one parent has been irradiated are listed in Table 24.) The largest study has been on the survivors of the Hiroshima and Nagasaki explosions. At first no statistically significant differences in sex ratio between

TABLE 24. Effect of parental irradiation on the sex-ratio of offspring (series with adequate control date only)

	Controls % of males	Irradiated: Estimated dose (R)	Irradiated: % of males
A. PATERNAL IRRADIATION			
American radiologists	52.4	? (small, repeated, occupational exposures)	51.4[a]
Atomb bomb survivors			
Children born 1948–53	52.1	c.8	51.6[a]
		c.60	53.3
		c.200	52.7
Children born 1956–62	51.4	c.8	51.8
		c.60	51.6
		c.200	51.0[a]
French patients	51.5	200–400	56.1
	52.7	2–20	46.0[a]
Japanese X-ray technicians	52.6	10–100	55.5[b]
Ditto (these two series may overlap)	50.9	290 ± 46	53.3
Dutch patients	47.0	300–600	52.5[b]
	52.3	1–10	53.3
B. MATERNAL IRRADIATION			
Atom bomb survivors			
Children born 1948–53	52.1	c.8	52.0
		c.75	51.4
		c.200	51.2
Children born 1956–62	51.4	c.8	51.6[a]
		c.75	50.8
		c.200	53.6[a]
French patients	54.6	200–400	44.7[b]
Dutch patients	54.1	300–600	48.5
Canadian patients	52.7	7–20	48.8

[a] Effect opposite to expected.
[b] Shift has a probability of less than 5%.

the irradiated and control groups were found, but the small observed differences were reasonably consistent – although the clearest difference (in the children of exposed mothers) showed a change from 52.1% boys in non-irradiated women to 51.2% boys born to women who received an estimated dose of 200 R, which is hardly catastrophic. Worse, a follow-up study of children born to the same group in later years showed a reduction in the *control* sex-ratio. The only possible conclusion from such confusing findings is that man is not abnormally radiosensitive.

Animal studies on chronically irradiated populations tend to give general support to this. For example, an area of the sea shore in Kerala, Southern India has a high proportion of thorium in the sand – enough to raise the background radiation 100-fold in parts, with an average 7 to 8-fold increase in one area where a canal parallel to the coast makes a long thin island and effectively isolates most of the animals living on it. The radiation received by animals living in close contact with the ground is enough to double the mutation if the backward extrapolation of dose-effect curves is valid. However rats living on this 'island' did not differ in either inherited skeletal traits or embryonic mortality from rats living ten miles away under normal background conditions (Grüneberg, Bains, Berry, Riles, Smith and Weiss, 1966).

Mutation in the service of man

Agriculturists have exploited mutations induced by various agencies (particularly radiation) to produce new variation in crop plants. The problem is that commercially valuable strains are selected for uniformity of expression of their food or harvesting characteristics. This usually means inbreeding, and eliminating traits which might otherwise be useful – such as resistance to particular diseases, or differences in rates of growth. Variation can be restored either by crossing to other strains or 'land races' (*i.e.* the unselected ancestors of the crop), or by selecting mutations occurring spontaneously or as a result of a mutagenic treatment. For example, the widely grown spring barley variety Bonneville used to be difficult to thresh properly because its awns tended to persist in the treated corn. In an attempt to overcome this, a sample of Bonneville seed was irradiated, and among the offspring were several plants with brittle awns that were easily broken during threshing. The 'brittle awn' character proved to be inherited, and seed carrying it (but otherwise identical with the old form) is now marketed as Bonneville 70.

At one time it seemed that the commercial Mitcham variety of the Peppermint Plant (*Mentha piperita*) from which peppermint oil is distilled, might become extinct because of the spread of a virulent form of (fungal) wilt. This problem was solved by irradiating Mitcham stolons, and allowing the crops grown from them to be infected by wilt. Only 1% of the six million or so plants grown survived, but some of these were found to combine the commercially important flavour of peppermint with a high degree of resistance to wilt.

One of the key factors underlying the successful introduction of high-yielding ('green revolution') wheat into India was the incorporation of a colour mutant in the strains developed in Mexico. The problem here was not so much the climate or diseases of India, but the Indians themselves: the Mexican wheat seeds were red, but Indians make their wheat into chapatis, which, as every Indian knows, is yellowish in colour, not red. A crash plant-breeding programme was set up at the Indian Agricultural Research Institute in which the Mexican seeds were irradiated. A few amber-coloured mutants were recovered from the progeny of the original seeds, and a special Indian variety was derived from them. It was readily accepted by Indian farmers.

ENVIRONMENTAL CHANGES

Although man is able to affect the genetical constitution of animals and plants directly by his control of physical and chemical mutagens, there is no doubt that his most important influence is indirect – on the environments which shape the organisms that live in them. Considered in this light, man is an extremely important evolutionary agent.

1. Intentional environmental activity

The pesticides and medicines used by man to protect himself from his natural enemies are broadcast so widely and with such accurate aim that they become a major factor in the environment of many species. We have already noted how an allele giving resistance to warfarin poisoning has spread in Norway Rats (page 118), and similar inherited resistance to the poison is found even more widely in House Mice. Even more intensive attempts at chemical control have been aimed at many insect species, notably the Mosquito species which are vectors of malaria. It is not surprising that resistance to commonly used poisons is found in virtually all exposed species. Brown (1967) listed 225 species which had been reported to be resistant to one or more poisons. The largest number showed resistance to the cyclodiene group (dieldrin, aldrin, lindane, etc.); next was the DDT group (DDT, DDD, methoxyclor, etc.); the organophosphates (malathion, fenthion, etc.) were third in importance; while only 20 species were resistant to compounds which fell into none of the 3 main groups. It was unusual to find cross-resistance between groups, although many species become resistant to insecticides of different groups as these are successively applied. The House-fly (*Musca domestica*) on Danish farms is now resistant to every insecticide which can safely be used, with the exception of pyrethrum in a very few places.

There can be no doubt that resistance is the result of selection for variants that increase survival. Mutations for resistance have been produced by radiation, but most insecticides themselves do not raise the mutation rate. The induction of enzymes to break down insecticides does sometimes occur, but seems to play a small part in natural resistance. Several workers have fed insects for several generations on amounts of poison just insufficient to kill them, without producing any increase in resistance. One *Drosophila* experiment involved raising the larvae in high DDT concentrations and went on for 50 generations, but the flies remained as susceptible to the poison at the end as they were at the beginning (Luers, 1953). In another experiment, Bennett (1960) produced a DDT resistant strain of *Drosophila melanogaster* entirely by breeding from the sibs (brothers and sisters) of individuals with the longest survival times when exposed. The resistant flies were the result of increasing the frequency of alleles already present in the base population since they themselves had not been in contact with DDT. A survey in the Mosquito *Anopheles gambiae* showed that between 0.4 and 6% of individuals are heterozygous for a DDT resistance gene in unsprayed villages in Nigeria, but selection by spraying produces a population with 90% homozygotes in areas which have been sprayed.

Since selection will affect any mechanisms increasing survival, it is not surprising to find several different mechanisms coexisting in heavily exposed species (Cook and Wood, 1976). For example, in House-flies there are at least three genes for DDT resistance and two for carbaryl (organophosphorus) resistance. Two of the DDT resistant loci are concerned with detoxification by different biochemical routes, and both interact with a third gene which increases resistance to a number of unrelated chemicals by reducing the rate of penetration through the cuticle. Another gene acts through an unknown mechanism, possibly by decreasing nerve sensitivity. In the Yellow Fever Mosquito (*Aedes aegypti*) there are alleles at two unlinked loci giving resistance to DDT, one increasing larval survival and the other adult survival. However the larval gene may interact with other genes to increase adult survival as well. Different flies homozygous for one of the resistance alleles may vary between 10 and 1000 times in their susceptibility to a given dose of poison: this shows the importance of the 'genetical background' of the individual fly.

Biological control of pests is often argued as a preferable alternative to chemical control. Most biological control methods involve the hopeful introduction of predators, but some use genetical knowledge, and are attempts to carry out genetical warfare. The strategy in these situations is to release so many artificially reared individuals of the pest carrying a deleterious gene that the natural pest population is swamped and becomes genetically debilitated. The favourite idea is to introduce 'sex-distorting' alleles, so that the progeny of matings between wild and released individuals are either sterile or only one sex. In this way it is hoped that the pest may be reduced to insignificant numbers. Some pilot schemes along these lines have been carried out with encouraging results, but it is a field where theory is well ahead of practical results (Whitten and Foster, 1975). An alternative approach which may have greater long-term importance is to attack a species genetically when it is vulnerable ecologically. For example, insecticide resistance in aphids rapidly increases during the summer months as resistant animals breed parthenogenetically. However selection in parthogenetic clones (page 96) is liable to reduce variation as unfit clones are eliminated through the short generations, and it is conceivable that variation could be reduced to a calamitously low level at the end of the breeding season, particularly if additional environmental pressures could be applied at a critical stage (Beranek and Berry, 1974).

Parallel but entirely different problems arise through the resistance of bacteria to drug action. Some big hospitals use valuable antibiotics like the penicillins only as a last resort because of the risk of producing resistant germs which could not be controlled. These dangers are highlighted in parts of the world where drugs are freely available and cheaper than doctors. A savage epidemic of typhoid in Mexico in 1972 revealed this arose through a widespread resistance of the disease bacillus to chloramphenicol, which is the mainstay of modern treatment.

Antibiotic resistance arises through the selection of pre-adapted variants by the drug in exactly the same way as pesticide resistance. Penicillin resistance may occur through the synthesis of a penicillinase enzyme which breaks down the antibiotic, or by a modification of the bacterial cell wall so that penicillin is not absorbed.

An additional problem with bacteria is that resistant individuals can transmit their resistance genes to members of different species, linked with other genes on an element of DNA during conjugation. The discovery of transferred resistance in Japan in 1956 caused a reappraisal of the use of antibiotics in animal husbandry, where they were used not only to cure and, hopefully, prevent infection, but also as food additives to promote animal growth. The urgency of the situation showed itself in Britain in the mid-1960s when a *Salmonella* epidemic in cattle proved to be resistant to seven well-established antibiotics, and resulted in an outbreak of food-poisoning among farmers and vets in which six people died, despite treatment with antibiotics which would usually have been effective (Anderson, 1968). The harmless gut bacteria of the affected cattle were also highly resistant to the same antibiotics. Walton (1966) investigated drug resistance in *Escherichia coli* from the guts of healthy pigs and calves, and found an association between the kinds of antibiotics in their food and *E. coli* strains resistant to these drugs. Even worse, 99 of 134 strains resistant to more than one antibiotic were able to transfer their resistance to sensitive strains of *Salmonella typhimurium* and *E. coli* in laboratory tests. As a result, a government committee recommended that no antibiotic with a therapeutic value for human disease should be used to prevent disease or promote growth in animals, although such drugs could be used for the treatment of animal disease (HMSO, 1969).

2. Incidental genetical results of man's activities

We have already seen that the smoke pollution resulting from the industrial revolution gave rise to the conditions for the spread of melanic forms of moths (pages 120–30). Melanism occurs also in other groups in industrial areas:

there is a black form of one Spider (*Salticus scenicus*) which apparently is confined to Stockport Gasworks, whilst another (of *Arctosa perita*) is found only on colliery spoil heaps in Warwickshire and Leicestershire.

Another species which shows a form of industrial melanism is the common Two-spot Lady-bird (*Adalia bipunctata*), which has dark forms (black with red spots) occurring at up to 97% around Liverpool and Glasgow, whilst the typical form (red with black spots) occurs most commonly in rural areas. The black forms are under the control of a single gene, and are genetically

FIG. 99. The three commonly encountered forms of the Two-Spot Ladybird (*Adalia bipunctata*). The specimen on the left is the typical red form with two prominent black spots; centre is *sexpustulata* and right *quadrimaculata* (from Creed, 1971*b*).

dominant over the typical forms. The polymorphism cannot be the result of predation against the more conspicuous forms, because all individuals of the species are distasteful to most vertebrates, and emit a distinctive scent. Only Redstarts (*Phoenicurus phoenicurus*) regularly eat lady-birds, and they are not common enough to exercise a major effect on frequencies. Moreover melanic frequencies of only about 10% are found in the London area, although they occur at 63% in Harrogate and 75% in Hexham – both towns in largely rural areas. On the other hand, melanics have decreased in the Birmingham area since smoke control legislation has improved the atmosphere (Creed, 1971*a*), showing that human activity has an important effect on the polymorphism. Creed (1971*b*) originally suggested that lady-bird melanics were less affected than the red form by a toxic component in smoke, but it now turns out that the frequencies are regulated in a much more subtle way – black lady-birds have an advantage over red ones in cold and dark conditions because of their greater capacity to absorb radiation (Benham, Lonsdale and Muggleton, 1974; Muggleton, Lonsdale and Benham, 1975). This means that lady-bird melanism is really a geographical phenomenon, but that local climatic conditions may change in favour of melanics if there is significant smoke cover reducing the amount of sunshine. The low frequency carries with it the implication that melanics will have an advantage during the summer months, and in fact melanics have been found to increase in frequency during the summer and decrease during the winter in a number of places. On this basis, selection in Birmingham is 24% against blacks in winter, and 9% against typical (red) homozygotes in summer (Creed, 1975). It explains also why the

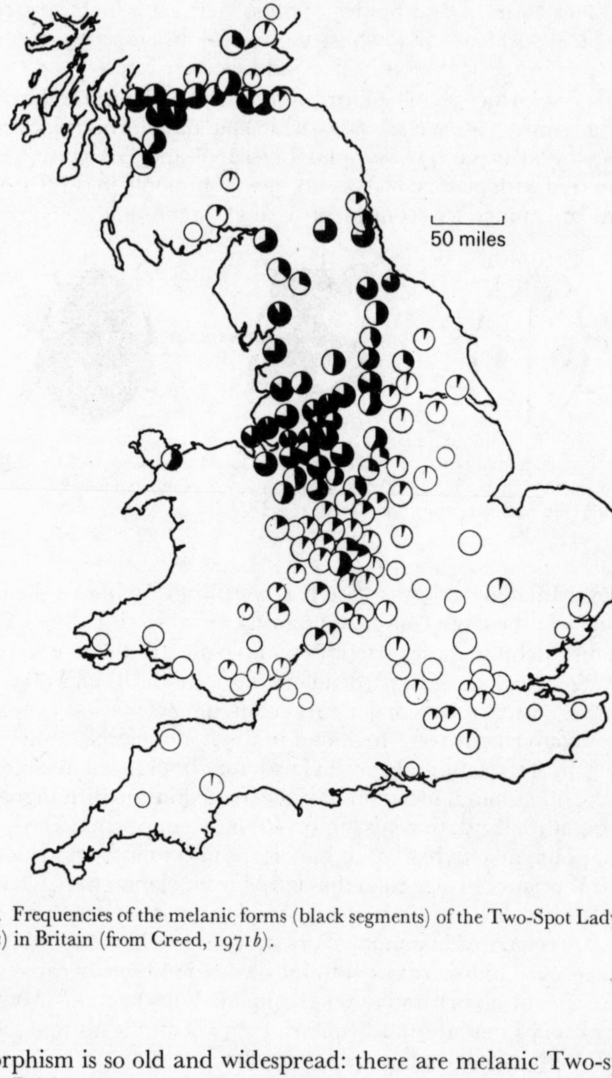

FIG. 100. Frequencies of the melanic forms (black segments) of the Two-Spot Ladybird (*Adalia bipunctata*) in Britain (from Creed, 1971*b*).

polymorphism is so old and widespread: there are melanic Two-spot Ladybirds in British Museum collections dating from 1696, 150 years before the first black Peppered Moth was recorded.

Another elegant series of investigations on genetical changes following man's activities has been carried out by Professor A. D. Bradshaw on the colonization by plants (mainly grasses) of polluted ground. The most extensive of these have involved the flora of spoil heaps from old metal mines in Wales. Antonovics, Bradshaw and Turner (1971) list 21 species of plants that

FIG. 101. Variability of copper tolerance in *Agrostis tenuis* plants and seeds collected from the waste heap of a copper mine, and seed produced from adults grown under experimental conditions. Isolation seed shows the same mean as adults (although a greater variance); seed produced on the mine has a lower mean tolerance, due to gene-flow from neighbouring non-tolerant populations (from McNeilly, 1968).

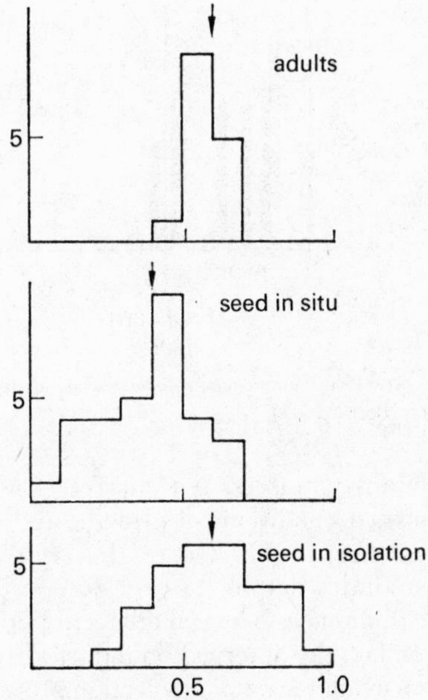

have evolved local races tolerant to soil concentrations of metals which are normally toxic.

Many of the waste heaps are small areas of a few hundred square feet containing up to 1% of copper, zinc, or lead (metals which normally prevent plant growth if they are in excess of 10 parts per million). The toxicity is due to the effect on root growth; an easy test for tolerance is to measure root growth of different plants in test-tubes containing solutions with different concentrations of metal ions. In a common grass like *Agrostis tenuis* all gradations of tolerance can be found, with no discontinuity between 'susceptible' and 'tolerant'. If an arbitrary level of metal-tolerant growth is chosen, one or two tolerant plants per thousand are found in most normal populations growing on uncontaminated soils (Walley, Khan and Bradshaw, 1974; Gartside and McNeilly, 1974). However only tolerant plants can grow on the extremely polluted soils: all other individuals will be selected against. As a result a tolerant population can evolve in only one or two generations.

A problem for grasses is that they are wind-pollinated, and most of the pollen received by an isolated population of tolerant plants will be from susceptible individuals (Gleaves, 1973). Consequently the seed produced by the tolerant population will be less well adapted to the mine environment than that of the parents. However, selection on the site of the mine will eliminate all but the tolerant individuals. In pastures on ordinary soils,

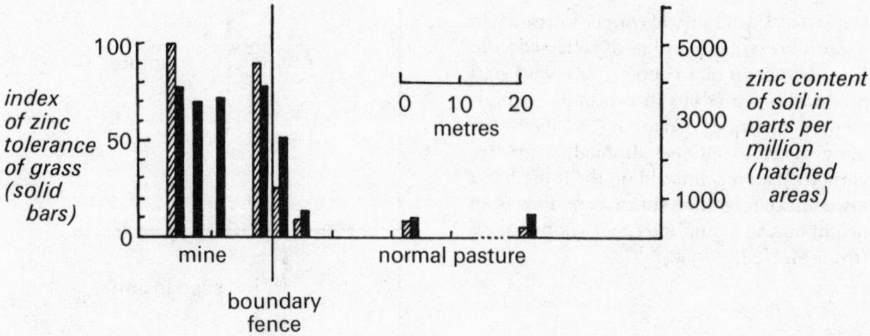

FIG. 102. Zinc tolerance in *Anthoxanthum odoratum* on, and adjacent to, Trelogan mine, Flintshire (from Berry, 1971).

tolerant plants do not survive as well as susceptible ones under normal crowded conditions of growth, and are rapidly eliminated away from the mine heaps. This means that very steep clines in tolerance exist at the boundaries, despite the gene flow occurring in both directions, tending to blur the differences between adjacent populations.

In fact the observed boundaries are even sharper than would be expected from available data on selection pressures and gene flow, presumably because of mechanisms to reduce gene exchange. Two such mechanisms that have been identified are differences in flowering-time between tolerant and non-tolerant populations, and higher levels of self-fertility around mine boundaries (Bradshaw, 1971).

Fall-out from aerial pollution can cause similar effects. Wu, Bradshaw and Thurman (1975) have described the situation around a metal refinery in Prescot. The refinery produces a smoke with high copper, plus lower lead and zinc contents. Grassland in the area has a total soil copper content of up to 4000 parts per million, and in some places the vegetation has been entirely destroyed. The maintenance of gardens in the immediate vicinity has usually depended on the surface soil being changed at frequent intervals. However lawns in the refinery grounds existed in excellent condition despite having no special treatment. Some of these were over 70 years old, and investigation showed that all the grass was highly tolerant to heavy metals. One side of one lawn had previously been a flowerbed, but this had not been successful, and 8 years previously had been sown with grass. Most of the new grass had failed to flourish, but the bare area was being progressively colonized by the few individuals which had survived (*i.e.* been selected) out of the original seed sown: in this case the new tolerant population was taking time to develop and give full cover because it was prevented from reseeding itself. Another lawn in the grounds was just achieving full cover after 15 years.

A parallel to the evolution of copper tolerant swards is copper tolerant algae on ships' bottoms. Antifouling paint is designed to release toxic compounds (usually copper, but more recently other metals) so as to prevent growth of marine organisms. As a result fouling by animals such as barnacles is now a minor problem. However green algae (*Ectocarpus* spp.) have continued to grow after antifouling treatment to the extent of causing economic problems because of the extra drag to the ship's movement through the water. It has been estimated that algal growth on a tanker's bottom may require the expenditure of an additional £50,000 of fuel at 1974 prices on a round trip between Britain and the Arabian Gulf, plus the cost of scraping and repainting the hull, which adds another £75,000 per ship to the annual cost (A. D. Bradshaw, *pers. comm.*). *Ectocarpus silculosus* collected from ships has been shown to be tolerant of 10 times the concentration of copper which a population growing on a normal rocky shore can stand (Russell and Morris, 1970), and also some *Ectocarpus* plants from ships may reach sexual maturity in half the normal time (Skelton, 1969). Once again we have evidence of rapid evolution in a special environment created by man.

3. Results from man's long-term activities

It is trite but true to state that virtually all the landscape in Britain is man-made. The large-scale removal of hedgerows is a fairly obvious example, and even this is merely producing a reversion to a state of open countryside characteristic of the time before the enclosures of the eighteenth and nineteenth centuries (Pollard, Hooper and Moore, 1974). What is relevant from the present point of view is that the changes produced in animal and plant communities by such activities automatically generate their own genetical pressures. This is most easily and spectacularly seen from the founder effect, some examples of which we have seen in Chapter 6. Other recorded examples relate to local introductions or extinctions (Hawksworth, 1974). In only a few cases have changes in genetical traits been recorded. Examples which have already been mentioned are changes in colour and banding pattern frequencies in *Cepaea nemoralis* and *C. hortensis* with vegetation between prehistoric times and the present (Cain, 1971), the spread of the Icelandic race of the Fulmar with the increase in deep-sea trawling (Fisher, 1952), movement of the hybrid zone of the Hooded and Carrion Crows in Scotland (Cook, 1975), change in skeletal traits in the St Kilda Field Mouse following human evacuation of the island (Berry, 1970*a*), and adjustments to the female wing spot distribution of the Meadow Brown butterfly in southern Britain during the late 1950s (Creed, Dowdeswell, Ford and McWhirter, 1962).

A rather intriguing example of local adjustment is that of the British race of the Large Copper butterfly (*Thersamonia dispar dispar*). This was first described from the Huntingdonshire fens in 1795, and its brilliant colour and restricted distribution soon attracted collectors. Nevertheless it was fairly widespread in the fens, and there can be little doubt that its extinction by 1851 was the result of over-collecting. Several attempts have been made to reintroduce European races of the species into England, but none of them have been entirely successful, probably due to the lack of suitable habitats following fen drainage. However, a colony at Woodwalton Fen founded from a Dutch race, *batavus*, in 1927 has survived with extensive human help – a stock is maintained artificially so that caterpillars and/or pupae can be released if required, and areas of food plant have been planted (Duffey, 1968). The interesting point about this English colony is that it has diverged considerably from the ancestral Dutch population, and many of the changes (especially in hind-wing pattern) are in the direction of the extinct British race, *dispar* (Bink, 1970). On the face of it, we are seeing a re-evolution of the old form.

A presumably similar but unequivocally deleterious situation is that in the British race of the Swallowtail butterfly (*Papilio machaon britannicus*) which became extinct at Wicken Fen in Cambridgeshire during the early 1950s, and has failed to recolonize there despite repeated attempts to re-introduce it. The main range of the race in Norfolk has also contracted, and the butterflies themselves are smaller than formerly. This may be an adaptation for reduced dispersal from the limited and decreasing patches of favourable habitat still available to them, but, whatever the true explanation, there is no doubt that an inherited recent change in body proportions has occurred at a time when the food plant of the race, Milk Parsley (*Peucedanum palustre*), has been on the decline (Dempster, King and Lakhani, 1976).

The difficulty about the Large Copper change is that we do not know how characteristic the butterflies introduced to Woodwalton Fen were of the Dutch *batavus* race. As we have had repeated cause to note, virtually all groups are likely to carry so much variation that any small isolate will almost inevitably differ from its ancestors (Berry & Warwick, 1974). The importance of studies like Ford's on Meadow Browns in the Scillies or my own on the Skokholm mice is that the amount of genetical change occurring from generation to generation can be recorded and its causes sought.

4. The genetics of domestication

One way in which man has consciously, drastically, and in some cases irrevocably affected the genetical constitution of species has been through his domestication activities pursued since neolithic times. Once our ancestors changed from a hunting to a settled life, and began to cultivate their own crops, they must quickly have begun to exercise selection for valuable traits, firstly in food plants, and then in their commensal animals (Ucko and Dimbleby, 1969; Brothwell and Brothwell, 1969).

As far as plants are concerned, it was first pointed out by the Russian N. I. Vavilov in the 1920s and confirmed by many since, that the origins of our cultivated plants are geographically spread unevenly, and the bulk of genetical diversity in our important agricultural crops is confined to relatively few centres (Zohary, 1970). In these centres, our ancestors exercised practices which exactly mimic the sequence of events in the natural evolution of local races or ecotypes – a reduction of the flow of genes from and to the bulk of the

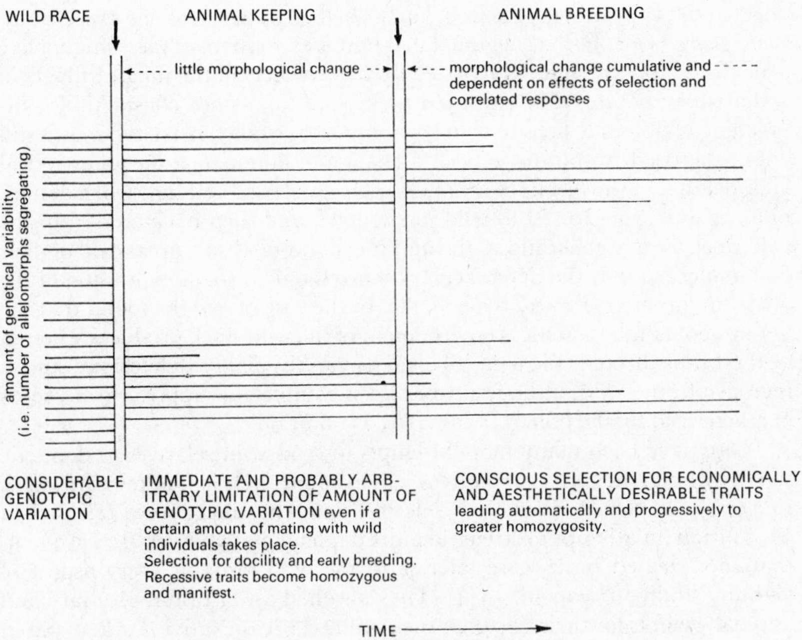

FIG. 103. Diagrammatic representation of the genetical changes undergone by a group of animals as they are domesticated (from Berry, 1969*b*).

original population; and adaptation to the specialized environment of the farmer, any response being dependent on the variation available in the founding members of the crop (Berry, 1969*b*).

Domestication of animals must have followed a similar pathway to that in plants, with a clearer distinction here between an early phase of animal *keeping* and a later one of animal *breeding*. The transition between the two stages was made in cattle many centuries ago, pigs more recently, and is only now beginning in an organized way in sheep.

Darwin was impressed with the ease with which distinct domesticated breeds can be produced. He filled notebooks with the fruits of his discussions with fanciers and commercial breeders. His description of the genetics of tame pigeons still provides fascinating reading.

Another example of domestication recalls one of the classical controversies of genetics. When Fisher was developing his theory of the evolution of dominance (page 188), one of the discrepancies that he had to explain was that a number of mutant conditions in hens were dominant to the wild type, whereas in virtually all other examples known at that time (the early 1930s), mutant phenotypes were recessive to the wild one. The conditions in question affected the appearance of hens in fairly obvious ways – *crest*, *rose-comb*, *barred*, *white*, *feathered feet*. Fisher suggested that minor expressions of these might have been distinctive enough to be prized by their owners in the jungle villages of India where poultry domestication took place, and since considerable out-breeding of domestic hens to wild cocks must have taken place then (as it still does), the result would have been a change in dominance (*i.e.* of genetical architecture, in the terminology of Chapter 5). He showed that when domestic hens were out-crossed to wild jungle fowl, and then back-crossed to the wild stock for five generations, the mutant conditions lost a great deal of their dominance, so that the heterozygotes showed only a small expression of the alleles in question (Fisher, 1935, 1938). In the case of *crest*, he found that the homozygote in the genetical environment of the wild stock produces a hernia of the brain through the skull. This is obviously highly deleterious, and is recessive in the wild, although it has been completely suppressed in breeds characterized by the homozygous crested condition.

There have been many more attempts to cross wild relatives to domestic stock, with a great variety of motives. For example, hybrid Turkeys (*Meleagris gallopavo*), in part derived from selected domestic stock, were released in Missouri in an attempt to strengthen the depleted native stock. It was found that the released birds were inferior to the wild birds in every aspect of viability studied (Leopold, 1944). They also had lower brain, pituitary and adrenal weights for their size than their wild relatives. Native Turkeys raised bigger broods than hybrids, and were more successful in rearing them. The commercial qualities of the domestic birds have been bought at a price of abandoning qualities favouring survival in the wild.

An illuminating study on domestication has been carried out on the grey or Norway Rat (*Rattus norvegicus*). This species was introduced to Britain about 1730 and to North America about 1775. It was probably taken into captivity around 1800. At that time rat-baiting was popular in England and France. Recently trapped rats were placed in a pit and a terrier let loose amongst them. The spectators bet on the time required by different terriers to kill the last rat. Large numbers of rats had to be caught and held in readiness for such spectacles. There are records of albinos being removed from such collections and retained for breeding and exhibition.

This was the beginning of the domestication of most modern strains of laboratory rats. In 1919, H. D. King began a 'repeat' of the domestication of the species. He took into his laboratory 16 male and 20 female wild-caught rats, and followed their change into servile commensals (King, 1939).

Early generations were marked by a high degree of female sterility (only 6 of the original 20 females bred) and infant mortality, and it was necessary to foster the first generation young on 'well-domesticated' mothers. It was suggested that the infertility of the wild females might have been due to nervous upset brought about by fear and confinement in cages. However the incidence of female sterility declined from 37% to 6% in the first eight generations, and all the females chosen for breeding from the thirteenth generation were fertile.

As the proportion of fertile females increased, so did the length of the reproductive period. Since litter size remained approximately constant (*c.* 6 young per litter), this had the effect that fecundity grew from an average of 23.1 young per female in the first generation to an average of 63.3 young in the twenty-fifth.

Mean adult weight increased fairly steadily to a gain of about 20% over the first 25 generations, so that the rats became slightly heavier on average than those of the long-established Wistar albino strain. The docility of the rats became more marked as the generations passed, but even at the 43rd generation they were more savage than the standard Wistar albino strain, and reacted to humans by jumping, squealing and biting. It may be that this difference between the strains reflected genetical differences in the founders of each.

A particularly interesting finding was that eight mutants appeared in the newly-captive strain, and sub-strains made by breeding from each differed from each other and the parental strain in a number of ways. Indeed the behaviour of the fully-domesticated Wistar strain could be mimicked by substituting three of the mutants (hooded, black, and albino) into the parental strain. The most important allele for this was the black one (for example, 73% of the blacks showed no anger when tickled on the nose with a fine brush, compared with only 15% of their grey sibs).

It is difficult to distinguish in this sort of experiment between a gene being important because of exerting more than one effect on its carriers, or because of being closely linked with loci affecting temperament and behaviour. Psychologists have been unable to detect any differences between coloured and albino members of the same rat family, although black animals in an F_2 are more timid than their grey sibs (Broadhurst, 1960). However 18 strains of laboratory rats are homozygous for black, which seems too many to have acquired the allele by chance. The segment of chromosome containing the black gene certainly seems important for domestication.

Black rabbits occur on Skokholm at a frequency of about 1%. They have a life expectancy of only one-third that of normal brown (agouti) animals, presumably because they are more easily seen and caught by birds, especially gulls. The allele persists because its bearers are less timid than normal animals and thus spend less time in their burrows. In conditions where food availability limits growth and survival (as on Skokholm and similar small islands), the disadvantage of conspicuousness is compensated by their extra feeding time.

Whilst acknowledging that domestication is not necessarily simple and may arise in different ways and at different speeds, one definite property of highly selected strains of both animals and plants is that they carry only a tiny proportion of inherited variation present in the wild species. This does not matter if the conditions for which the strain was derived remain constant, but diseases, husbandry and cultivation practices, climate, and soil are never fixed. Moreover successful crops and breeds are now spread much more widely than in past times, and may completely replace older strains. Thus the variation available to a farmer or breeder may be insufficient for his needs. Bee keepers frequently argue whether the native British black bee was superior to the imported Italian bees which have largely replaced them. We have seen that one response to the need for new variation is the production of new mutations, usually by radiation (page 259), but this amounts to no more than a 'fine tuning' of an already existing genome. For sound economic reasons, it is essential to be able to draw upon a wider pool of variants.

This need is now internationally recognized, and from the plant side one of the major efforts of the International Biological Programme was to record and collect seeds (or other propagable elements) from as many 'land races' (*i.e.* ancestral or unimproved varieties) of crop plants as possible (Frankel and Bennett, 1970). Preservation of animal variation is more difficult, but many zoos now see themselves as repositories of rare breeds (Rowlands, 1964). The success of the Severn Wildfowl Trust in breeding and releasing back into the wild the rare Hawaiian Ne-Ne Goose is well known. A determined effort at genetical conservation is being mounted in Britain by the Rare Breeds Survival Trust. Among other projects, they have established a flock of primitive North Ronaldsay sheep on a small Orkney island (Lamba Holm) lest disaster befalls the main stock of this unique seaweed-eating isolate on North Ronaldsay itself.

Pools of 'unimproved' animal and plant varieties provide essential material for research into the possibilities of better strains. Plant-breeders are more and more turning to wild races for genes which increase resistance to disease (Watson, 1970), while the extensive study of Soay sheep on St Kilda carried out by the Nature Conservancy and the Hill Breeding Research Organization (Jewell, Milner and Boyd, 1974) will provide invaluable data for the future on the hazards and reactions of an unshepherded flock. Frankel (1970) has called the loss of variation through concentration on a few commercially successful strains 'genetic erosion'. One of the more important applications for the future of man of the subjects discussed in this book will be the sensible containment of genetical erosion – not as an exercise in sentimental preservation, but as a resource to be husbanded.

GUIDELINES FOR GENETICAL CONSERVATION

It is impossible in the present state of knowledge to lay down exact rules about conserving gene pools. Already practical questions about the numbers of individuals that need to be taken into cultivation (or captivity) are becoming important, but we are still too ignorant about the amount and relevance of variation carried by a small group of individuals compared with that in the species as a whole to be able to give useful answers. The best we can do is to state some guidelines to be borne in mind when undertaking the care of an animal or plant population considered to have important inherited characteristics (Berry, 1971). These guidelines are less important for botanists than zoologists, because of the ease with which most seeds can be kept dormant over many years.

1. Any environmental change (be it natural or part of 'management') within the normal tolerance of a species will result in genetical adaptation in the affected population; if the changes exceed the species tolerance, the population will, of course, become extinct. There is a slight paradox here, because 'tolerance' is a genetical property. However this should not be a serious problem in practice, since management is effectively directed towards maintaining or restoring a *status quo*.
2. Adaptation is rapid and precise, and has to be based on variation existing within the population. The chances of random – and potentially deleterious – changes taking place are slight and can probably be neglected.
3. The amount of variation (at least in a normally outbred, diploid species) is extremely large and resistant to loss. Even in a population subjected to rigorous and long-continued directional selection, considerable variation remains unused (Lerner, 1954; Carson, 1967). This generalization is not contradicted by forms subject to a varying environment at the edge of their range which lose genetical markers such as inversions, and hence appear to have a poverty of variation (Carson, 1958; Zohary and Imber, 1963). It is extremely unlikely that any natural management procedures could significantly affect the amount of variation in a local population to the extent of making that population unable to respond to environmental change.
4. Intermittent drift (page 56) could lead to a change in a local form – but so could straightforward natural selection. Local forms have ecological interest and taxonomic kudos, but no eternal value. Consequently 'gene-banks' should be seen as maintaining particular alleles, and not preserving whole genomes.
5. There is no reason to think that particular genetical problems will arise through inbreeding necessary to rescue 'threatened species' from extinction. The difficulties are much more likely to be ecological and related to the reason why the species became rare in the first place.

5. Man looking after man

Finally we must consider what effect we are having and can have on the genetical constitution of our own species. We are repeatedly exhorted by optimistic humanists to take care of our posterity by controlling our genes. We shall see in the next chapter that an awful lot of nonsense has been talked about the possibility of man's extinction through his heavy 'load' of mutations, dragging him down as Christian in Bunyan's *Pilgrim Progress* was weighed down by his load of sin. In this section we must consider some of the speculation and inanities that have been used about *Homo sapiens* from at least the time of Plato (He said, 'If we want to prevent the human race from degenerating, we shall take care to encourage union between the better specimens of both sexes, and limit that of the worse'). Frederick II of Prussia is said to have given the best girls to his best soldiers, to promote the formation of an elite.

There is a story of Bernard Shaw being approached by the dancer Isadora Duncan with the request that they should have a child, because her body and his brain would make a wonderful combination. 'Yes' replied Shaw, 'the snag would be if it had your brain and my body'. This sums up the two problems which face any attempt of human 'improvement': what are the characters we choose, and, given the complicated inheritance of traits like beauty and brain power, how do we arrange for them to increase in frequency? A stock of animals or plants – or, for that matter, a tool – can be 'improved' for a particular purpose so that we have a perfect instrument for our purpose. For example, a Jersey cow is an excellent mechanism for producing quantities of creamy milk – but it is only excellent on good pasture and will not function as well as other breeds on cold, wet hill pastures. What traits should we look for in a perfect man? It is repeatedly argued that modern medicine weakens mankind by allowing the weak to live and breed; in other words, that the nature of medicine is to work against natural selection. This is a fallacy: natural selection is a product of the physical and biological environment in which we live. Everything we do modifies this environment, and hence changes selective pressures. But survival and not perfection is the acid test. This means that 'improvement' always has to be related to some ideal; human perfection has an absolute philosophical or spiritual meaning, but a biological one only in particular environmental and cultural contexts. This is important in any debate about racial differences, because claims about individual superiority or inferiority must always be qualified by an understanding of the value of the person for the community or culture in which he lives. A high level of intelligence is valuable in western-style communities, but is much less important than health and strength in a harsh, marginal existence.

The history of the human species has been like that of virtually all other species: a succession of small, exploring populations with contact and conflict between them through the generations. As we have seen, men still tend to marry the girl next door (page 107). In every age, the problem has been survival and reproduction. The fact that the problems of survival and rearing a family in Britain are now vastly different from the ones that existed in the times of Boadicea or Bede, and that the qualities needed to solve them have also changed, only emphasizes the dynamic nature of adaptation to the environment faced by all species.

Human Races

The fossil record of man is probably as good as that of any species. The genus *Homo* emerged from the earlier *Australopithecus* during the Pleistocene. Australopithecines have been found in China, Indonesia, Asia Minor, and East and South Africa. Different specimens have been given different names (*Telanthropus*, *Plesianthropus*, *Paranthropus*, and *Zinjanthropus*) but all the forms are similar, and fairly closely related. The most abundant fossils known so far come from the Olduvai Gorge in Kenya, dated about $1\frac{3}{4}$ million years ago.

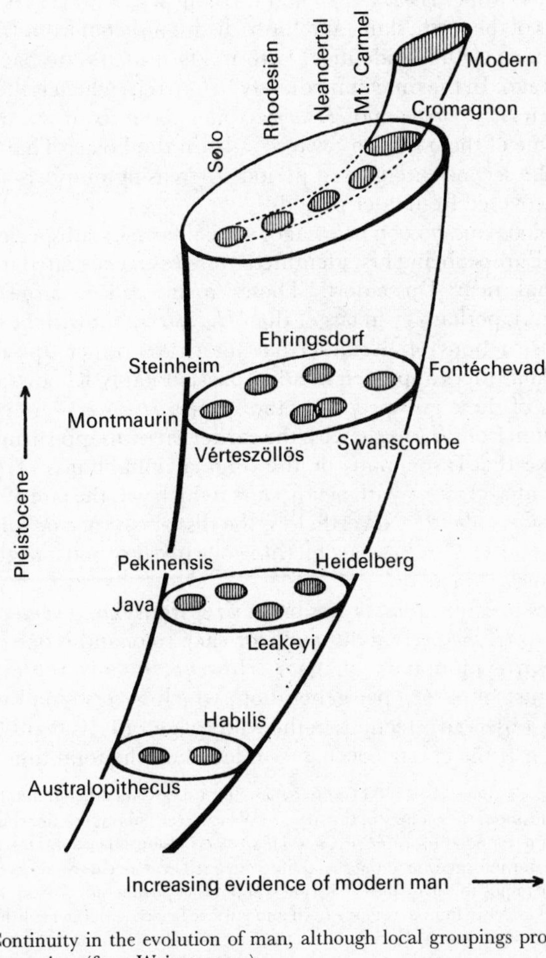

FIG. 104. Continuity in the evolution of man, although local groupings produce a variety of forms at any one time (from Weiner, 1971).

Some at least of the Olduvai hominids (*i.e.* man-like apes) made stone tools, but they cannot be regarded as true men in any normal sense.* *Australopithecus* probably walked on two feet, but its brain size was only about 450 to 600 c.c. This is not much larger than that of the living apes and less than half the size of modern man.

The first specimens generally agreed to belong to the genus *Homo* appeared about 500,000 years ago. These have been put in a species called *erectus*, and once again remains are known from the Far East ('Pekin' and 'Java' man, previously called *Sinanthropus* and *Pithecanthropus* respectively), from the Olduvai Gorge, from Ternifine in Algeria (previously *Atlanthropus*), and from Heidelberg in Europe. *Homo erectus* had a brain size of about 1000 c.c., but characteristics of the teeth, skull, and limbs distinguished it from *Homo sapiens*.

Fossil skeletons virtually identical with modern man date back to about 300,000 years ago. In the time immediately before this, which is the transition period between *Homo erectus* and *H. sapiens*, only about six or seven specimens are known. One of these is from Swanscombe in the Lower Thames Valley, dating from the second interglacial period. Increasing numbers of skeletons have been uncovered from later periods.

Between 30,000 and 70,000 years ago, a local form or subspecies occurred in north-west Europe, being first identified from fossils excavated in the valley of Neanderthal near Dusseldorf. These 'men' had a large skull size (1300–1600 c.c.), perhaps even larger than *H. sapiens*, but with heavier bones so that reconstructions of 'Neanderthal man' are rather ape-like in appearance. Similar races appeared in other parts, notably Rhodesia. The best interpretation of these groups is that the species *Homo sapiens* evolved as a single continuum from *H. erectus*, with the earlier races disappearing, either by extinction (like the Tasmanians or the original inhabitants of Taiwan or Japan), or by inter-crossing with neighbours (which was the fate of the Viking settlements in Greenland). Nevertheless, the disappearance or submergence of archaic varieties of *H. sapiens* is something of a mystery and a neglected field of study (Weiner, 1971).

The changes in skull capacity (= brain size) from *Australopithecus* to *Homo erectus* and on to *H. sapiens* is about 50% for each step, and is one of the most rapid of known evolutionary changes. However it only represents a net selection pressure of 0.04% per generation, which is very small when compared to many observed selections in the wild (page 130) (Cavalli-Sforza and Bodmer, 1971). If the change occurred in less than the total time available,

*In all the discussions about early and primitive man, it must be remembered that fossils can tell us only a very little about the biology of the species. We can learn about relationships with other species from remains found at the same site, and at later periods burial customs tell us something about beliefs. But our information about social life from fossil evidence must always be fragmentary, and it is quite impossible to do more than guess about the spiritual side of man. Indeed one of the most drastic changes in human organization and culture began with the neolithic revolution about 10,000 years ago, which dates modern man at roughly the same time as the Biblical creation of man! (Berry, 1975*b*).

selection would have had to be stronger. De Beer (1958) has argued that the key factor in human evolution was not a gradual trend but a somewhat spectacular change in growth rates. If so, man is really an ape with a greatly extended immature stage. On this view, adults are really sexually mature children, with the body proportions (and possibilities for brain growth) of a young person.

None of this helps to explain the amount of racial variation that exists in modern human populations. Throughout the Pleistocene, stone tools changed in shape and sophistication, with similar but distinct traditions throughout the world. Men experienced a greater variety of habitats than ever before, particularly in the northern hemisphere where they had to endure the cold conditions of the late glacial stages. These early men were hunters, as the debris in their caves shows. The hunting groups must have had all the possibilities of differing widely from each other through the repeated operation of the founder effect, and would have different conditions to respond to in different parts of their range. Unfortunately we have to depend upon a few unhelpful skeletons for most of our knowledge of the origin of different races, and there is insufficient material to confirm or deny suggestions that, for example, the Negroes originated from Rhodesia man, or that Pekin man is the ancestor of the Oriental races.

Coon (1962) has argued that the development of the human races was a process taking a million years or more, and that *Homo sapiens* represents a number of different lines established during the *H. erectus* time. He divides *H. sapiens* into five subspecies (Mongoloid, Australoid, Negroid, Capoid, and Caucasoid), corresponding roughly to the earliest inhabitants of the five continents, and implies that their development has been parallel for many generations.

Completely parallel evolution is unlikely in populations living under widely different conditions, and an interpretation like this lays the way open for suggestions that one race may be more 'advanced' or 'primitive' than others. In fact there is not much evidence for Coon's ideas (he himself cites mainly dental variations) and one intriguing bit of counter-evidence in the frequencies of mutant proteins.

We have seen that change in the DNA 'code' for a protein is likely to alter the amino-acid sequence of the protein formed (page 35). If most of these mutations do not affect the function of the protein, they will accumulate in the population as time passes. The structure and variation of haemoglobin is well known, but instead of there being a large number of variants at most amino-acid positions, there is a clear distinction between common substitutions (presumably affected by selection, like the one producing sickle-cell haemoglobin), and very rare ones. This suggests that the human species has not 'had time' to gather chemical detritus, and it seems likely that the species went through a bottleneck in numbers a few thousand years ago which eliminated the past substitutions (Smith, 1970). This is rather a roundabout argument, but it supports the traditional view that racial differences have arisen relatively recently.

But we still do not know what the differences *mean*. Valiant attempts are periodically made to identify adaptations that have been important in man's past history. The most determined of these are speculations about skin colour, and particularly the possibility that dark skin might protect against tumours induced by ultraviolet radiation from strong sunlight and prevent the for-

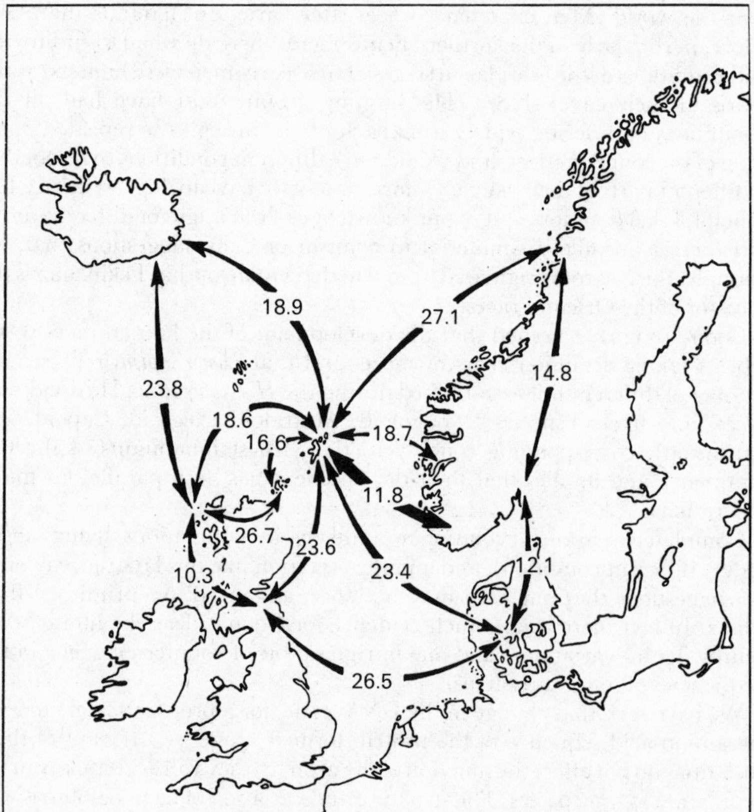

FIG. 105. Genetical 'distances' calculated from non-metrical variant frequencies in human skulls. The population of the Shetland Isles is particularly close to southern Norway. Evidence such as this can be used to reconstruct human population movements in the same way as for mice (see figure 77). Comparisons based on multigenically determined traits are often more useful than those derived from blood group or enzyme variations, because many more genes are involved (from Berry and Muir, 1975).

mation of excessive amounts of vitamin D. More plausible are suggestions that the distribution of blood group and enzyme variant frequencies are the result of epidemic diseases in the past, complicated by subsequent population movements (Haldane, 1949; Sheppard, 1959). For example, people with blood groups A and AB are most likely to get smallpox during an epidemic, while people of groups B and O tend to have a mild attack and survive (Vogel and Chakravartti, 1966). The discovery that people with certain blood groups have a higher chance than others of acquiring certain degenerative diseases (for example, group O people are 40% more likely to get a duodenal ulcer than other people; group A people are 20% more likely to get stomach cancer than others) does not affect selective pressures much, but does demonstrate that the blood group substances are exercising previously unsuspected physiological effects.

Notwithstanding we still do not know the reason why the racial groups split in the first place, nor how much of their differentiation has been the result of adaptation in partially isolated groups. It is not difficult to show natural selection at work in human population (Bajema, 1971); indeed we have described examples of selection acting in both multifactorial and unifactorial systems (on birth weight, page 131, and sickle cell haemoglobin, page 69). But the problem with humans is exactly the same as with other species: without a longitudinal study of genetical traits through many generations, we cannot tell how much the genetical composition of a group is the result of the largely chance assortment of genes carried by its founders, and how much the consequence of subsequent adaptation. The situation is exactly comparable with the taxonomic diversity that exists in the island field mice described in Chapter 6. Indeed a study of the North Atlantic people descended from the Vikings using similar skull characters to those used in the mouse study,

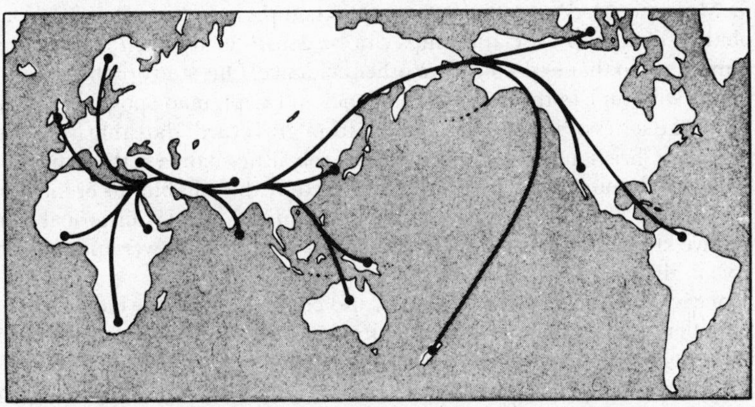

FIG. 106. An attempted reconstruction of world-wide human migrations based on population relationships (from Edwards and Cavalli-Sforza, 1964).

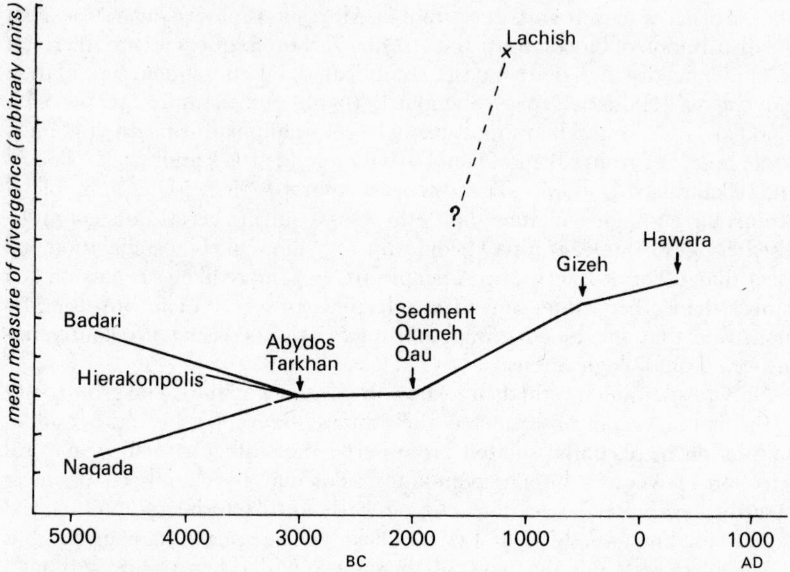

FIG. 107. Changes in the genetical constitution of the population of ancient Egypt, based on measures of divergence calculated from non-metrical skull variant frequencies. If the population did not change at all, the graph would be horizontal (after Berry and Berry, 1972).

produced a series of putative relationships which could be interpreted in just the same way as that outlined for the mouse races (A. C. Berry, 1974; Berry and Muir, 1975). When the same sort of comparisons are attempted on a global scale, a map of relationships can be constructed which bears a good resemblance to that expected from other evidence. One study of the genetical changes with time in the population of ancient Egypt, made possible because of the large series of skeletons excavated from graveyards dateable from 3000 BC to early Christian times, showed remarkably little change in the indigenous population despite all the vicissitudes suffered by the inhabitants of that part of the world. The only significant changes occurred in early historical times when we know from written records that ships and traders were particularly active in the East Mediterranean.

Race is a loaded term for sociologists and politicians. For the biologist there is no difference in principle between human and animal races, stocks, strains or sub-species. To him, racial conflict is no more surprising than any event of intra-specific competition wherever it occurs. To return to the discussion with which we began this section, perhaps the biologist's main contribution to the hot air that is talked about racial differences (particularly in intelligence) is to emphasize that notions of racial inferiority or superiority are invalid extrapolations from the demonstration of differences *per se*. This makes a positive contribution, whilst in no way lessening his personal compassion nor his commitment to reduce the casualties of racial tension.

Eugenics

In the heroic days of Victorian enterprise, Francis Galton advocated careful breeding to improve the human race and eliminate its defects. He called this *eugenics*. So-called positive eugenics has had a short and untriumphant history, for reasons already touched on (page 274). One of its most influential propagandists was H. J. Muller, who was awarded a Nobel Prize for his work on radiation-induced mutations. He planned a sperm bank where the sperm of prize men could be stored (like that of prize bulls wanted for artificial insemination). This bank would be available for any woman who hankered for children of a higher quality than those likely to be sired by her no doubt desirable but otherwise unselected husband (Muller, 1965). Muller saw no difficulty in selecting the characters which would qualify one to be a high-quality mate: 'practically all people venerate creativity, wisdom, brotherliness, loving-kindness, perceptivity, expressivity, joy of life, fortitude, vigor, longevity'. The problem is that many of these traits may exist only in the eye of the beholder (never mind the contribution of genes to their inheritance): after spending some time in Russia (during which he gave up his doctrinaire communism), Muller crossed Lenin and Marx off his list of suitable sperm donors.

The possibility of actively improving the human race genetically is impractical. However, we are still faced with a problem of negative eugenics. The burden of genetical disease on the community is substantial: 6% of the population has a readily detectable genetical defect at birth (Table 25); 25% of children admitted to hospital will be there because of some inherited anomaly. Furthermore, genetical disease is becoming relatively more important. In 1922 37% of blindness cases in Britain had a genetical origin; by 1952 this had almost doubled to 68%. Deaths at the Hospital for Sick Children, Great Ormond Street, with an environmental cause (pneumonia, tuberculosis, etc.) constituted 68% of the total in 1914, but only 14% in 1954. The incidence of congenital malformations has remained more or less constant throughout this century, but they now account for a quarter of the deaths in the birth period, whereas in 1900 they were the cause of only one in 50 deaths. It seems unlikely that congenital malformations will be significantly reduced in frequency by any further increases in maternal care.

The old-fashioned sledge-hammer approach to the prevention of genetical disease was to sterilize anyone thought unfit to bear children. A number of countries still have laws for compulsory sterilization, but they have never found widespread favour because of their ineffectiveness: only people suffering from clear-cut single factor conditions (heterozygotes for dominant

TABLE 25. Incidence of genetical disease in man (after Berry, 1972)

Type of inheritance		*Incidence per 1000 births*
Single gene effects		
Autosomal dominant		9·5
Autosomal recessive		1.3
Sex-linked recessive		0.4
All single locus effects		11.2
Chromosome anomalies		
Trisomy-21		1·4
Trisomy-13		0.1
Trisomy-18		0.2
Cri du chat		>0.1
Klinefelter's syndrome, etc.	in males	1.7
XYY syndrome	in males	c.0.1
XXX syndrome	in females	1.0
Turner's syndrome	in females	0.2
Total clinical chromosomal anomalies		5·4
Congenital malformations		14·1
Serious 'constitutional' disorders (diabetes mellitus, idiopathic epilepsy, schizophrenia, etc.)		14.8

conditions or homozygotes for recessives) are sure candidates for sterilization, but these are the very people least likely to pass on their genes because they are severely handicapped themselves. Moreover the frequency of deleterious dominants is largely maintained by recurrent mutation, while recessive alleles are almost all carried by healthy and unidentified heterozygotes. Since we all carry four or five deleterious alleles in the heterozygous state (page 288), it would be delightfully impractical to attempt any lowering of recessive gene frequencies by sterilization.

The most important recent advance in the reduction of genetical disease has undoubtedly been the ability to diagnose diseased foetuses in the first four months of pregnancy, giving the possibility of eliminating the abnormal by selective abortion. The procedure involves taking a small amount of amniotic fluid with a syringe (*amniocentesis*), and examining the chromosomes or metabolic activity of the foetal cells floating in the fluid (or in some cases the constituents of the fluid itself). Amniocentesis involves a small risk (less than 1%) of causing a miscarriage, and its cost in money and skilled manpower will probably rule it out from becoming a routine test for all pregnant women. Nevertheless it will play an increasing part in preventing abnormal children being born to women at risk and hence lower the burden both on individual families and on the community as a whole (Harris, 1974):

1. When a woman has had a child with a malformation of the central nervous system (anencephalus or spina bifida), her risk of having another is 25 times higher than before. Defective foetuses in future pregnancies can be identified at about the eighteenth week of preganancy from the concentration of 'α-foeto-protein' in the amniotic fluid.
2. Older women (aged 38 years or more) have a high chance of producing a trisomic child (most importantly one with Down's syndrome or mongolism – page 77). These can be easily identified in chromosome culture. Similarly some parents are carriers of translocated chromosomes, and have a raised chance of having a child with unbalanced chromosome complement (*i.e.* partially trisomic).
3. A couple who are both heterozygous for a deleterious recessive condition have a one in four chance of producing an abnormal child. More than 40 recessive conditions can now be identified in cell culture, making it worthwhile to monitor all pregnancies of such couples in order to identify homozygous foetuses.

Genetic counselling and amniocentesis have become more or less routine medical procedures in medically advanced countries. But there is another part of man's attempts to control his own heredity which borders on science fiction.

1. *Advances in reproductive physiology.*

Although not strictly genetics, the possibility of transplanting a fertilized ovum into a woman who might be sterile, or who might undertake the chore of pregnancy for another will obviously affect the genetical structure of the population (Edwards, 1969). The ability to separate X- and Y-bearing sperm outside the body is probably near, allowing the sex of future children to be chosen. Most families claim to want at least one boy, so this procedure would lead to an excess of males: if a male is the first-born in a family, a couple would not have recourse to artificial methods for the next child, and have an equal chance of having a boy or a girl; if, on the other hand, the first child was a girl, families able to choose are likely almost always to opt for a boy. On the other hand, the possibility of producing children of the desired sex would presumably lead to fewer children being born.

2. *Algeny or chemical control of the genotype.*

This is the fancy which tends to be called 'genetical engineering'. In 1967, Kornberg synthesised DNA in a test-tube, and pointed out that 'harmless DNA from a virus attached to a necessary gene and then used as a vehicle for delivering a gene to the cells of a patient suffering from a hereditary disease could thereby cure him'. Tatum (1965) carried the speculation a stage further. His plan was to incorporate the 'good' gene into a tissue culture of the patient's own cells; allow them to grow in culture; and then inject them into the patient. This would prevent any immunological rejection of the graft cells.

Sinsheimer (1969) has pointed out that some sort of procedure like this will have to be seriously considered for diabetes. There are now over four million sufferers in the United States alone, and the usual source of insulin from slaughter-houses is not keeping pace with the increase of the disease.

These speculations are not as wild as might appear. In one experiment cells cultured from a galactosaemic patient (*i.e.* an individual homozygous for a recessive gene which prevents the formation of the enzyme catalysing the production of glucose from galactose; this leads to mental deficiency) were infected with a bacterial virus carrying the gene for normal enzyme production in *E. coli*. This led to a 10- to 75-fold increase in enzyme activity in the deficient cells (Merril, Geier and Petricciani, 1971).

Repairing genes in this way has the merit that all the individual stages are possible. The difficult decision is the incorporation of transformed cells into the body, since they may have lost the ability to control their growth and as a consequence be cancer-producing. However, if the choice is between saving a patient's life in the short-term whilst storing up unknown future problems, it is likely that such techniques will be tried in the foreseeable future.

Other suggestions for algeny have been made, such as using repressor molecules to block the expression of certain characteristics, or, even more radically, direct gene surgery involving the manipulation of chromosomes by laser beams, etc. Essentially these techniques would lead to the early Mendelian grail of directed mutation. The problem always remains of damaging other genes than the one being 'operated on'.

3. *Clonal proliferation* (or vegetative reproduction, as with potatoes or roses). This story began in 1952 with the successful development of an amphibian embryo from an egg whose nucleus was transplanted from a blastula cell by Briggs and King. In 1961 John Gurdon in Oxford achieved normal growth of a tadpole from an egg whose nucleus had come from the intestinal wall of an adult. Since the nucleus of all cells in the body has an identical genetical complement, 'if a superior individual – and presumably genotype – is identified, why not copy it directly, rather than suffer all the risks, including those of sex determination, involved in the disruption of recombination?' (Lederberg, 1966). In other words, take a body cell and use it as a zygote by implanting it in the uterus. Since we can choose the individual from which the cell came:

a. There will be no risk of an inferior phenotype, which can only be recognized 15 to 20 years after the original experiment.
b. No problem with gene segregation or sex as in the normal recombination lottery.
c. All the individuals produced will be genetically identical and hence have interchangeable parts.

d. There will be academic interest in comparing identical twins of different generations: will the second Einstein or Sinatra be better than the original?

Clonal proliferation ought to be one of the goals of women's lib., since it does away with the need for males; it has been described as auto-adultery. Indeed it is the realization of the process described by Aldous Huxley in the *Brave New World* where appropriate genotypes were bred for each job – epsilon demimorons for lift operators, and so on.

Rattray Taylor (1969) stated his belief in the *Biological Timebomb* that 'The power to interfere in the processes of heredity is the most serious of all the human problems raised by biological research'. Inevitably it involves a host of ethical questions which cannot be dodged (*q.v.* Ramsey, 1970; Hilton *et al.*, 1973). Nevertheless it may be unwise to be too concerned about the future. Speaking about 'test-tube babies', Dame Honor Fell has pointed out that 'The cost of providing the necessary plant, materials and staff for the production of population on even a limited scale would be simply prohibitive. I cannot imagine that even the most irresponsible government would be prepared to finance such a project, especially in view of the fact that a superabundance of quite adequate babies can be produced at very little expense to the Treasury, by the old-fashioned methods'. A similar note of pessimistic realism comes from the complaint of J. B. S. Haldane: 'If we ever want a race of angels, we shall probably have to have new mutations, for the present supply of genes includes neither those for the wings nor for the requisite moral perfection'.

CHAPTER 9

WHY BE VARIABLE?

'THE mass of genetically distinct forms which make up the Linnaean species do not distribute themselves indiscriminately over an area comprising different types of localities, but, on the contrary, are found in nature to be grouped into different types, each confined to a definite habitat. Further, these "ecotypes" do not originate through sporadic variation preserved by chance isolation; they are, on the contrary, to be considered as products arising through the sorting and controlling effect of the habitat-factors upon the heterogeneous species-population'. So wrote Turesson (1922) in one of a series of papers in which he set out to synthesize genetical, ecological and taxonomic approaches to plant communities, an endeavour which was so successful that it changed the face of botany in succeeding decades (Heslop-Harrison, 1953, 1964). However our interpretations are still changing: Gregor and Watson (1961) have suggested that a better understanding of natural processes 'could be achieved through the accumulation of records in which the emphasis has been transferred from the discreet ecotype to the trends of ecotypic differentiation'. To this less static approach we must add the understanding of intra-population variation which followed Ford's definition of polymorphism (*q.v.* page 18), and which has been the particular contribution of animal workers to natural populations. The time has now come to link concepts developed separately by geneticists and ecologists for their own purposes.

Here laboratory results are relevant. Hitherto we have not referred much to laboratory work in population genetics and ecology because our main concern has been – and is – the real situation in nature. Nevertheless it is fair to say that field and laboratory conclusions complement each other, particularly in revealing the large pool of variation available for response to selection in outbreeding animals and plants. In general, laboratory work provides stringent models for which field results are examples of particular cases. Unfortunately in the years after being established by the early work of Fisher, Haldane, and Wright, theoretical population genetics moved into the lower floors of an ivory tower of unreality. There grew up an attitude among theoreticians that all the worthwhile genetical problems had been solved, just as physics was supposed to have ended with Newton – until it became complicated again by atoms, relativity and the like. This patronizing of practical geneticists has changed dramatically as a result of two related events: the calculation that there is a maximum to the variation which any population can carry, and the finding that most populations have too much variation for

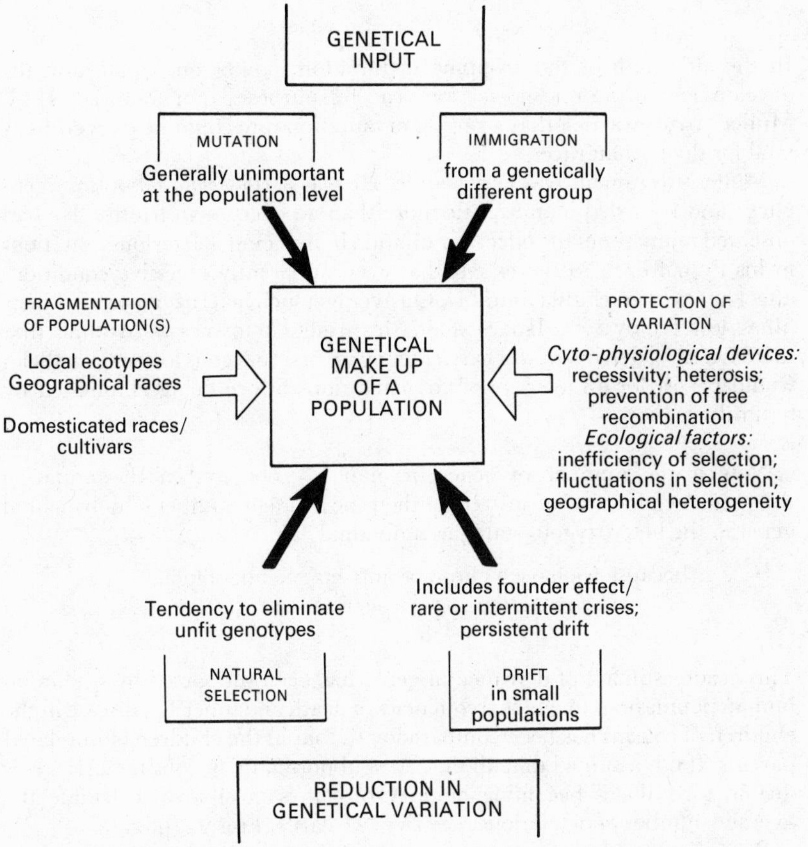

FIG. 108. Summary of the factors contributing to the genetical make-up of a population or group of populations (after Berry, 1974*b*).

their own survival. This discovery upset the theoretical apple-cart as decisively as Copernicus' observations on the solar system destroyed the elaborate artificiality of the mediaeval cosmos. Since Lewontin, Hubby and Harris showed in 1966 that both *Drosophila pseudoobscura* and man did not fit the previously accepted theoretical dogma, theory and observation have come much closer together to the potential benefit of both, but with field observation now leading theory.

GENETIC LOAD

In the aftermath of the dropping of the atom bombs on Japan, and the development of atomic power for peaceful purposes that followed, H. J. Muller (1950) warned that a raising of mutation rate could be dangerous or fatal for the human race.

Muller's argument was very simple. He was a *Drosophila* laboratory geneticist and regarded man as little more than an overgrown fruitfly. He had observed many times the effect of radiation in inducing deleterious mutations in his fly cultures, to the extent that even apparently recessive conditions affect heterozygotes unfavourably. On average, such heterozygotes have their fitness lowered by $2\frac{1}{2}\%$. If one extends this argument to man and assumes that there is a $2\frac{1}{2}\%$ detriment to a carrier for every heterozygous locus, such alleles would persist for an average of 40 generations before being eliminated by natural selection.

Now if the number of genes in man is 5,000 and if the mutation rate/gene/generation is 1 in 50,000, then the average number of detrimental genes in the heterozygous state per individual

$$= 2 \text{ (because we have 2 chromosome sets)} \times \text{no. of loci} \times \text{mutation rate} \times \text{persistance} = 8$$

This crude estimate of detrimental genes has been supported by studies on human populations in which frequencies of death and inherited disease in the children of cousins has been compared with that in the children of unrelated parents. If it is assumed that all the excess abnormality in cousin marriages is due to rare alleles becoming homozygous, it is possible to calculate the average number of deleterious recessives we carry (Fraser, 1962).

Returning to Muller's argument, he maintained that the mean fitness in man is reduced by ($8 \times 2\frac{1}{2}\% =$) 20% as compared to the maximum fitness of a person carrying no mutants. He called this the *genetic load*. It means that the average number of surviving children born to a couple is only 80% of those that might have been born, and hence at least 2.5 children per couple must be conceived if the population size is to remain constant.

Muller believed this genetic load was effectively the result of recurrent mutation. It follows that if the mutation rate is doubled, only 60% of the children conceived would survive. This would mean that each couple would conceive 2.5 children, but only produce 1.5 living adults for the next breeding generation. The population would not replace itself, and would rapidly become extinct.

This argument attained the status of a prophecy, and resulted in large amounts of government conscience money being given for genetical research. However it involves two fallacies. Family size is limited by many factors, of which the least important in most communities is the possibility of only siring 2.5 children. However much more far-reaching is Muller's assumption that most genetical disease is the result of recurrent mutation. We have seen that stabilising selection, where the extreme expressions of a trait have a lower fitness than the mean, is common (page 131). The simplest genetical mechanism for producing stabilizing selection is where the heterozygote is favoured over the homozygotes (or multiple homozygotes for a multigenically determined trait). In such a situation some genotypes will be less fit than others, and therefore impose a load upon the population. This is a *segregational* load in contrast to the *mutational* load discussed primarily by Muller. About half the zygotes formed in human populations fail to survive and reproduce in the next generation, even under the best medical conditions; it has been suggested that half this failure is due to stabilising selection (Penrose, 1955).

There are other components of the total genetic load. J. B. S. Haldane (1957) called attention to the *substitutional* load, which he called 'the cost' of natural selection. This occurs whenever one allele spreads in a population. The allele that is being replaced will confer a lower fitness on its carriers than the allele that is spreading. If an allele spreads to fixation (*i.e.* if the former allele is completely eliminated in a population whose size remains constant), the number of individuals that die through natural selection will be equal to nearly two-thirds of those alive in any one generation. The advantage produced by the new allele does not matter: an allele that spreads rapidly (*i.e.* has a high selective advantage) will be responsible for a higher proportion of selective deaths in a few generations when compared with an allele spreading more slowly. Clearly, however, a population can only tolerate a limited number of substitutions going on at the same time, or too many individuals will be dying to maintain a viable population. Haldane calculated that one allele will be substituted every 30 generations on average, and supported this conclusion by deductions about the rate of evolutionary change in fossil lineages.

Other processes which may contribute to the total genetic load are: recombination (*q.v.* page 183), environmental heterogeneity (since a genotype may have a higher fitness in one part of the population range than in another), meiotic drive (*q.v.* page 228), maternal-foetal incompatibility in mammals, population size (since ill-adapted phenotypes may arise through genetic drift), and migration (Brues, 1969; Crow, 1970). From the point of view of the genetical hazards of mutation-causing agents such as radiation, it is important to know how much of the load is due to mutation. Considerable ingenuity has been expounded in trying to apportion the contribution of each process to the total, but without much success (Wallace, 1970). Notwithstanding, the implication of the genetic load argument is that there is an upper limit to the amount of variation that is possible.

This can be illustrated by a human example. A population which segregates for both sickle cell and normal haemoglobin possesses a segregational load, and it is possible to measure the relative fitnesses of the three genotypes, by counting the number of children produced by each (Allison, 1954). In a typical Central African population where the frequency of the sickle cell allele (*S*) is 28.6%:

	AA	*AS*	*SS*
Fitness (*f*)	0.8	1.0	0.5
Proportion of each genotype (*r*)	0.510	0.408	0.082
Proportion dying from genetical causes ($r(1-f)$)	0.102	0	0.041

Hence (0.102 + 0.041 =) 14.3% of the population die from causes which it would not do if every individual possessed the optimum genotype (*AS*).

The mean fitness of a population like this will be $(1-0.143) = 0.86$. If the population were to support 10 such polymorphisms, the average fitness would be $(0.86)^{10}$ or about 0.15. This means that only one zygote in seven would survive to become a reproductively effective member of the population. The human species could just about survive in a precarious way, provided that most of the genetical loss took effect before implantation. If selection acted later, an intolerable amount of available pregnancy time would be wasted. *Drosophila* flies lay many eggs; they could support 10 such segregating systems fairly easily, but would be strained with 20 – in this case only one individual from every 50 zygotes would survive to be a parent in the next generation.

It can now be seen why the estimates of the numbers of segregating loci obtained by electrophoresis were so startling. The original finding apparently showed that a third of both *Drosophila pseudoobscura* and human gene loci had more than one allele – which meant several thousand loci in both species. Theoretically both species were extinct; obviously both were happily thriving.

EXPECTATIONS AND EXPLANATIONS

Ironically, the amount of variation revealed by electrophoretic techniques is exactly what was expected by classical population geneticists. It is worth emphasizing this; problems about over-abundant variation affect only those seduced by abstract reasoning.

1. On a previous occasion when genetical expectation seemed remote from observed evolutionary changes, it will be recalled that Fisher put forward his theory of the evolution of dominance. The essence of his theory was that the expression of a gene could be modified by other genes, called modifiers. Experimental work on a variety of organisms has shown that such modifying genes exist wherever they have been looked for.

2. The genetical relationships between many species of *Drosophila* have been worked out in great detail in the United States, but in no case do species

differences involve only genes with major effects. Interspecific crosses and comparisons are easier to make in plants than in animals. Polyploidy has been a major factor in plant evolution (page 75), but with this exception apart, genetical differences between plant species are similar to those between animal (Renner, 1929; Dobzhansky, 1970). Sometimes speciation may arise through a chromosomal change which prevents successful meiosis, but most differences involve many small inherited differences between the groups compared.

3. Some *Drosophila* species are highly polymorphic for chromosomal inversions. In *Drosophila pseudoobscura*, for example, the frequencies of different inversions are controlled by fine selective pressures, and change seasonally, with altitude, food, competition, and other factors of fly ecology (page 87). In a particular environment, two inversions may be heterotic, *i.e.* they may occur together much more commonly than expected, and the fitness of the heterozygotes and homozygotes can be measured. However – and this is the important point – inversions from different parts of the species range do not usually show heterosis when their bearers are crossed. Indeed the heterozygotes may have a lower fitness than the homozygotes (Dobzhansky, 1970). This is because inversions have become locally adapted by incorporating locally-arising variants: the chromosomes show variation within each inversion. Direct evidence for such intra-inversion variation has come from determining the electrophoretic variants carried in inverted sequences: the same inversion may include different alleles in different populations (Prakash and Lewontin, 1968).

A similar conclusion comes from comparing the effects of genes in different genetical backgrounds. For example, Wallace (1956,1963) carried out a long series of experiments to test if radiation-induced mutations in *Drosophila melanogaster* were invariably deleterious. He found that mutations induced in flies made homozygous by artificial breeding were often heterotic, although mutations occurring in flies with a heterozygous background were usually deleterious to their carriers – since the gene expression here has been adjusted by ordinary selection. The converse of the same problem was studied by Jones and Yamazaki (1974). It is known that different environments affect the frequency of alleles of genes determining esterase functions in *Drosophila pseudoobscura*, suggesting that they are controlled by selection. Jones and Yamazaki found the equilibrium frequencies of the alleles concerned are affected by genes linked to the *Est* locus, showing once again the effect and interdependence of the many genes that normally contribute to an individual's development and survival.

These experiments are similar to those carried out on the modification of dominance (*q.v.* pages 189–93). The advantage of experiments with *Drosophila* is that many species are so well known genetically that small differences in fitness and survival can be identified and analysed.

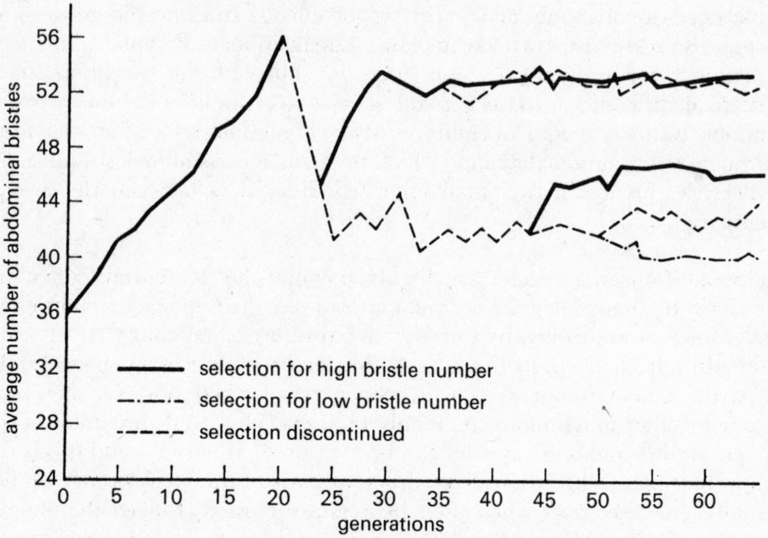

FIG. 109. Summary of selection experiments by Mather and his colleagues in *Drosophila melanogaster* for change in the number of abdominal bristles (from Purdom, 1963).

4. Both commercial and experimental selection is directed towards the greater (or lesser) expression of some inherited trait. Sometimes this may involve the incorporation of a characteristic not present in the initial strain, but it always includes change in allele frequency and arrangement on the chromosomes. Once again we have to turn to *Drosophila* for most information on the genetical effects following selection. The classical work of Mather and his colleagues on selection for the number of bristles at the rear end of *Drosophila melanogaster* has already been mentioned (Mather, 1943, 1953, 1974; also Thoday, 1961, 1972). Its relevance in the present context is that:

a. Genes affecting bristle number are on all the chromosomes, mixed with ones for controlling other characters. This means that selection for bristles produces changes in other traits, and upsets developmental processes.
b. After several generations response to selection stops. However there is still ample variation for the character hidden in the genotype, because after some time further progress becomes possible. This change is almost certainly due to recombination restoring the balance of linked processes, and permitting further selection for bristle number.
c. Selected stocks often remain stable for bristle number, even if selection is discontinued. This is because the arrangement of bristle genes is 'held' by the linked genes affecting general viability.

The conclusion from these results is that a tremendous amount of variation is concealed in the genotype, which can thus respond to environmental change by a rearrangement of the many genes of small effect. Mather calls these *polygenes* to distinguish them from those with a larger effect (*oligogenes*).

We can now face the paradox presented by the almost unlimited amount of inherited variation expected by the experiment and arguments described above, and the firm limits to variation demanded by theory. When species after species was shown by electrophoresis to contain a high proportion of segregating loci (Tables 2, 26), a considerable rethink had to be done by

TABLE 26. Heterozygosity in some rodent species (after Selander and Kaufman, 1973*b*)

	Mean % heterozygosity/individual
Dipodomys ordii	0.8
D. merriami	5·1
Eutamias panamintinus	6.1
Mus musculus (Denmark)	7·8
Sigmodon hispidus	1.7–2.6
S. arizonae	2.9
Thomomys bottae	4·3–10.6
T. umbrinus	3·3
Peromyscus polionotus	5·1–8.1
P. floridanus	5·3

the thinkers. This is an excellent example of the way science advances – hypothesis continually modified by relevant observations. We now have a much clearer understanding of the significance of genetical variation for individuals and populations.

1. Both Haldane and Muller effectively thought of every gene acting independently on its carrier. This is patently not true. The difference in fitness between an individual heterozygous at (say) 20% of his loci and one heterozygous at 10% is unlikely to be great, and extreme multiple heterozygotes will be so rare that they will not contribute significantly to a normal population (Sved, Reed and Bodmer, 1967). Selection is likely to pick out individuals with relatively high frequencies of heterozygous loci, and eliminate the tail of predominantly homozygous individuals. For a given level of mortality, this will produce typical stabilizing selection. In other words, the entire phenotype will be selected, not independent aspects of it (King, 1967; Milkman, 1967; Smith, 1968).

2. We have seen (pages 135–9) that selection pressures change in time and

space. Looked at from the point of view of the individual this means that fitness is not a constant, but depends on both the physical and biological environment. Sheppard (1975) has argued that the 'segregational load' ought to be called the 'environmental load' since it depends on selection exerted against the homozygotes, and the strength of this selection is directly related to environmental pressures. In the classical example of the polymorphism for sickle cell haemoglobin, the advantage of the heterozygote disappeared in negro slaves taken to North America where there was no malaria. The frequency of the sickle cell allele is now declining there entirely through selection against the anaemic homozygotes.

In formal terms, selection pressures may be density and frequency dependent or independent. Wallace (1968, 1970) has suggested a further distinction between 'hard' and 'soft' selection. Under hard selection, an individual has to meet certain requirements to survive. For example, a marine organism is unable to cope with the osmotic problems of living in fresh-water; most plants are poisoned by the soil concentrations of heavy metals on mine spoil heaps; unselected mosquitoes cannot tolerate a high concentration of insecticide; and so on. In contrast 'soft' selection begins to operate only when the normal carrying capacity of an environment is filled – a low density of a prey species attracts neither prey nor intra-specific pressures, and a struggle for survival only intensifies when numbers increase.

Indeed when we examine the situations where genetically caused deaths occur, we are faced with the enigma that a genetically invariable population has no 'load', but may be faced with extinction if the environment deteriorates without new variants appearing. This is true of the classical case of the spread of the melanic form of the Peppered Moth which Haldane (1957) used as his example of the cost of gene substitution. The species only survived in polluted areas *because* of the occurrence of the melanic mutants. The Rosy Minor Moth (*Miana literosa*) was extinct for many years in the Sheffield area because no melanic form appeared. Only in the mid-1940s was it able to recolonize the centre of the city through a newly arisen melanic.

In situations where an environment changes so as to reduce the chance of survival of species living in it, a 'load' is placed upon the species which has nothing to do with genetics. A gene replacement process offers an escape from this external load. If the replacement is rapid enough, the gradual decrease in environmentally caused deaths will enable the population to survive, while another population without an alternative allele may succumb to the changed environment.

Although different genetical causes of death all contribute to 'genetic load' and can all be described by the same algebra, they differ considerably in their contribution to the species. Mutational load – the original Mullerian load – is bad. It is the result of actual deterioration of the genetical material, and the elimination of mutants is a maintenance job to keep the population in working condition. Substitutional load is definitely favourable to the population, although it may coincide with a time of increased mortality if it occurs in response to an environmental crisis. Segregational load is a paradox. It can result in clearly identifiable genetical deaths, yet its overall effect on the species is to increase its tolerance to environmental variation (pages 72, 303). Perhaps it is offensive to our sense of fairness: it would be more democratic to have everyone slightly sickly than to have a few die outright.

Much of the difficulty of the load concept is due to the fact that it wrongly combines the mean and variance of a population's viability through an invalid assumption that increase in the number of segregating genes must be associated with decreased mean fitness. It would be an advantage to keep the mean fitness and variance of a population distinct: the mean fitness involves biological and ecological factors of considerable complexity, while the statistical variance of fitness can be dealt with unambiguously if the question of population 'success' can be omitted (Brues, 1969; Fraser, 1962; Fraser and Mayo, 1974).

3. Population composition cannot be separated from population history. In many ways the composition of a population is a record of its history through past environments and vicissitudes (Nei *et al.*, 1975). Some alleles will respond to current environmental pressures but others will reflect past events. For example, it has been suggested that human blood group frequencies were determined by the major epidemic diseases of the past (page 279). Once a change has taken place, the whole genetical environment of the population is that of the post-change situation. Examples of this are the skeletal variant frequencies in the St. Kilda field mouse, which were stable for many years, changed over the period when the human population left the island and the house mice became extinct, but have remained the same since (page 267); and the wing pattern in the Marsh Fritillary Butterfly which became extremely variable at a time when the population size increased, and settled to a new pattern when numbers again decreased (page 147). Attempts to detect selective processes may fail if the pressures are not constant. Changes in the frequency of alleles of the *Hbb* locus in the Skokholm mouse population were described in Chapter 7; in the winter of 1972–3 there was no changes. This was a very mild winter, the mice on the island are very susceptible to low temperatures and apparently there was no selective death, and hence no change in *Hbb* frequencies.

This means that the genetical forces maintaining (or lessening) variation in a population will vary with time. For example heterosis is usually more marked in a poor environment than an easy one (*e.g.* larval survival and longevity in *Drosophila melanogaster*; growth rates in Wall Cress (*Arabidopsis thaliana*) and maize; breeding success in laboratory mice; etc.) In classical experiments with laboratory populations of *Drosophila pseudoobscura*, Wright and Dobzhansky (1946) found great differences in fitness between inversion genotypes (karyotypes) which led to heterozygous advantage and a stable polymorphism at the unpleasantly high temperature (for the flies) of 25°C, but not at 16.5°C, with an intermediate situation at 22°C. Examples can be multiplied (Parsons, 1971). We have described intermittent drift as being much more important in the life of a population than persistent drift; likewise we must recognize the existence of intermittent selection, operating only at particular times in the life-cycle, in particular parts of the species range, or even at particular rare occasions in the species history.

The danger about using past events to explain a present situation is that it is only too easy to cloak ignorance by untestable speculation (Birch and Ehrlich, 1967). Too often 'drift' has been claimed merely as a confession of not recognizing the true nature of the genetical process moulding a population; and there is a danger of using selection in a similar unthinking way. Notwithstanding it would be a curious biologist who believed that all the variable traits of an organism are always of advantage or always of disadvantage to it. McWhirter (1957) has suggested that there may be a difference between ancient and modern gene systems (*palaeogenes* and *neogenes* respectively), with the former so inbuilt that they resist minor changes in the environment and often participate in supergenes (pages 184–5). This idea has never been worked out in detail, but it may relate to some of the considerations set out in this section.

4. Finally, there may be many genes with only a small primary effect on the phenotype, but which exercise an important role as modifiers of major gene effects (Cook, 1971). This would mean that the selection pressures exercised on such polymorphic loci would be nearer 1% than the 10% indicated by studies on major polymorphic genes (Karlin and McGregor, 1974). Indeed the simple ideas of genic polymorphism and its maintenance which have been the inspiration and mainstay of ecological genetics are far too naive when we consider gene interactions in development and behaviour. At the level of coadaptation it would be presumptuous to pontificate that a particular enzyme or protein change had no effect on cell functioning: a high proportion of electrophoretically separable proteins are being found to differ in their biochemical properties, and this implies they will be effective in fine 'metabolic tuning'.

NEUTRAL GENES

Looking at the evidence as a whole, there is clearly no reason to believe that populations have an inexplicable excess of variation. Rather, the evidence from electrophoretic investigations supports the expectations of naturalists. We need, therefore, spend little time on the idea that the majority of enzyme differences detected by electrophoresis are 'neutral', *i.e.* that they do not affect the phenotype at all, and hence are not subject to selection. Indeed, if this was so, evolution would not proceed by adaptation, but would be 'non-Darwinian' (King and Jukes, 1969; Clarke, 1970 *a*, *b*). Such a neutralist view regards different alleles as no more than biochemical mistakes which occur from time to time, but since they do no harm, there is nothing to stop them accumulating. This would be philosophically interesting, since the chief alternative to 'Darwinian' evolution is special creation, but it would be biologically untenable. It would be surprising if there were not a few perpetually neutral variants, but it is probably significant that most people who think they are important and common are mathematicians or chemists.

The main positive evidence cited by protagonists of neutral mutations is not data from enzyme and protein polymorphism within species, but the alleged accuracy of the 'protein clock', *i.e.* apparent similarities in the rate of substitution of amino acids in the same protein in different species. The amino acid sequence of certain proteins like cytochrome C, haemoglobin, and insulin have now been determined in many animals. From the number of amino acid differences between the same protein in different species it is possible to construct genealogical trees to show common descent and degree of relationship. A timescale for the branching points on such trees can be obtained from palaeontological knowledge. When this is done, there appears to be a remarkable uniformity in the rates of substitution along completely different evolutionary lines. For example, the ancestors of mammals and fish diverged 350 to 400 million years ago. When the differences in the haemoglobin molecule of carp and man are added up, it can be calculated that the average rate of change since the existence of a common ancestor has been 8.9×10^{-10} per amino acid site per year (*i.e.* about one substitution every ten million years). Comparisons among mammals from different orders which radiated about 80 million years ago produce a substitution rate of 8.8×10^{-10}. The β chain in the globin molecule is probably the result of a duplication of the α chain. Both carp and man have β chains, so the duplication must have occurred at least 450 million years ago. When the α and β chains of human haemoglobin are compared, a rate of change of 8.9×10^{-10} can be obtained. Thus the rate at which the α and β chains have evolved over the last 450 million years seems to be about the same as the rate of change in the α chain over both the last 350 million years and the last 80 million years (Kimura, 1968 *b*; Kimura and Ohta, 1971).

Although these rates of change are impressively constant, they disguise many differences. For example, the calculations are based on years and not generation time. Mutation rates seem to be approximately the same in all organisms, but different species have very different generation times (and opportunities for substitution). Secondly, once calibrated the protein clock can be used to 'date' the time of separation between groups. When this is done, some groups are claimed to have split at a period significantly different from that accepted by palaeontologists on direct fossil evidence. In the light of this it is not surprising that the rate of substitution in different species is not as constant as first appeared. For example, the rate of change of the haemo*globin* α chain of the early vertebrates was faster than in mammals. The β chain evolved faster than the α one, and both α and β chain rate of change decreased in the line giving rise to the apes (Goodman, Barnabas, Matsuda and Moore, 1971). As we have noted, moreover, the extreme 'neutral mutation' argument allows no place for adaptive changes to occur, despite the fact that (for example) the oxygen-binding requirements and characteristics of human and carp blood are quite different. The most likely history of an 'inoffensive' mutation is that it may find itself affecting its carrier for good or ill at some time during its evolutionary life, and therefore be exposed to selection. The mutations 'waiting' to confer insecticide resistance on unsprayed flies or heavy metal tolerance on plants growing on normal soil come into this category. One of the important theoretical contributions of R. A. Fisher was to show how small selection pressures need to be to produce gene frequency changes (Fisher, 1930).

Another problem for the neutralists is the remarkably similar degree of heterozygosity in different species and populations (*e.g.* Table 26). If mutations are neutral, they should be more common in more numerous forms, A similar argument is that the same allele frequencies over a large geographical area are evidence for selective control, since if drift is important, fluctuations would occur between localities (Ayala, 1972). The difficulty (for both neutralists and selectionists) is that even a modest amount of migration between populations (a few individuals in each generation) will prevent significant divergence between them due to drift (neutral effects), and thus produce similar frequencies even if selection is not acting.

A more important line of reasoning is the distinction observed between less variable and more variable proteins: enzymes involved in glucose metabolism and energy production tend to be less variable (*i.e.* have fewer segregating alleles) than non-specific enzymes such as esterases, and alcohol and aldehyde dehydrogenases. This difference is in fact between enzymes which use a single substrate, usually produced and broken down within a cell, and enzymes capable of reacting with a variety of substrates which reflect the diversity of environments in which the organisms live (Gillespie and Langley, 1974). Furthermore the proportion of polymorphic 'non-specific' enzymes is much higher in *Drosophila* than in either mouse or man. *Drosophila* is more dependent on environmental variability than is a mammal; the difference between the two classes of enzymes makes sense if their variants are subject to selection, but not otherwise.

Looked at rather harshly, much of the support for the neutral mutation point of view comes from the elegance of the mathematical arguments for it, rather than their necessary correctness (Kimura, 1968 *a*; Kimura and Ohta, 1972). This does not mean that the neutralist theory is wrong (Van Valen, 1974). Actually the most likely reality is that *both* extreme neutralists and *extreme* selectionists are over-persuaded. Some mutations will have no effect on fitness when they first occur and thus be uninfluenced by selection; the crucial point is that the neutralism of such alleles is unlikely to persist in the long term.

NICHES AND GENES

If an organism can manage to live in more than one environment, it is likely to be more successful (in the sense of having greater numbers) than if it was restricted to a single environment. For example trunk sitting moths that rely on camouflage for their survival have many more individuals if they can exist in both a polluted and an unpolluted situation. When we speak of 'environment' or 'situation' in this context, we are describing what ecologists call a *niche* (Miller, 1967; Whittaker, Levin and Root, 1973). The 'fundamental niche' of a species can be defined by the limits of all the environmental variables which allow it to survive and breed. In a situation where the species is competing with other species, these limits will usually be reduced (Hutchinson, 1944, 1957). The relevance of this in the present context is that different genotypes within a species also have limits, and ideas developed by ecologists for competition between species can equally well be used for intra-specific competition.

Consider a population inhabiting two niches (these may be different food plants, different nest preferences, different soil acidities, etc.), and assume that one genotype *AA* is fitter in one niche and another *aa* is fitter in the other. If selection occurs among young individuals in each niche, followed by migration of surviving adults and random mating with random settlement of zygotes, each sub-population will contribute its genes in proportion to its number of survivors. A stable equilibrium will be established if the fitness differences are reasonably large, and density is controlled independently in the two niches (Levene, 1953; Smith, 1966). This is the situation of disruptive selection (page 133).

However 'niche' is a complicated concept which has caused many to stumble, and at this point we must distinguish the 'niche' of an animal or plant from both its habitat and its survival strategy. Habitat is probably the easiest idea to grasp, because it refers to a place in space which can be fully described. Clearly, different habitats vary in their stability (*i.e.* their permanence and predictability) – ranging from temporary pools, carrion, or dung patches to the changelessness of placid tropical forest. From an inhabitant's point of view, this stability depends on his generation time and on the length of time the habitat remains favourable for sustaining him. If generation time is significantly shorter than the period of habitat favour, one generation may influence the resources available for the next and population numbers will affect future survival for better or worse. Under these circumstances density dependent regulating mechanisms will be important (Roughgarden, 1973; Charlesworth, 1971). In contrast a population of longer lived organisms (relative to habitat change) cannot affect the food, shelter, and so on necessary for the next generation, since the environment will have changed by the time the next generation appears. This means there will be no penalty for rapid population growth and the attainment of 'too large' numbers.

This leads us to the ways organisms cope with situations of different predictability. In good conditions, the numbers of any population will increase exponentially at a rate determined only by its net reproductive rate (or Malthusian parameter) (*r*). This rapid increase will not last long, and individuals will soon begin to react with one another and compete for resources, and increase in numbers will slow to a point where the population size remains constant. This is termed the carrying capacity (*K*) of the environment.

The values of *r* and *K* for a population in a particular environment are determined by the characteristics of the individuals in the population, and as such are inherited traits subject to natural selection. Now species differ in their *r* and *K* traits: some species are highly mobile and colonizing, with high reproductive rates but low competitive abilities so that their populations tend to be ephemeral; whilst others fluctuate comparatively little in numbers, and put their energy into competition and maintenance. These alternatives have been called respectively *r* and *K* strategies, produced by *r* and *K* selection (MacArthur and Wilson, 1967). An *r*-strategist lives in a variable or an unpredictable environment and is subject to recurring heavy mortality; selection will favour rapid development and early reproduction, high reproductive rates, and small competitive abilities. In contrast, selection for a *K*-strategist leads to slower development, a greater body size, and the ability to withstand both inter- and intra-specific competition, since such species will usually live in complex communities near their equilibrium numbers. Presumably the Loch Ness monsters are *K*-specialists. Since they are monsters, they are by definition large; since no corpses have ever been found, it is likely they are long-lived and reproduce rarely; and there can be little doubt that the depths of Loch Ness provide a stable environment with a constant food supply (Sheldon and Kerr, 1972). However *r* and *K* strategies are not really alternatives; species will range in their survival stratagems as widely as the stability of habitats, although in general *r*-strategists will be opportunists, and *K*-strategists will be equilibrium species in constant environments (MacArthur, 1960; Pianka, 1970).

Two closely related species which follow different strategies are the Field and House Mouse (*Apodemus sylvaticus* and *Mus musculus*); in breeding seasons of similar length, Field Mice increase their population size about four-fold, whilst House Mice increase ten to twenty fold. However House Mice respond to competition by moving into new areas, and if this is not possible, may become locally extinct. This is what happened on St Kilda; when men lived on the island, House Mice were commensals in their houses; after the evacuation of the human population in 1930, Field Mice invaded the deserted village, and within 18 months the House Mice were all dead – not through inter-specific savagery, but because their breeding was upset and births were insufficient to replace adult animals dying from old age (Berry and Tricker, 1969).

When the ways animals and plants adapt to the relative stabilities of their habitats are recognized, certain implications follow about their 'niche' (*i.e.* their set of interactions with the physical and biological environments).

1. *Niches vary in their 'width'*. If a form (meaning either a species or a genotype) has little competition in a particular niche, it may be able to occupy more of its 'fundamental' niche than would be possible if competition was more intense. This is another way of saying that density dependent factors will begin to operate at higher densities than usual. If density dependent selection is important in controlling allele frequencies at a locus, these frequencies may be different in a 'wide' niche. For example, the consequences of territorial behaviour will be increased if inter-specific competition is added to intra-specific. Moore (1964) has recorded that inter-specific encounters are more frequent and effective between Dragonflies that resemble each other, *e.g.* he found that the Emperor (*Anax imperator*) invariably drove the Southern Aeschna (*Aeschna juncea*) from small ponds, although the presence of other, less similar species, did not provoke attacks. The motivation for inter-specific encounters among Dragonflies may be aggressive or sexual, but the result is the same in that only one male remains within the territory.

One place where niche width is increased are small islands, which have fewer species than mainland areas. For example, House Mice on the Isle of May have occupied the niche normally filled by voles, and have added the behaviour shown by voles when they meet each other to their normal mouse behaviour (J. Mackintosh, *pers. comm.*). Comparison of island and mainland variability is complicated because island races will tend to have less variation than their mainland relatives because of the founder effects and the relative harshness of directed selection. However some apparently meaningful tests of island variability have been made. For example, bill length in small birds (Chiff-chaff, *Phylloscopus collybita*; Blue Tit, *Parus caeruleus*; Chaffinch, *Fringilla coelebs*) is more variable in the Canaries than on the mainland of Europe (Van Valen, 1965; Rothstein, 1973). On the islands these birds eat a greater variety of food than on the mainland. The real problem is that there is no good way of measuring niche 'size' to determine its effect on variation.

2. *'Experience' of a niche is reduced by mobility*. Men, mice and cockroaches may live in the same area but they will have a totally different 'experience' of their environment. This is largely a result of their sizes and abilities to move; clearly a man rides rough-shod over many features which are major obstacles to a mouse. One way of looking at this difference is to think of an environment as having 'roughness' or 'grain'. Small, relatively immobile creatures tend to experience environmental factors as sets of alternatives. Their environment is 'coarse-grained', and even if the overall environment for the species is constant, it will be uncertain for the individual. Such species will be mainly *r*-strategist opportunists (Southwood, May, Hassell and Conway, 1974).

With greater mobility, environmental differences become increasingly fine-grained and are experienced as a succession of differences with a similar average for all members of the population. The effective environment is therefore more certain. Vertebrates with generally larger body size, greater mobility and better homeostatic control than invertebrates 'need' less ability to cope with environmental alternatives. They tend to be *K*-strategists.

The idea of grain includes not only the sizes of patches of habitat relative to the size and mobility of the organism, but also variation of temperature, food, parasites, etc. in time and space (Levins, 1968). Environmental grain may be different at different stages of the life-cycle. For example, larval insects have a coarse-grained environment, while winged adults have a finer-grained one. Social dominance may affect grain, since a dominant animal has less restriction than a subordinate one. It is therefore entirely in accord with expectation that in general heterozygosity decreases with size (Selander & Kaufman, 1973 *b*) (Table 2).

These considerations apply almost exclusively to animals. It is not yet possible to discuss levels of intra-population variation in different plant species in the same way. Sufficient data on heterozygosity are not yet available, while the factors influencing variability are likely to be very different because of: the stronger tendency of gametes and seeds to be dispersed randomly with respect to environmental patchiness; the great degree of phenotypic flexibility; and greater diversity of breeding systems (Bradshaw, 1972). Available data suggest that some plant species at least have a heterozygosity similar to small mammals.

3. *A population may never attain ecological, never mind genetical, equilibrium.* The notion of an *r*-strategy implies that some populations will rarely, if ever, come into equilibrium with their food-supply. It follows that the same populations are likely to show frequent and large reversals of selection pressures. The classical studies of Dobzhansky on seasonal adjustment of chromosomal inversion frequencies in *Drosophila pseudoobscura* are outstanding examples of this. Another example is the cyclical change in *Hbb* allele frequencies in the Skokholm mouse population. In contrast a population which fluctuates little in numbers is likely to show less volatility in genetical constitution, although density dependent stabilizing selection will be strong. Moreover, *K*-selection tends to produce larger size and long life. These characteristics are related, and both tend to dampen effects of seasonal breeding and mortality (Bonner, 1965; Southwood, 1971). The low genetical variation in larger animals describes their environmental experience (or grain), and also, therefore, correlates with their ecological strategies.

Genetical equilibrium takes much longer to attain than ecological equilibrium, since it is affected by rare events like mutation, specific recombinations, and uncommon migrants. A dedicated *r*-strategist may never reach a genetical steady state, and even a *K*-strategist may take a longer time to equilibrium than its period of habitat stability (Anderson, 1970). House Mice are *r*-strategists but on Skokholm they are living in a *K*-situation where intra-specific pressure are strong and colonizing ability unimportant – yet they are still changing progressively after seventy generations of isolation.

Habitat, niche, and survival strategy are ecological ideas which overlap each other to large extents. They are helpful for understanding genetical processes because:

1. We have already seen that it is wrong and frequently misleading to speak of *the* coefficient of natural selection acting on a population or species. The main reason for this is the changing nature of the habitat or niche in time and space. This immediately shows why a population benefits from being variable: as soon as one adaptive genotype increases in frequency in response to an environmental constraint, the environment changes (either with time, or the movement of the organism), and genotype frequencies respond as individuals die or fail to reproduce.

2. The recognition of different ecological strategies helps to distinguish the importance of different types of selection acting at different times during the life-cycle, and the needs of different organisms to maintain variation. Indeed one of the more profitable aspects of the neutralist gene controversy has been a search for and discovery of associations between environmental and genetical changes. For example, under strict laboratory conditions, *Drosophila* cultures subjected to environmental change in time and space retained more enzymic variability than cultures of the same origin kept in constant environments (Powell, 1971); radiation induced mutations increased the competititive ability (*i.e.* 'niche-width') in one species of *Drosophila* when cultured in the presence of another (Shugart and Blaylock, 1973). In natural populations, inherited variation (both total heterozygosity and the number of alleles per locus) decreases with depth in six bivalve molluscs, including the Common Mussel (*Mytilus edulis*) (Levinton, 1973). Conditions in deeper water are more constant for a variety of ecological factors than they are nearer the surface. Interestingly enough, another American shallow water species, the Horseshoe Crab (*Limulus polyphemus*) is also highly variable (it has as much population variability as, say, a rodent species) (Selander, Yang, Lewontin and Johnson, 1970) despite being a 'living fossil' which has not changed superficially for millions of years. Its morphological constancy is clearly not due to lack of variability, and the existence of so much variation implies a use for it.

Another method of evaluating the importance of inherited variation is to examine important ecological variables in the life of a species, and test if fluctuation in them is correlated with genetical variance. This is a crude approach, because it ignores the properties and reactions of specific loci. Nevertheless many studies have found significant correlations. For example, between eighty and ninety per cent of enzymic variation in populations as distinct as *Drosophila*, ants, and mice can be accounted for (in the statistical sense) by variation in mean climatic variables recorded in the places they were caught (Bryant, 1974). Similar conclusions have resulted from studies on fish (Koehn, 1972) and cereal species (Marshall and Allard, 1972). These findings do not prove that there is a causal correlation between heterozygosity and environmental variation (although at least one cline in enzyme variation in the fish *Catostomus clarkii* has been related to temperature differences (Smith and Koehn, 1969)), but they imply strongly that the majority of polymorphisms are adaptively important in relation to time and space.

3. Ecological studies may provide more accurate and relevant predictions about genetical variables than specifically genetical ones. For example, studies on lizards living on islands have shown that the smaller (which usually means more ecologically distinct) and older is the island, the lower is the heterozygosity of the lizard population. This implies that directional selection tends to prevail over stabilizing selection (Soulé and Yang, 1974; Gorman, Soulé, Yang and Nevo, 1975).

Nevertheless there are cases known where populations have far less variation than average, but this does not seem to matter in the short term. For example, the Archatinid Land Snail *Rumina decollata* was introduced to North America (probably from Europe) before 1822. It is usually a self-fertilizing hermaphrodite, although individuals can cross. It is one of the most successful snail colonizers of America and has achieved an extensive distribution in the temperate United States and Northern Mexico attaining high numbers in both agricultural and natural habitats. Yet it has few segregating loci, and only one of the 25 studied had more than one common allele (Selander and Kaufman, 1973 *a*). This genetical poverty is not simply the consequence of introducing a limited number of founders from elsewhere: the species has a similar low variability in its native range of Europe and North Africa. In contrast two other snail species introduced into the United States (*Helix aspersa* and *Otala lactea*) have been shown to have about 18% of their loci heterozygous per individual, which is a similar variability to native North America snails of the genera *Mesodon* and *Rabdotus*.

4. A species which has evolved a predominantly *r* or *K* 'method' of survival, may increase its success by incorporating adaptations more characteristic of the other strategy. For example, insects are primarily short-lived opportunists, but industrial melanic moths have locally reduced their mortality through the spread of alleles and reorganizing their genome in traits affecting dominance and behaviour in order to improve crypsis; Arctic Skuas are *K*-strategists which have adopted a polymorphism for varying the reproductive success of different morphs; the spectacular spreads of Fulmars and Collared Doves may be the consequence of a genetical change for more efficient exploitation of resources in the short-term; melanic ladybirds survive better than typical ones during the unfavourable season, and hence damp down

fluctuations in total population size; *Drosophila* species in the Hawaiian Islands have increased in size, longevity, territoriality, and decreased in reproductive rate in the process of adapting to an existence in isolated 'kipukas' (copses) (Carson, Hardy, Spieth and Stone, 1970); and so on. Unless a population is very specialized, the potentialities of extending its habitat tolerance (or reducing the interactions within its niche) are likely to be highly advantageous.

5. Finally, a recognition of the differences between *r*- and *K*-strategies resolves much of the disagreement about the evolution of reproductive rates (*q.v.* Chapter 3). Wynne-Edwards' claim that clutch size, breeding age, behaviour, and similar attributes have evolved to adjust the reproductive rate to available resources is largely based on evidence from species which have been strongly *K*-selected (petrels, albatrosses, man); while Lack's contention that 'each species has evolved those adaptations which enable it to reproduce as rapidly as it can in its natural habitat' (*i.e.* that a clutch or litter size has been evolved such that, on average, most young are raised), derives mainly from those species with an *r*-strategy (Wynne-Edwards, 1962; Lack, 1966; Cody, 1966; Southwood, May, Hassell and Conway, 1974). The belief of Andrewartha and Birch (1954) that density-dependence is unimportant in controlling population size comes mainly from opportunist *r*-type species (especially insect pests) which colonize empty habitats and rarely reach equilibrium numbers.

VARIATION MUST BE UNDERSTOOD BIOLOGICALLY

The whole of the theory behind genetical variation and environmental heterogeneity is the same as that behind inter-specific competition (Clarke, 1973). There is a cautionary tale here. As a result of the theoretical work of Volterra and Lotka, and experimental work by Gause, there has grown up a dogma that complete competitors cannot coexist, usually known as Gause's principle. For a long time experimental ecologists were faced with the paradox that many species seemed to coexist in nature but not in laboratory experiments: Gause (1934) put different species of micro-organisms into media in bottles, and only one species survived; Park (summarized 1962) put different species of beetles or moths in flour in bottles, and nearly always only one species survived. The answer came from some sensible experiments performed by Crombie (1946). He put two flour beetles (*Tribolium confusum* and *Oryzaephilus suranimensis*) together, and one (*Oryzaephilus*) became extinct. However when he put the species together in a mixed environment, in flour containing wheat husks or pieces of glass tubing, both coexisted indefinitely. In other words, a diversified habitat provided conditions which allowed species diversity to exist, and almost certainly mirrored the natural habitat of the beetles more accurately than pure flour.

Silly conclusions about genetical questions come from over-simplified assumptions about genes, genotypes, and environments. Variation is an integral part of the survival strategy of most species, and cannot be isolated from an understanding of the whole biology of the organisms concerned (Dickinson and Antonovics, 1973). It is not surprising that the discussion in this chapter on the causes and maintenance of variation has called upon many of the ideas and results of the earlier chapters, as well as introducing a series of relevant ecological concepts. The message for geneticists is that their studies will only be complete if they are conceived as part of population natural history. Equally, and perhaps more seriously, ecologists will repeatedly fail if they persist in treating all members of a species as equivalent to each other. They will be falling into the same error as those mathematicians who recognize genetical diversity but ignore the biology represented in their sums. To take one example, it was only when differences in mating behaviour between the colour phases of the Arctic Skua were recognized that a clear picture emerged of the breeding biology of the species (page 144). There is an obvious visible difference between the Skua morphs, and such clear-cut polymorphisms invite investigation. The classical work of E. B. Ford was based upon exploiting polymorphisms of this type (Ford, 1971).

However there is a hazard in concentrating too hard upon classical polymorphisms. Genetical variation is the normal situation for virtually every species, and not an esoteric exception. Furthermore segregating genes affect the whole of the biology of an animal or plant. It is not enough to restrict attention to particular convenient loci. The earlier geneticists did this when they treated the genetical constitution of an organism as a bag of beans which was regularly shaken and sorted (Haldane, 1964). Such an approach is a valid beginning to understanding genetical mechanisms, but becomes a hindrance when these same mechanisms are dissected, and shown to lie behind a variety of population strategies (Berry, 1976).

There is a form of intellectual arrogance which treats molecular biology as the key to all biology (Ayala, 1974). It is not. A living organism is greater than the sum of its parts, and it is the phenotype not the genome which lives and dies, thrives and breeds. The extent to which this is not fully appreciated is highlighted by the surprise – almost alarm – that greeted the apparently objective measures of total inherited variation made possible by electrophoretic techniques. The time should by now have arrived when every naturalist, be he professional or amateur, is conditioned to seek differences between members of the population he studies as an essential to a full understanding of population processes.

The United Nations Environmental Programme was launched at the Stockholm Conference in 1972. The 'keynote' book for it was *Only One World*, written jointly by Barbara Ward and René Dubos. Dubos suffered a reaction to writing this emphasis on unity, and as soon as he finished his involvement with it, he began another book to emphasize the diversity and uniqueness of all men (*A God Within*, 1973). In this he explored the differences that persist between the inhabitants of different parts of a country *despite* their genetical similarity, and he affirmed how background and upbringing make us all different. Individuals of an animal or plant population are not as distinct as humans, but we will learn more as we look for differences rather than try to force all diversity into the same dreary mould.

APPENDIX

MITOSIS AND MEIOSIS

An organism receives an equal number of chromosomes from both parents (except for sex chromosome differences, page 91). This number is referred to as the *haploid* number. In the cells of the body of an individual there are twice the haploid number of chromosomes (the *diploid* number), made up of two haploid sets. In animals a cell division reducing the chromosomes to the haploid number takes place in the formation of gametes. This reduction division is called *meiosis*. The diploid number is restored when two gametes fuse to form a zygote. Most plants (indeed, all 'higher' plants) have an 'alternation of generations' involving a diploid sporophyte (which is the 'plant' in the normal sense) and male and female haploid gametophytes. Consequently meiosis in plants produces spores (pollen and ovule) which give rise to a haploid stage. The two stages are distinct in 'lower' plants (mosses and liverworts) but in flowering plants both gametophytes are reduced to a few cells.

Meiosis can be regarded as two divisions of the cell nucleus, in the course of which the chromosomes only divide once. Presumably it arose through the modification of two 'normal' divisions (mitoses – which produce virtually identical cells and involve a single longitudinal replication of the chromosomes) (*q.v.* John and Lewis, 1973).

The first meiotic division begins with a lengthy *prophase* during which the chromosomes become visible and the nuclear membrane disappears. It is divided into five stages which have names indicating the appearance of the chromosomes as they undergo various transformations: *leptotene*, *zygotene*, *pachytene*, *diplotene*, and *diakinesis*.

Leptotene (= thin thread)
The chromosomes become visible as long, thin, apparently undivided threads.

Zygotene (= yoked thread)
Pairs of *homologous* chromosomes (*i.e.* ones derived from different haploid sets) come together (*synapsis*) from their ends, proceeding like a zip-fastener along their length to give very intimate and strict pairing.

Pachytene (= thick thread)
As soon as synapsis is complete, the degree of condensation of the chromosomes increases, so that at the end of pachytene the chromosomes are relatively thick threads. As a result of synapsis there appears to be only the haploid number of chromosomes in the nucleus, each pair (or *bivalent*) ap-

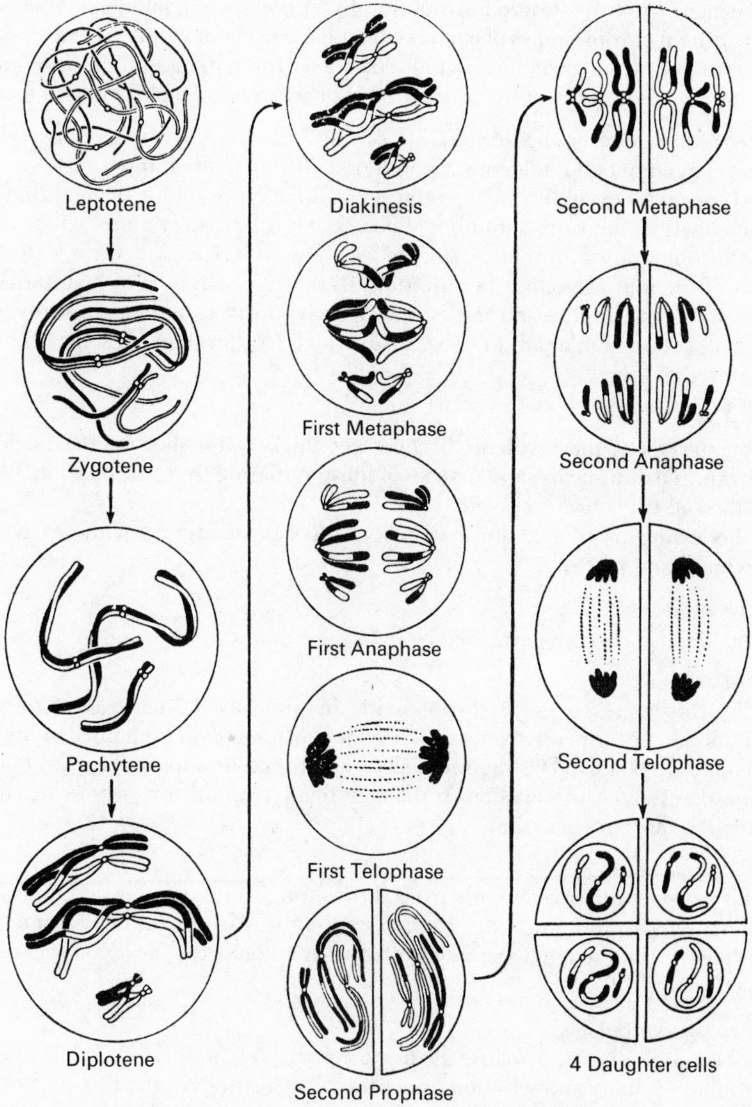

FIG. 110. Diagram of cell division by meiosis illustrated by a cell nucleus containing three pairs of chromosomes. Chromosomes of paternal origin are shown in white, those derived from the mother in black. The first five stages (leptotene to diakinesis) constitute the prophase of the first meiotic division. The chromosomes are double by diplotene, but the centromeres do not split until the end of metaphase of the second division. The four daughter nuclei finally produced all contain the haploid number of three (from Berry, 1965*a*).

pearing as a two-stranded structure. By the end of pachytene, the two constituent chromosomes of each bivalent have become visibly two stranded, so that the whole structure is four-stranded. Interchange of chromosomal material between strands (*crossing-over*) is believed to take place at this stage.

Diplotene (= double thread)
The attraction of homologous chromosomes for each other apparently ceases and the members of the pairs would segregate were it not for the fact that in each bivalent there are a number of places (2, 3, or 4; rarely more) where two of the four strands form an X-shaped association (a *chiasma*). At these places two of the four *chromatids* (a chromatid is one of the results of longitudinal chromosomal duplication and splitting) have broken and reciprocally re-joined, resulting in a pair of chromatids which have exchanged some of their material.

Diakinesis
The strands of the bivalents become yet thicker and shorter; the nuclear membrane disappears, and a series of fibres radiating from the opposite ends of the cell (the *spindle*) appears.

Following prophase, the first meiotic division continues with *metaphase*, *anaphase*, and *telophase*.

First metaphase
The bivalents become attached close to the equator of the spindle.

First anaphase
The centromeres (page 78) do not divide; in this meiosis differs from mitosis in which the centromeres always divide. The centromeres in each bivalent move to opposite poles, each dragging a pair of chromatids after it. This forces the chiasmata along the bivalent until they finally slip off the ends as the half bivalents are torn asunder.

First telophase
This does not differ essentially from the last phase of mitosis, except that the 'daughter chromosomes' are each composed of two dissimilar chromatids (following the exchange of material between homologous chromosomes during diplotene).

Second meiotic division
A second division may follow the first immediately, or after some delay. The prophase is short and without any of the complications of the first prophase. Metaphase differs from the metaphase of ordinary mitosis in that there are only half the diploid number of chromosomes. Also, no DNA synthesis precedes this division. The first division of centromeres during meiosis takes place in second anaphase, and the four daughter nuclei which result from the original cell in second telophase all contain the haploid number of daughter chromosomes.

BIBLIOGRAPHY and AUTHOR INDEX

The page number(s) after each citation refers to the main text of the book

ACKEFORS, H. (1971). Mercury pollution in Sweden with special reference to conditions in the water habitat. Proc. roy. Soc. B, *177*: 365–387. (p. 256).

ACTON, A. B. (1957). Chromosome inversions in natural populations of *Chironomus tentans*. J. Genet., *55*: 61–94. (p. 87).

ALLARD, R. W., BABBEL, G. R., CLEGG, M. T. & KAHLER, A. L. (1972). Evidence for coadaptation in *Avena barbata*. Proc. Nat. Acad. Sci., U.S.A., *69*: 3043–3048. (p. 184).

ALLISON, A. C. (1954). Notes on sickle-cell polymorphism. Ann. hum. Genet., *19*: 39–57. (pp. 69, 290).

AMES, B. N., KAMMEN, H. O. & YAMASAKI, E. (1975). Hair dyes are mutagenic: identification of a variety of mutagenic ingredients. Proc. Nat. Acad. Sci., U.S.A., *72*: 2423–2427. (p. 256).

ANDERSON, E. S. (1968). Drug resistance in *Salmonella typhimurium* and its implications. Br. med. J., *iii*: 333–339. (p. 262).

ANDERSON, P. K. (1970). Ecological structure and gene flow in small mammals. In *Variation in Mammalian Populations*: 299–325. Berry, R. J. & Southern, H. N. (Eds.). London: Academic. (p. 228).

ANDERSON, P. K., DUNN, L. C. & BEASLEY, A. B. (1964). Introduction of a lethal allele into a feral house mouse population. Am. Nat., *98*: 57–64. (p. 228).

ANDREWARTHA, H. G. & BIRCH, L. C. (1954). *The Distribution and Abundance of Animals*. Chicago: University Press. (pp. 113, 305).

ANGSEESING, J. P. A. (1974). Selective eating of the acyanogenic form of *Trifolium repens*. Heredity, *32*: 73–83. (p. 160).

ANTONOVICS, J. (1969). The heterogeneous environment: its effect on the genetics of natural populations. Paper read at the meeting of the British Association for the Advancement of Science in Exeter, September 1969. (p. 130).

ANTONOVICS, J., BRADSHAW, A. D. & TURNER, R. G. (1971). Heavy metal tolerance in plants. Adv. ecol. Res., *7*: 1–85. (p. 265).

ARNOLD, R. W. (1970). A comparison of populations of the polymorphic land snail *Cepaea nemoralis* (L.) living in a lowland district in France with those in a similar district in England. Genetics, *64*: 589–604. (p. 149).

ARNOLD, R. W. (1971). *Cepaea nemoralis* on the east Sussex South Downs, and the nature of area effects. Heredity, *26*: 277–298. (p. 152).

ASKEW, R. R., COOK, L. M. & BISHOP, J. A. (1971). Atmospheric pollution and melanic moths in Manchester and its environs. J. appl. Ecol., *8*: 247–256. (p. 164).

AUERBACH, C. (1962). *Mutation*. Edinburgh & London: Oliver & Boyd. (p. 253).

AYALA, F. J. (1972). Darwinian versus non-Darwinian evolution in natural populations of *Drosophila*. Proc. 6th Berkeley Symp. Math. Stat. Prob., *5*: 211–236. (p. 298).

AYALA, F. J. (1974). Introduction. In *Studies in the Philosophy of Biology*: vii–xvi. Ayala, F. J. & Dobzhansky, T. (Eds.). London: Macmillan. (p. 306).

BACOT, A. W. (1905). On *Triphaena comes*. Proc. ent. Soc., London., for 1905: 67–71. (p. 192).

BAJEMA, C. J. (1971). *Natural Selection in Human Populations*. New York & London: Wiley. (p. 279).

BAKER, E. C. S. (1942). *Cuckoo Problems*. London: Witherby. (p. 115).

BALFOUR-LYNN, S. (1956). Parthenogenesis in human beings. Lancet, *i*: 1071–1072. (p. 97).

BALLANTINE, W. J. (1961). A biologically-defined exposure scale for the comparative description of rocky shores. Fld Stud., *1*(3): 1–19. (p. 136).

BANTOCK, C. R. (1974*a*). Experimental evidence for non-visual selection in *Cepaea nemoralis*. Heredity, *33*: 409–412. (p. 155).

BANTOCK, C. R. (1974*b*). *Cepaea nemoralis* (L.) on Steep Holm. Proc. malac. Soc., Lond., *41*: 223–232. (p. 155).

BANTOCK, C. R. & BAYLEY, J. A. B. (1973). Visual selection for shell size in *Cepaea*. J. Anim. Ecol., *42*: 247–262. (p. 156).

BANTOCK, C. R. & COCKAYNE, W. C. (1975). Chromosomal polymorphism in *Nucella lapillus*. Heredity, *34*: 231–245. (pp. 79, 80).

BANTOCK, C. R. & NOBLE, K. (1973). Variation with altitude and habitat in *Cepaea hortensis* (Mull.). Zool. J. Linn. Soc., *53*: 237–252. (pp. 155, 184).

BANTOCK, C. R. & PRICE, D. J. (1975). Marginal populations of *Cepaea nemoralis* on the Brendon Hills, England. I Ecology and ecogenetics. Evolution, Lancaster, Pa, *29*: 267–277. (p. 155).

BARRETT-HAMILTON, G. E. H. (1900). On geographical and individual variation in *Mus sylvaticus* and its allies. Proc. zool. Soc., Lond., 397–428. (p. 201).

BATEMAN, A. J. (1950). Is gene dispersion normal? Heredity, *4*: 353–363. (p. 109).

BATESON, W. (1894). *Materials for the Study of Variation, treated with especial regard to discontinuity in the origin of species*. London & New York: Macmillan. (p. 27).

BATTAGLIA, E. (1963). In *Recent Advances in the Embryology of Angiosperms*: 221–264. Maheshwari, P. (Ed.). Delhi: University Press. (p. 94).

BATTEN, C. A. & BERRY, R. J. (1967). Prenatal mortality in wild-caught house mice. J. Anim. Ecol., *36*: 453–463. (pp. 112, 241).

BATTEN, J. L. & THODAY, J. F. (1969). Identifying recombinational lethals in *Drosophila melanogaster*. Heredity, *24*: 445–456. (p. 195).

BEECHEY, C. V., GREEN, D., HUMPHREYS, E. R. & SEARLE, A. G. (1975). Cytogenetic effects of plutonium-239 in male mice. Nature, Lond., *256*: 577–578. (p. 256).

BEIRNE, B. P. (1952). *The Origin and History of the British Fauna*. London: Methuen. (p. 213).

BELLAMY, D., BERRY, R. J., JAKOBSON, M. E., LIDICKER, W. Z., MORGAN, J. & MURPHY, H. M. (1973). Ageing in an island population of the house mouse. Age and Ageing, *2*: 235–250. (p. 248).

BENHAM, B. R., LONSDALE, D. & MUGGLETON, J. (1974). Is polymorphism in two-spot ladybirds an example of non-industrial melanism? Nature, Lond., *249*: 179–180. (p. 263).

BENNETT, D., BRUCK, R., DUNN, L. C., KYLDE, B., SHUTSKY, F. & SMITH, L.

J. (1967). Persistence of an introduced lethal in a feral house mouse population. Am. Nat., *101*: 538–539. (p. 228).

BENNETT, J. (1960). A comparison of selective methods and a test of the pre-adaptation hypothesis. Heredity, *15*: 65–77. (p. 261).

BENNETT, J. (1965). *Gregor Mendel's Experiments in Plant Hybridisation*. Edinburgh & London: Oliver & Boyd. (p. 41).

BENSON, R. B. (1950). An introduction to the natural history of British sawflies. Trans. Soc. Brit. Entomol., *10*: 45–142. (p. 96).

BERANEK, A. P. & BERRY, R. J. (1974). Inherited changes in enzyme patterns within parthenogenetic clones of *Aphis fabae*. J. Ent., *48A*: 141–147. (p. 262).

BERRY, A. C. (1974). The use of non-metrical variations of the cranium in the study of Scandinavian population movements. Am. J. phys. Anthrop., *40*: 345–358. (p. 280).

BERRY, A. C. & BERRY, R. J. (1967). Epigenetic variation in the human cranium. J. Anat., *101*: 361–379. (p. 208).

BERRY, A. C. & BERRY, R. J. (1971). Epigenetic polymorphism in the primate skeleton. In *Comparative Genetics in Monkeys, Apes and Man*: 13–41. Chiarelli, B. (Ed.). London & New York: Academic. (p. 179).

BERRY, A. C. & BERRY, R. J. (1972). Origins and relationships of the ancient Egyptians. J. hum. Evolution, *1*: 199–208. (pp. 232, 280).

BERRY, A. C., BERRY, R. J. & UCKO, P. J. (1967). Genetical change in ancient Egypt. Man, *2*: 551–568. (p. 47).

BERRY, R. J. (1963). Epigenetic polymorphism in wild populations of *Mus musculus*. Genet. Res., *4*: 193–220. (p. 231).

BERRY, R. J. (1964). The evolution of an island population of the house mouse. Evolution, Lancaster, Pa, *18*: 468–483. (p. 237).

BERRY, R. J. (1965*a*). *Teach Yourself Genetics*. London: English Universities Press. (pp. 14, 24, 86, 292, 309).

BERRY, R. J. (1965*b*). The evolution of the Skokholm mouse. Nature Wales, *9*: 110–115. (p. 236).

BERRY, R. J. (1967). Genetical changes in mice and men. Eug. Rev., *59*: 78–96. (p. 235).

BERRY, R. J. (1968*a*). The biology of non-metrical variation in mice and men. In *The Skeletal Biology of Earlier Human Populations*: 103–133. Brothwell, D. R. (Ed.). London: Pergamon. (pp. 207, 208).

BERRY, R. J. (1968*b*). The ecology of an island population of the house mouse. J. Anim. Ecol., *37*: 445–470. (pp. 239, 242).

BERRY, R. J. (1969*a*). History in the evolution of *Apodemus sylvaticus* (Mammalia) at one edge of its range. J. Zool., Lond., *159*: 311–328. (pp. 202 ff.).

BERRY, R. J. (1969*b*). The genetical implications of domestication in animals. In *The Domestication and Exploitation of Plants and Animals*: 207–217. Ucko, P. J. & Dimbleby, G. W. (Eds.). London: Duckworth. (p. 269).

BERRY, R. J. (1970*a*). Covert and overt variation, as exemplified by British mouse populations. In *Variation in Mammalian Populations*: 3–26. Berry, R. J. & Southern, H. N. (Eds.). London: Academic. (pp. 200, 267).

BERRY, R. J. (1970*b*). Viking mice. Listener, *84*: 147–148. (p. 206).

BERRY, R. J. (1970*c*). The natural history of the house mouse. Fld Stud., *3*: 219–262. (pp. 239, 240).

BERRY, R. J. (1971). Conservation aspects of the genetical constitution of populations. In *The Scientific Management of Animal and Plant Communities for Conservation*: 177–206. Duffey, E. & Watt, A. S. (Eds.). Oxford: Blackwell. (pp. 130, 131, 216, 272).

BERRY, R. J. (1972). Genetical effects of radiation on populations. Atomic Energy Rev., *10*: 67–100. (pp. 257, 282).

BERRY, R. J. (1973). Chance and change in British Long-tailed field mice. J. Zool., Lond., *170*: 351–366. (p. 208).

BERRY, R. J. (1974*a*). The Shetland fauna, its significance or lack thereof. In *The Natural Environment of Shetland*: 151–163. Goodier, R. (Ed.). Edinburgh: Nature Conservancy Council. (p. 171).

BERRY, R. J. (1974*b*). Conserving genetical variety. In *Conservation in Practice*: 99–115. Warren, A. & Goldsmith, F. B. (Eds.). London: Wiley. (p. 287).

BERRY, R. J. (1975*a*). On the nature of genetical distance and island races of *Apodemus sylvaticus*. J. Zool., Lond., *176*: 293–296. (p. 59).

BERRY, R. J. (1975*b*). *Adam and the Ape*. London: Falcon. (p. 276).

BERRY, R. J. (1976). Variability in mammals – concepts and complications. Proc. 1st intern. Congr. Theriol., Moscow. (p. 306).

BERRY, R. J. & CROTHERS, J. H. (1968). Stabilizing selection in the Dog-whelk (*Nucella lapillus*). J. Zool., Lond., *155*: 5–17. (pp. 136, 137).

BERRY, R. J. & CROTHERS, J. H. (1974). Visible variation in the Dog-whelk, *Nucella lapillus*. J. Zool., Lond., *174*: 123–148. (p. 222).

BERRY, R. J. & DAVIS, P. E. (1970). Polymorphism and behaviour in the Arctic Skua (*Stercorarius parasiticus* (L.)). Proc. roy. Soc. B, *175*: 255–267 (pp. 143, 144).

BERRY, R. J., EVANS, I. M. & SENNITT, B. F. C. (1967). The relationships and ecology of *Apodemus sylvaticus* from the Small Isles of the Inner Hebrides, Scotland. J. Zool., Lond., *152*: 333–346. (pp. 208, 209).

BERRY, R. J. & JAKOBSON, M. E. (1971). Life and death in an island population of the house mouse. Exp. Geront., *6*: 187–197. (p. 241).

BERRY, R. J. & JAKOBSON, M. E. (1974). Vagility in an island population of the house mouse. J. Zool., Lond., *173*: 341–354. (p. 240).

BERRY, R. J. & JAKOBSON, M. E. (1975*a*). Ecological genetics of an island population of the house mouse. J. Zool., Lond., *175*: 523–540. (pp. 238, 246, 247).

BERRY, R. J. & JAKOBSON, M. E. (1975*b*). Adaptation and adaptability in wild-living house mice. J. Zool., Lond., *176*: 391–402. (p. 250).

BERRY, R. J., JAKOBSON, M. E. & TRIGGS, G. S. (1973). Survival in wild-living mice. Mammal Rev., *3*: 46–57. (p. 249).

BERRY, R. J. & MUIR, V. M. L. (1975). Natural history of man in Shetland. J. biosoc. Sci., *7*: 319–344. (pp. 278, 280).

BERRY, R. J. & MURPHY, H. M. (1970). Biochemical genetics of an island population of the house mouse. Proc. roy. Soc. B, *176*: 87–103. (p. 244).

BERRY, R. J. & PETERS, J. (1975). Macquarie Island house mice: a genetical isolate on a sub-Antarctic island. J. Zool., Lond., *176*: 375–389. (p. 246).

BERRY, R. J. & ROSE, F. E. N. (1975). Islands and the evolution of *Microtus arvalis* (Microtinae). J. Zool., Lond., *177*: 395–409. (pp. 214, 215).

BERRY, R. J. & RUDGE, P. J. (1973). Natural selection in Antarctic limpets. Br. Antarct. Surv. Bull., *35*: 73–81. (p. 139).

Berry, R. J. & Searle, A. G. (1963). Epigenetic polymorphism of the rodent skeleton. Proc. zool. Soc., Lond., *140*: 577–615. (p. 208).

Berry, R. J. & Tricker, B. J. K. (1969). Competition and extinction: the mice of Foula, with notes on those of Fair Isle and St. Kilda. J. Zool., Lond., *158*: 247–265. (p. 300).

Berry, R. J. & Warwick, T. (1974). Field mice (*Apodemus sylvaticus*) on the Castle Rock, Edinburgh: an isolated population. J. Zool., Lond., *174*: 325–331. (p. 268).

Bink, F. A. (1970). A review of the introductions of *Thersamonia dispar* Haw. (Lep., Lycaenidae) and the speciation problem. Entomol. Ber., *30*: 179–183. (p. 268).

Birch, L. C. & Ehrlich, P. R. (1967). Evolutionary history and population biology. Nature, Lond., *214*: 349–352. (pp. 114, 296).

Bishop, J. A. (1969). Changes in genetic constitution of a population of *Sphaeroma rugicauda* (Crustacea, Isopoda). Evolution, Lancaster, Pa, *23*: 589–601. (p. 139).

Bishop, J. A. (1972). An experimental study of the cline of industrial melanism in *Biston betularia* (L.) (Lepidoptera) between urban Liverpool and rural North Wales. J. Anim. Ecol., *41*: 209–243. (pp. 108, 125, 127, 164).

Bishop, J. A. & Cook, L. M. (1975). Moths, melanism and clean air. Sci. Am., *232*: 90–99. (p. 128).

Bishop, J. A., Cook, L. M., Muggleton, J. & Seaward, M. R. D. (1975). Moths, lichens, and air pollution along a transect from Manchester to North Wales. J. appl. Ecol., *12*: 83–98. (pp. 124, 125).

Blaylock, B. G. (1965). Chromosomal aberrations in a natural population of *Chironomus tentans* exposed to chronic low level radiation. Evolution, Lancaster, Pa, *19*: 421–429. (p. 87).

Boardman, M., Askew, R. R. & Cook, L. M. (1974). Experiments on resting site selection by nocturnal moths. J. Zool., Lond., *172*: 343–355. (p. 127).

Böcher, T. W. (1949). Racial divergences in *Prunella vulgaris* in relation to habitat and climate. New Phytol., *48*: 285–314. (p. 167).

Bodmer, W. F. (1960). The genetics of homostyly in populations of *Primula vulgaris*. Phil. Trans. roy. Soc. Ser. B, *242*: 517–549. (p. 187).

Bodmer, W. F. & Parsons, P. A. (1962). Linkage and recombination in evolution. Adv. Genet., *11*: 1–100. (p. 183).

Bonner, J. T. (1965). *Size and Cycle: an essay on the structure of biology*. Princeton: University Press. (p. 302).

Boyce, A. C., Küchemann, C. F. & Harrison, G. A. (1967). Neighbourhood knowledge and the distribution of marriage distances. Ann. hum. Genet., *30*: 335–338. (p. 106).

Boyle, C. M. (1960). Case of apparent resistance of *Rattus norvegicus* Berkenhout to anticoagulant poisons. Nature, Lond., *188*: 517. (p. 117).

Bradshaw, A. D. (1971). Plant evolution in extreme environments. In *Ecological Genetics and Evolution*: 20–50. Creed, E. R. (Ed.). Oxford: Blackwell. (p. 266).

Bradshaw, A. D. (1972). Some of the evolutionary consequences of being a plant. Evol. Biol., *5*: 25–47. (pp. 166, 229. 302).

Breese, E. L. & Mather, K. (1960). The organization of polygenic activity within a chromosome in *Drosophila*. II Viability. Heredity, *14*: 375–399. (p. 183).

Brewen, J. G. & Preston, R. J. (1974). Cytogenetic effects of environmental mutagens in mammalian cells and the extrapolation to man. Mutation Res., *26*: 297–305. (p. 255).

BRIDGES, B. (1975). The mutagenicity of captan and related fungicides. Mutation Res., *32*: 3–34. (p. 256).

BRIGGS, D. & WALTERS, S. M. (1969). *Plant Variation and Evolution*. London: Weidenfeld & Nicolson. (p. 31).

BROADHURST, P. L. (1960). Experiments in psychogenetics. Applications of biochemical genetics to the inheritance of behaviour. In *Experiments in Personality, Psychogenetics and Psychopharmacology*: 1–102. Eysenck, H. J. (Ed.). London: Routledge & Kegan Paul. (p. 271).

BROTHWELL, D. & BROTHWELL, P. (1969). *Food in Antiquity*. London: Thames & Hudson. (p. 268).

BROWN, A. W. A. (1967). Insecticide resistance – genetic implications and applications. World Rev. pest Control, *6*: 104–114 (p. 260).

BROWN, J. L. (1966). Types of group selection. Nature, Lond., *211*: 870. (p. 230).

BROWN, L. N. (1965). Selection in a population of house mice containing mutant individuals. J. Mammal., *46*: 461–465. (p. 61).

BRUÈRE, A. N. (1974). The segregation patterns and fertility of sheep heterozygous and homozygous for three different Robertsonian translocations. J. Reprod. Fert., *41*: 453–464. (p. 85).

BRUES, A. M. (1969). Genetic load and its varieties. Science, N. Y., *164*: 1130–1136. (pp. 289, 295).

BRYANT, A. (1940). *English Saga (1840–1940)*. London: Collins. (p. 120).

BRYANT, E. H. (1974). On the adaptive significance of enzyme polymorphisms in relation to environmental variability. Am. Nat., *108*: 1–19. (p. 304).

BUMPUS, H. C. (1899). The elimination of the unfit as illustrated by the introduced sparrow. Biol. Lect., Woods Hole for 1898: 209–226. (p. 132).

BURNET, M. & WHITE, D. O. (1972). *Natural History of Infectious Disease*, 4th Edition. Cambridge: University Press. (p. 146).

CAIN, A. J. (1953). Visual selection by tone in *Cepaea nemoralis* L. J. Conch., *23*: 333–336. (p. 149).

CAIN, A. J. (1971). Undescribed polymorphisms in two British snails. J. Conch., *26*: 410–416. (pp. 150, 153, 267).

CAIN, A. J. & CURREY, J. D. (1963*a*). Area effects in *Cepaea*. Phil. Trans. roy. Soc. Ser. B, *246*: 1–81. (pp. 152, 153, 193).

CAIN, A. J. & CURREY, J. D. (1963*b*). The causes of area effects. Heredity, *18*: 467–471. (pp. 153, 193).

CAIN, A. J., KING, J. M. B. & SHEPPARD, P. M. (1960). New data on the genetics of polymorphism in the snail *Cepaea nemoralis* L. Genetics, *45*: 393–411. (p. 187).

CAIN, A. J. & SHEPPARD, P. M. (1950). Selection in the polymorphic land snail *Cepaea nemoralis* (L.). Heredity, *4*: 275–294. (p. 149).

CAIN, A. J. & SHEPPARD, P. M. (1954). Natural selection in *Cepaea*. Genetics, *39*: 89–116. (p. 149).

CAMIN, J. H. & EHRLICH, P. R. (1958). Natural selection in water snakes (*Natrix sipedon* L.) on islands in Lake Erie. Evolution, Lancaster, Pa, *12*: 504–511. (pp. 62, 63).

CARLQUIST, S. (1974). *Island Biology*. New York & London: Columbia. (p. 198).

CARSON, H. L. (1955). The genetic characteristics of marginal populations of *Drosophila*. Cold Spring Harb. Symp. quant. Biol., *20*: 276–287. (p. 225).

CARSON, H. L. (1958). Response to selection under different conditions of recombination in *Drosophila*. Cold Spring Harb. Symp. quant. Biol., *23*: 291–306. (p. 273).

CARSON, H. L. (1959). Genetic conditions which promote or retard the formation of species. Cold Spring Harb. Symp. quant. Biol., *24*: 87–105. (p. 225).

CARSON, H. L. (1967). Permanent heterozygosity. Evol. Biol., *1*: 143–168. (p. 273).

CARSON, H. L., HARDY, D. E., SPIETH, H. T. & STONE, W. S. (1970). The evolutionary biology of Hawaiian Drosophilidae. In *Essays in Evolution and Genetics*: 437–543. Hecht, M. K. & Steere, W. C. (Eds.). Amsterdam: North-Holland. (p. 305).

CARTER, M. A. (1968). Studies on *Cepaea*. II Area effects and visual selection in *Cepaea nemoralis* (L.) and *Cepaea hortensis*. Phil. Trans. roy. Soc. Ser. B, *253*: 397–446. (pp. 149, 152).

CAVALLI-SFORZA, L. L. & BODMER, W. F. (1971). *The Genetics of Human Populations*. San Francisco: Freeman. (pp. 14, 195, 276).

CHANCE, E. (1922). *The Cuckoo's Secret*. London: Sidgwick & Jackson. (p. 115).

CHARLESWORTH, B. (1971). Selection in density-regulated populations. Ecology, *52*, 469–474. (p. 299).

CHITTY, D. (1952). Mortality among voles (*Microtus agrestis*) at Lake Vyrnwy, Montgomeryshire in 1936–9. Phil. Trans. roy. Soc. Ser. B, *236*: 505–552. (p. 226).

CHITTY, D. (1967). The natural selection of self-regulatory behaviour in animal populations. Proc. Ecol. Soc. Aust., *2*: 51–78. (p. 226).

CLARK, J. M. (1975). The effects of selection and human preference on coat colour gene frequencies in urban cats. Heredity, *35*: 195–210. (p. 52).

CLARKE, B. C. (1960). Divergent effects of natural selection on two closely related polymorphic snails. Heredity, *14*: 423–443. (p. 152).

CLARKE, B. C. (1962). Balanced polymorphism and the diversity of sympatric species. In *Taxonomy and Geography*: 47–70. Nichols, D. (Ed.). London: Systematics Association. (pp. 135, 152).

CLARKE, B. C. (1964). Frequency-dependent selection for the dominance of rare polymorphic genes. Evolution, Lancaster, Pa, *18*: 364–369. (p. 140).

CLARKE, B. C. (1966). The evolution of morph-ratio clines. Am. Nat., *100*: 389–402. (pp. 193, 194).

CLARKE, B. C. (1968). Balanced polymorphism and regional differentiation in land snails. In *Evolution and Environment*: 351–368. Drake, E. T. (Ed.). New Haven & London: Yale University Press. (p. 194).

CLARKE, B. C. (1970*a*). Darwinian evolution of proteins. Science, N.Y., *168*: 1009–1011. (p. 296).

CLARKE, B. C. (1970*b*). Selective constraints on amino-acid substitutions during the evolution of proteins. Nature, Lond., *228*: 159–160. (p. 296).

CLARKE, B. C. (1972). Density-dependent selection. Am. Nat., *106*: 1–13. (p. 146).

CLARKE, B. C. (1973). Mutation and population size. Heredity, *31*: 367–379. (p. 305).

CLARKE, B. C. & MURRAY, J. J. (1962). Changes of gene frequency in *Cepaea nemoralis*. Heredity, *17*: 445–465. (pp. 150, 151).

CLARKE, B. C. & MURRAY, J. J. (1969). Ecological genetics and speciation in land snails of the genus *Partula*. Biol. J. Linn. Soc., *1*: 31–42. (p. 54).

CLARKE, C. A. & SHEPPARD, P. M. (1960). Super-genes and mimicry. Heredity, *14*: 175–185. (p. 185).

CLARKE, C. A. & SHEPPARD, P. M. (1966). A local survey of the distribution of industrial melanic forms in the moth *Biston betularia* and estimates of the selective values of these in an industrial environment. Proc. roy. Soc. B, *165*: 424–439. (p. 125).

CLARKE, C. A., SHEPPARD, P. M. & THORNTON, I. W. B. (1968). The genetics of the mimetic butterfly *Papilio memnon*. Phil. Trans. roy. Soc. Ser. B, *254*: 37–89. (pp. 13, 185).

CLAUSEN, J. & HIESEY, W. M. (1958). Experimental studies on the nature of the species. IV Genetic structure of ecological races. Carnegie Inst., Washington, publ. no. *615*: 1–312. (pp. 179–81).

CODY, M. L. (1966). A general theory of clutch size. Evolution, Lancaster, Pa, *20*: 174–184. (p. 305).

COLEMAN, D. A. (1973). Marriage movements in British cities. In *Genetic Variation in Britain*: 33–57. Roberts, D. F. (Ed.). London: Taylor & Francis. (p. 106).

COLWELL, R. N. (1951). The use of radioactive isotopes in determining spore distribution patterns. Am. J. Bot., *38*: 511–523. (p. 107).

COOK, A. (1975). Changes in the Carrion/Hooded Crow hybrid zone and the possible importance of climate. Bird Study, *22*: 165–168. (p. 177, 267).

COOK, L. M. (1971). *Coefficients of Natural Selection*. London: Hutchinson. (p. 296).

COOK, L. M., ASKEW, R. R. & BISHOP, J. A. (1970). Increasing frequency of the typical form of the Peppered Moth in Manchester. Nature, Lond., *227*: 1155. (p. 125).

COOK, L. M. & WOOD, R. J. (1976). Genetic effects of pollutants. Biologist, 23: 129–139. (p. 261).

COOKE, A. H. (1895). *Mollusca*. London: Macmillan. (pp. 222, 223).

COOMBE, D. E. (1961). *Trifolium occidentale*, a new species related to *T. repens* L. Watsonia, *5*: 68–87. (p. 75).

COOMBS, V. A. (1973). Dessication and age as factors in the vertical distribution of the Dog-whelk, *Nucella lapillus*. J. Zool., Lond., *171*: 57–66. (p. 221).

COON, C. S. (1962). *The Origin of Races*. London: Cape. (p. 277).

CORBET, G. B. (1963). An isolated population of the bank vole (*Clethrionomys glareolus*) with aberrant dental pattern. Proc. zool. Soc., Lond., *140*: 316–319. (p. 59).

CORBET, G. B. (1964). *The Identification of British Mammals*. London: British Museum (Natural History). (p. 200).

CORBET, G. B. (1970). Patterns of subspecific variation. In *Variation In Mammalian Populations*: 105–116. Berry, R. J. & Southern, H. N. (Eds.). London: Academic. (p. 28).

CORBET, G. B. (1975). Examples of short- and long-term changes of dental pattern in Scottish voles (Rodentia; Microtinae). Mammal Rev., *5*: 17–21. (p. 59).

CRANE, M. B. & BROWN, A. C. (1937). Incompatibility in the sweet cherry *Prunus avium* L. J. Genet., *15*: 86–117. (p. 65).

CREED, E. R. (1971*a*). Industrial melanism in the two-spot ladybird and smoke abatement. Evolution, Lancaster, Pa, *25*: 290–293. (p. 263).

CREED, E. R. (1971*b*). Melanism in the two-spot ladybird in Great Britain. In *Ecological Genetics and Evolution*: 134–151. Creed, E. R. (Ed.). Oxford: Blackwell. (p. 263).

CREED, E. R. (1975). Melanism in the two-spot ladybird: the nature and intensity of selection. Proc. roy. Soc. B, *190*: 135–148. (p. 264).

CREED, E. R., DOWDESWELL, W. H., FORD, E.B. & MCWHIRTER, K. G. (1962). Evolutionary studies on *Maniola jurtina*: the English mainland, 1958–60. Heredity, *17*: 237–265. (p. 267).

CREED, E. R., DOWDESWELL, W. H., FORD, E.B. & MCWHIRTER, K. G. (1970). Evolutionary studies on *Maniola jurtina* (Lepidoptera, Satyridae): The "boundary phenomenon" in southern England 1961 to 1968. In *Essays in Evolution and Genetics in honor of Theodosius Dobzhansky*: 263–287. Hecht, M. K. & Steere, W. C. (Eds.). Amsterdam: North-Holland. (p. 216).

CROMBIE, A. C. (1946). Further experiments on insect competition. Proc. roy. Soc. B, *133*: 76–109. (p. 305).

CROSBY, J. L. (1949). Selection of an unfavourable gene-complex. Heredity, *13*: 127–131. (p. 187).

CROW, J. F. (1970). Genetic loads and the cost of natural selection. In *Mathematical Topics in Population Genetics*: 128–177. Kojima, K. I. (Ed.). Heidelberg: Springer Verlag. (p. 289).

CROW, J. F. & KIMURA, M. (1970). *An Introduction to Population Genetics Theory*. New York, Evanston & London: Harper & Row. (pp. 14, 167, 195).

CROW, J. F., SCHULL, W. J. & NEEL, J. V. (1956). *A Clinical, Pathological and Genetic Study of Multiple Neurofibromatosis*. Springfield, Illinois: Thomas. (p. 62).

CULBERSON, W. L. & CULBERSON, C. F. (1967). Habitat selection by chemically differentiated lichens. Science, N.Y., *158*: 1195–1197. (p. 229).

CURTIS, H. J. (1971). Genetic factors in ageing. Adv. Genet., *16*: 305–324. (p. 252).

DADAY, H. (1954). Gene frequencies in wild populations of *Trifolium repens*. Heredity, *8*: 61–78; 377–384. (p. 160).

DAMIAN, R. T. (1964). Molecular mimicry: antigen sharing by parasite and host and its consequences. Am. Nat., *98*: 129–149. (p. 145).

DARLINGTON, C. D. (1971). The evolution of polymorphic systems. In *Ecological Genetics and Evolution*: 1–19. Creed, E. R. (Ed.). Oxford: Blackwell. (p. 186).

DE BEER, G. R. (1958). *Embryos and Ancestors*, 3rd Edition. Oxford: Clarendon. (p. 277).

DEGOS, L. & DAUSSET, J. (1974). Human migrations and linkage disequilibrium of *HL-A* system. Immunogenetics, *3*: 195–210. (p. 195).

DELANY, M. J. (1961). The ecological distribution of small mammals in north-west Scotland. Proc. zool. Soc., Lond., *137*: 107–126. (p. 201).

DELANY, M. J. (1964). Variation in the Long-tailed field-mouse (*Apodemus sylvaticus* (L)) in north-west Scotland. I Comparison of individual characters. Proc. roy. Soc. B, *161*: 191–199. (pp. 201, 204).

DELANY, M. J. & DAVIS, P. E. (1961). Observations on the ecology and life history of the Fair Isle field mouse *Apodemus sylvaticus fridariensis* (Kinnear). Proc. zool. Soc., Lond., *136*: 439–452. (p. 204).

DEMPSTER, J. P., KING, M. L. & LAKHANI, K. H. (1976). The status of the swallowtail butterfly in Britain. Ecol. Ent., *1*: 71–84. (p. 268).

DE SERRES, F. J. (1974). AF-2 – food preservative or genetic hazard? Mutation Res., *26*: 1–2. (p. 257).

De Winton, W. E. (1895). The Long-tailed field mouse of the Outer Hebrides: a proposed new species. Zoologist (3) *19*: 369–371, 426. (p. 201).

Dickinson, H. & Antonovics, J. (1973). Theoretical considerations of sympatric divergence. Am. Nat., *107*: 256–274. (p. 306).

Dobson, T. (1974). Studies on the biology of the kelp-fly *Coelopa* in Great Britain. J. nat. Hist., *8*: 155–177. (p. 167).

Dobzhansky, Th. (1946). Genetics of natural populations. XIII Recombination variability in populations of *Drosophila pseudoobscura*. Genetics, *31*: 269–290. (p. 195).

Dobzhansky, Th. (1961). On the dynamics of chromosomal polymorphism in *Drosophila*. Symp. R. ent. Soc., Lond., *1*: 30–42. (p. 87).

Dobzhansky, Th. (1970). *Genetics of the Evolutionary Process*. New York & London: Columbia. (pp. 26, 185, 291).

Dowdeswell, W. H. (1961). Experimental studies on natural selection in the butterfly, *Maniola jurtina*. Heredity, *16*: 39–52. (p. 217).

Dowdeswell, W. H. & Ford, E. B. (1953). The influence of isolation on variability in the butterfly *Maniola jurtina* L. Symp. Soc. exp. Biol., *7*: 254–273. (p. 217).

Dowrick, V. P. J. (1956). Heterostyly and homostyly in *Primula obconica*. Heredity, *10*: 219–236. (p. 186).

Drummond, D. C. (1970). Variation in rodent populations in response to control measures. In *Variation in Mammalian Populations*: 351–367. Berry, R. J. & Southern, H. N. (Eds.). London: Academic. (p. 119).

Dubos, R. (1973). *A God Within*. London & Sydney: Angus & Robertson. (p. 307).

Dubos, R. & Ward, B. (1972). *Only One Earth: The Care and Maintenance of Our Small Planet*. London: Penguin. (p. 306).

Duffey, E. (1968). Ecological studies on the large copper butterfly *Lycaena dispar* Haw. *batavus*. Obth. at Woodwalton Fen National Nature Reserve, Huntingdonshire. J. appl. Ecol., *5*: 69–96. (p. 268).

Dunn, L. C. (1960). Variations in the transmission ratios of alleles through egg and sperm in *Mus musculus*. Am. Nat., *94*: 385–393. (p. 228).

East, E. M. (1913). Inheritance of flower size in crosses between species of *Nicotiana*. Bot. Gaz., *55*: 177–188. (p. 42).

Edwards, A. W. F. & Cavalli-Sforza, L. L. (1964). Reconstruction of evolutionary trees. In *Phenetic and Phylogenetic Classification*: 67–76. Heywood, V. H. & McNeill, J. (Eds.). London: Systematics Association. (p. 279).

Edwards, R. G. (1969). Reproduction: chance and choice. In *Genetic Engineering*: 25–32. Paterson, D. (Ed.). London: B.B.C. (p. 283).

Ehrlich, P. R. & Holm, R. W. (1963). *The Process of Evolution*. New York: McGraw-Hill. (p. 53).

Ehrlich, P. R. & Raven, P. H. (1969). Differentiation of populations. Science, N.Y., *165*: 1228–1232. (p. 228).

Elton, C. (1930). *Animal Ecology and Evolution*. London: Oxford University Press. (pp. 56, 238).

Endler, J. A. (1973). Gene flow and population differentiation. Science, N.Y., *179*: 243–250. (p. 160).

Falconer, D. S. (1960). *An Introduction to Quantitative Genetics*. Edinburgh & London: Oliver & Boyd. (pp. 14, 159).

FISHER, J. (1952). *The Fulmar*. London: Collins. (p. 169, 267).
FISHER, R. A. (1918). The correlation between relatives on the supposition of Mendelian inheritance. Trans. roy. Soc., Edinb., *52*: 399–433. (p. 158).
FISHER, R. A. (1922). On the dominance ratio. Proc. roy. Soc., Edinb., *42*: 321–341. (p. 72).
FISHER, R. A. (1928). The possible modification of the response of the wild-type to recurrent mutations. Am. Nat., *62*: 115–126. (p. 188).
FISHER, R. A. (1930). *The Genetical Theory of Natural Selection*. Oxford: Clarendon. (pp. 195, 298).
FISHER, R. A. (1935). Dominance in poultry. Phil. Trans. roy. Soc. Ser. B, *225*: 197–226. (pp. 189, 270).
FISHER, R. A. (1936). Has Mendel's work been rediscovered? Ann. Sci., *1*: 115–137. (p. 41).
FISHER, R. A. (1938). Dominance in poultry: feathered feet, rose comb, internal pigment and pile. Proc. roy. Soc. B, *125*: 25–48. (pp. 189, 270).
FISHER, R. A. (1939). Selective forces in wild populations of *Paratettix texanus*. Ann. Eugen., *9*: 109–122. (p. 186).
FISHER, R. A. (1949). *Theory of Inbreeding*. Edinburgh & London: Oliver & Boyd. (p. 229).
FISHER, R. A. (1950). Gene frequencies in a cline determined by selection and diffusion. Biometrics, *6*: 353–361. (p. 164).
FISHER, R. A. & FORD, E. B. (1947). The spread of a gene in natural conditions in a colony of the moth *Panaxia dominula* L. Heredity, *1*: 143–174. (pp. 56, 224).
FORD, C. E. & HAMERTON, J. L. (1970). Chromosome polymorphism in the common shrew, *Sorex araneus*. In *Variation in Mammalian Populations*: 223–236. Berry, R. J. & Southern, H. N. (Eds.). London: Academic. (p. 88).
FORD, E. B. (1937). Problems of heredity in the Lepidoptera. Biol. Rev., *12*: 461–503. (pp. 121, 126).
FORD, E. B. (1940). Genetic research in the Lepidoptera. Ann. Eugen., *10*: 227–252. (pp. 18, 121, 126, 189).
FORD, E. B. (1945). *Butterflies*. London: Collins. (pp. 27, 127).
FORD, E. B. (1955). Polymorphism and taxonomy. Heredity, *9*: 255–264. (pp. 193, 216).
FORD, E. B. (1964). *Ecological Genetics*. London: Methuen. (pp. 216, 218).
FORD, E. B. (1967). *Moths*, 2nd Edition. London: Collins. (p. 27).
FORD, E. B. (1971). *Ecological Genetics*, 3rd Edition. London: Chapman & Hall. (pp. 218, 306).
FORD, E. B. & SHEPPARD, P. M. (1969). The *medionigra* polymorphism of *Panaxia dominula*. Heredity, *24*: 561–569. (p. 224).
FORD, H. B. & FORD, E. B. (1930). Fluctuation in numbers and its influence on variation in *Melitaea aurinia*. Trans. R. ent. Soc., Lond., *78*: 345–351. (p. 147).
FRANKEL, O. H. (1970). Genetic conservation in perspective. In *Genetic Resources in Plants*: 469–489. Frankel, O. H. & Bennett, E. (Eds.). Oxford: Blackwell. (p. 272).
FRANKEL, O. H. & BENNETT, E. (1970). *Genetic Resources in Plants*. I.B.P. Handbook, No. 11. Oxford: Blackwell. (p. 272).
FRANKLIN, I. & LEWONTIN, R. C. (1970). Is the gene the unit of selection? Genetics, *65*: 707–734. (p. 184).

FRASER, B. D. (1972). Population dynamics and recognition of biotypes in the pea aphid. Can. Ent., *104*: 1729–1733. (p. 99).

FRASER, G. R. (1962). Our genetical 'load'. A review of some aspects of genetical variation. Ann. hum. Genet., *25*: 387–415. (p. 295).

FRASER, G. R. & MAYO, O. (1974). Genetical load in man. Humangenetik, *23*: 83–110. (p. 295).

FREDGA, K. (1968). Idiogram and trisomy of the water vole (*Arvicola terrestris* L.), a favourable animal for cytogenetic research. Chromosoma, *25*: 75–89. (p. 77).

FROGGATT, P. & NEVIN, N. C. (1971). The 'Law of Ancestral Heredity' and the Mendelian-Ancestrian controversy in England, 1889–1906. J. med. Genet., *8*: 1–36. (p. 136).

FRÖST, S. (1958). The geographical distribution of accessory chromosomes in *Centaurea scabiosa*. Hereditas, *44*: 75–111. (p. 83).

GADGIL, M. & BOSSERT, W. H. (1970). Life history consequences of natural selection. Am. Nat., *104*: 1–24. (p. 145).

GARTSIDE, D. W. & MCNEILLY, T. (1974). The potential for evolution of heavy metal tolerance in plants. II Copper tolerance in normal populations of different plant species. Heredity, *32*: 335–348. (p. 265).

GAUSE, G. F. (1934). *The Struggle for Existence*. Baltimore: Williams & Wilkins. (p. 305).

GILLESPIE, J. H. & LANGLEY, C. H. (1974). A general model to account for enzyme variation in natural populations. Genetics, *76*: 837–848. (p. 298).

GILLETT, J. D. (1967). Natural selection and feeding speed in a blood-sucking insect. Proc. roy. Soc. B, *167*: 316–329. (p. 116).

GLEAVES, J. T. (1973). Gene flow mediated by wind born pollen. Heredity, *31*: 355–366. (p. 265).

GOODHART, C. B. (1962). Variation in a colony of the snail *Cepaea nemoralis* (L.). J. Anim. Ecol., *31*: 207–237. (p. 57).

GOODHART, C. B. (1963). "Area effects" and non-adaptive variation between populations of *Cepaea* (Mollusca). Heredity, *18*: 459–465. (p. 193).

GOODHART, C. B. (1973). A 16-year survey of *Cepaea* on the Hundred Foot Bank. Malacologia, *14*: 327–331. (p. 57).

GOODMAN, G. T. & GILLHAM, M. E. (1954). Ecology of the Pembrokeshire islands. II Skokholm, environment and vegetation. J. Ecol., *42*: 296–327. (p. 234).

GOODMAN, M., BARNABAS, J., MATSUDA, G. & MOORE, G. W. (1971). Molecular evolution in the descent of man. Nature, Lond., *233*: 604–613. (p. 298).

GORDON, C. (1935). An experiment on a released population of *D. melanogaster*. Am. Nat., *69*: 381. (p. 195).

GORMAN, G. C., SOULÉ, M., YANG, S. Y. & NEVO, E. (1975).Evolutionary genetics of insular Adriatic lizards. Evolution, Lancaster, Pa, *29*: 52–71. (p. 304).

GRANT, V. (1963). *The Origin of Adaptations*. London & New York: Columbia University Press. (p. 53).

GREAVES, J. H. (1973). Warfarin resistance in the rat. U. Lond.: Ph.D. (p. 117).

GREAVES, J. H. & RENNISON, B. D. (1973). Population aspects of warfarin resistance in the brown rat, *Rattus norvegicus*. Mammal Rev., *3*: 27–29. (pp. 117, 118, 119).

GREENWOOD, J. J. D. (1974). Visual and other selection in *Cepaea*: a further example. Heredity, *33*: 17–31. (p. 149).

GREGOR, J. W. (1938). Experimental taxonomy. II Initial population differentiation in *Plantago maritima* L. of Britain. New Phytol., *37*: 15–49. (p. 30).

GREGOR, J. W. (1946). Ecotypic differentiation. New Phytol., *45*: 254–270. (pp. 30, 31).

GREGOR, J. W. & WATSON, P. J. (1961). Ecotypic differentiation. Evolution, Lancaster, Pa, *15*: 166–173. (p. 286).

GRÜNEBERG, H. (1963). *The Pathology of Development*. Oxford: Blackwell. (p. 207).

GRÜNEBERG, H., BAINS, G. S., BERRY, R. J., RILES, L. E., SMITH, C. A. B. & WEISS, R. A. (1966). A search for genetic effects of high natural radioactivity in South India. Spec. Rep. Ser. med. Res. Coun. No. 307: 1–59. (p. 259).

GUSTAFFSON, A. (1946). The effect of heterozygosity on variability and vigour. Hereditas, *32*: 263–286. (p. 94).

HALDANE, J. B. S. (1924). A mathematical theory of natural and artificial selection. Trans. Camb. phil. Soc., *23*: 19–40. (pp. 125, 129, 136).

HALDANE, J. B. S. (1932). *Causes of Evolution*. London: Longmans, Green. (p. 195).

HALDANE, J. B. S. (1948). The theory of a cline. J. Genet., *48*: 277–284. (p. 171).

HALDANE, J. B. S. (1949). Disease and evolution. Ricerca Sci., Suppl., *19*: 68–76. (pp. 145, 279).

HALDANE, J. B. S. (1955). Population genetics. New Biology, *18*: 34–51. (p. 229).

HALDANE, J. B. S. (1956). The theory of selection for melanism in Lepidoptera. Proc. roy. Soc. B, *145*: 303–306. (p. 125).

HALDANE, J. B. S. (1957). The cost of natural selection. J. Genet., *55*: 511–524. (pp. 289, 294).

HALDANE, J. B. S. (1959). Natural selection. In *Darwin's Biological Work*: 101–149. Bell, P. R. (Ed.). Cambridge: University Press. (p. 136).

HALDANE, J. B. S. (1964). A defense of beanbag genetics. Persp. Biol. Med., *7*: 343–360. (p. 306).

HALKKA, O. & HALKKA, L. (1974). Partial population transfers as a means of estimating the effectiveness of natural selection. Hereditas, *78*: 314–315. (p. 220).

HALKKA, O., HALKKA, L., RAATIKAINEN, M. & HOVINEN, R. (1973). The genetic basis of balanced polymorphism in *Philaenus* (Homoptera). Hereditas, *74*: 69–80. (p. 219).

HALKKA, O., RAATIKAINEN, M. & HALKKA, L. (1974*a*). Radial and peripheral clines in northern polymorphic populations of *Philaenus spumarius*. Hereditas, *78*: 85–96. (p. 219).

HALKKA, O., RAATIKAINEN, M. & HALKKA, L. (1974*b*). The founder principle, founder selection, and evolutionary divergence and convergence in natural populations of *Philaenus*. Hereditas, *78*: 73–84. (p. 220).

HAMILTON, W. D. (1964). The genetical evolution of social behaviour. J. theor. Biol., *7*: 1–16, 17–51. (p. 230).

HAMILTON, W. D. (1972). Altruism and related phenomena, mainly in social insects. Ann. Rev. Ecol. Syst., *3*: 193–232. (p. 230).

HAMILTON, W. D. (1975). Innate social aptitudes of man: an approach from evolutionary genetics. In *Biosocial Anthropology*: 133–155. Fox, R. (Ed.). London: Malaby. (p. 230).

HAMMOND, J. (1971). *Farm Animals*, 4th edition. London: Arnold. (pp. 157, 158).

HARBERD, D. L. (1961). Observations on population structure and longevity of *Festuca rubra* L. New Phytol., *60*: 184–206. (p. 105).

HARRIS, H. (1966). Enzyme polymorphisms in man. Proc. roy. Soc. B, *164*: 298–310. (p. 287).

HARRIS, H. (1970). *The Principles of Human Biochemical Genetics*. Amsterdam & London: North-Holland. (p. 162).

HARRIS, H. (1974). *Prenatal Diagnosis and Selective Abortion*. London: Nuffield Provincial Hospitals Trust. (p. 282).

HARRIS, H., HOPKINSON, D. A. & ROBSON, E. B. (1973). The incidence of rare alleles determining electrophoretic variants: data on 43 enzyme loci in man. Ann. hum. Genet., *37*: 237–253. (p. 19).

HARRISON, J. W. H. (1956). Melanism in the Lepidoptera. Entomologist's Rec., *68*: 172–181. (p. 121).

HARVEY, P. H., JORDAN, C. A. & ALLEN, J. A. (1974). Selection behaviour of wild blackbirds at high prey densities. Heredity, *32*: 401–404. (p. 73).

HAWKSWORTH, D. L. (1974). *The Changing Flora and Fauna of Britain*. London & New York: Academic. (p. 267).

HAYNE, D. W. (1949). Two methods for estimating population from trapping records. J. Mammal., *30*: 399–411. (p. 239).

HEATH, D. J. (1974). Seasonal changes in frequency of the "yellow" morph of the isopod, *Sphaeroma rugicauda*. Heredity, *32*: 299–307. (pp. 139, 140).

HEATH, D. J. (1975*a*). Geographical variation in populations of the polymorphic isopod, *Sphaeroma rugicauda*. Heredity, *35*: 99–107. (p. 139).

HEATH, D. J. (1975*b*). Colour, sunlight and internal temperatures in the land-snail *Cepaea nemoralis* (L.). Oecologia, *19*: 29–38. (p. 154).

HEATH, J. (1974). A century of change in the Lepidoptera. In *The Changing Flora and Fauna of Britain*: 275–292. Hawksworth, D. L. (Ed.). London & New York: Academic. (p. 90).

HEBERT, P. D. N. (1974). Ecological differences between genotypes in a natural population of *Daphnia magna*. Heredity, *33*: 327–337. (p. 99).

HEBERT, P. D. N., WARD, R. N. & GIBSON, J. B. (1972). Natural selection for enzyme variants among parthenogenetic *Daphnia magna*. Genet. Res., *19*: 173–176. (p. 99).

HELLER, J. (1975*a*). The taxonomy of some British *Littorina* species with notes on their reproduction (Mollusca: Prosobranchia). Zool. J. Linn. Soc., *56*: 131–151. (p. 150).

HELLER, J. (1975*b*). Visual selection of shell colour in two littoral prosobranchs. Zool. J. Linn. Soc., *56*: 153–170. (p. 151).

HESLOP-HARRISON, J. (1953). *New Concepts in Flowering Plant Taxonomy*. London: Heinemann. (p. 286).

HESLOP-HARRISON, J. (1964). Forty years of genecology. Adv. ecol. Res., *2*: 159–247. (pp. 166, 286).

HEWITT, G. M. & BROWN, F. M. (1970). The B-chromosome system of *Myrmeleotettix maculatus*. V A steep cline in East Anglia. Heredity, *25*: 363–371. (pp. 80, 81, 82).

HILTON, B., CALLAHAN, D., HARRIS, M., CONDLIFFE, P. & BERKLEY, B. (1973). *Ethical Issues in Human Genetics*. New York: Plenum. (p. 285).

HINTON, M. A. C. (1914). Notes on the British forms of *Apodemus*. Ann. Mag. nat. Hist., (9) *14*: 117–134. (p. 201).

HINTON, M. A. C. (1919). The field mouse of Foula. Scott. Nat., 177–181. (p. 201).

HMSO (1969). *Report of the Joint Committee on the Use of Antibiotics in Animal Husbandry and Veterinary Medicine*. Cmnd 4190. London: HMSO. (p. 262).

HOWE, W. L. & PARSONS, P. A. (1967). Genotype and environment in the determination of minor skeletal variants and body weight in mice. J. Embryol. exp. Morph., *17*: 283–292. (p. 207).

HOWELLS, R. (1968). *The Sounds Between*. Llandysul: Gomerian. (p. 237).

HUBBARD, C. J. E. (1965). *Spartina* marshes in Southern England. 6 Pattern of invasion in Poole Harbour. J. Ecol., *53*: 799–813. (p. 76).

HUGHES, A. W. MCK. (1932). Induced melanism in the Lepidoptera. Proc. roy. Soc. B, *110*: 378–402. (p. 121).

HUNT, W. G. & SELANDER, R. K. (1973). Biochemical genetics of hybridisation in European house mice. Heredity, *31*: 11–33. (p. 178).

HUTCHINSON, G. E. (1944). Limnological studies in Connecticut. VII A critical examination of the supposed relationship between phytoplankton periodicity and chemical changes in lake waters. Ecology, *25*: 3–26. (p. 299).

HUTCHINSON, G. E. (1957). Concluding remarks. Cold Spring Harb. Symp. quant. Biol., *22*: 415–427. (p. 299).

HUXLEY, J. S. (1942). *Evolution, the Modern Synthesis*. London: Allen & Unwin. (p. 27).

JACQUARD, A. (1974). *The Genetic Structure of Populations*. Heidelberg & New York: Springer-Verlag. (p. 195).

JAYANT, K. (1966). Birth weight and survival: a hospital survey repeated after 15 years. Ann. hum. Genet., *29*: 367–375. (p. 135).

JEFFERIES, D. J. & PARSLOW, J. L. F. (1976). The genetics of bridling in guillemots from a study of hand-reared birds. J. Zool., Lond., *179*: 411–420. (p. 168).

JENSEN, R. A. C. (1966). Genetics of cuckoo egg polymorphism. Nature, Lond., *209*: 827. (p. 116).

JEWELL, P. A., MILNER, C. & BOYD, J. M. (1974). *Island Survivors: the Ecology of the Soay Sheep of St. Kilda*. London: Athlone. (p. 272).

JOHN, B. & HEWITT, G. M. (1969). Parallel polymorphism for supernumerary segments in *Chorthippus parallelus* (Zelterstedt). Chromosoma, *28*: 73–84. (p. 228).

JOHN, B. & LEWIS, K. R. (1973). *The Meiotic Mechanism*. Oxford Biological Reader, No. 65. London: Oxford University Press. (p. 308).

JOHNSON, M. S. (1976). Allozymes and area effects in *Cepaea nemoralis* on the Lambourn Downs. Heredity, *36*: 105–121 (p. 194).

JONES, D. A. (1967). Polymorphism, plants and natural populations. Sci. Prog., *55*: 379–400. (pp. 160, 161).

JONES, D. A. & WILKINS, D. A. (1971). *Variation and Adaptation in Plant Species*. London: Heinemann. (p. 105).

JONES, J. S. (1973*a*). Ecological genetics and natural selection in molluscs. Science, N.Y., *182*: 546–552. (pp. 151, 153, 154).

JONES, J. S. (1973*b*). Ecological genetics of a population of the snail *Cepaea nemoralis* at the northern limit of its range. Heredity, *31*: 201–211. (pp. 154, 225).

JONES, J. S. (1973*c*). The genetic structure of a southern peripheral population of the snail *Cepaea nemoralis*. Proc. roy. Soc. B, *183*: 371–384. (p. 154).

JONES, J. S., BRISCOE, D. A. & CLARKE, B. C. (1974). Natural selection on the polymorphic snail *Hygromia striolata*. Heredity, *33*: 102–106. (p. 149).

JONES, J. S. & YAMAZAKI, T. (1974). Genetic background and the fitness of allozymes. Genetics, *78*: 1185–1189. (p. 291).

JONES, M. E. (1971). The population genetics of *Arabidopsis thaliana*. Heredity, *27*: 39–72. (p. 95).

KARLIN, S. & MCGREGOR, J. (1974). Towards a theory of the evolution of modifier genes. Theor. Pop. Biol., *5*: 59–103. (p. 296).

KARN, M. N. & PENROSE, L. S. (1951). Birth weight and gestation time in relation to maternal age, parity and infant survival. Ann. Eugen., *16*: 147–164. (pp. 131, 135).

KERR, W. E. & WRIGHT, S. (1954). Experimental studies of the distribution of gene frequencies in very small populations of *Drosophila melanogaster*. Evolution, Lancaster, Pa, *8*: 172–177; 225–240; 293–302. (p. 54).

KETTLEWELL, H. B. D. (1955*a*). Selection experiments on industrial melanism in the Lepidoptera. Heredity, *9*: 323–342. (p. 18).

KETTLEWELL, H. B. D. (1955*b*). Recognition of appropriate backgrounds by pale and dark phases of Lepidoptera. Nature, Lond., *175*: 943–944. (p. 126).

KETTLEWELL, H. B. D. (1956). Further selection experiments on industrial melanism in the Lepidoptera. Heredity, *10*: 287–301. (p. 122).

KETTLEWELL, H. B. D. (1961*a*). Selection experiments on melanism in *Amathes glareosa* Esp. Heredity, *16*: 415–434. (p. 170).

KETTLEWELL, H. B. D. (1961*b*). Geographical melanism in the Lepidoptera of Shetland. Heredity, *16*: 393–402. (p. 173).

KETTLEWELL, H. B. D. (1965). Insect survival and selection for pattern. Science, N.Y., *148*: 1290–1296. (p. 190).

KETTLEWELL, H. B. D. (1973). *The Evolution of Melanism*. Oxford: Clarendon. (pp. 121, 123, 126, 128, 129, 175, 192).

KETTLEWELL, H. B. D. & BERRY, R. J. (1961). The study of a cline. Heredity, *16*: 403–414. (p. 172).

KETTLEWELL, H. B. D. & BERRY, R. J. (1969). Gene flow in a cline. Heredity, *24*: 1–14. (p. 172).

KETTLEWELL, H. B. D., BERRY, R. J., CADBURY, C. J. & PHILLIPS, G. C. (1969). Differences in behaviour, dominance and survival within a cline. Heredity, *24*: 15–25. (p. 174).

KETTLEWELL, H. B. D. & CADBURY, C. J. (1963). Investigations on the origins of non-industrial melanism. Entomologist's Rec., *75*: 149–160. (p. 170).

KETTLEWELL, H. B. D., CADBURY, C. J. & LEES, D. R. (1971). Recessive melanism in the moth *Lasiocampa quercus* L. In *Ecological Genetics and Evolution*: 175–201. Creed, E. R. (Ed.). Oxford: Blackwell. (p. 164).

KIMURA, M. (1968*a*). Genetic variability maintained in a finite population due to mutational production of neutral and nearly neutral isoalleles. Genet. Res., *11*: 247–269. (p. 298).

KIMURA, M. (1968*b*). Evolutionary rate at the molecular level. Nature, Lond., *217*: 624–626. (p. 297).

KIMURA, M. & OHTA, T. (1971). Protein polymorphism as a phase of molecular evolution. Nature, Lond., *229*: 467–469. (p. 297).

KIMURA, M. & OHTA, T. (1972). Population genetics, molecular biometry, and evolution. Proc. 6th Berkeley Symp. Math. Stat. Prob., *5*: 43–68. (p. 298).

KING, C. E. (1972). Adaptation of rotifers to seasonal variation. Ecology, *53*: 408–418. (p. 99).

KING, H. D. (1939). Life processes in gray Norway rats during fourteen years in captivity. Am. anat. Mem., No. 17. (p. 270).

KING, J. L. (1967). Continuously distributed factors affecting fitness. Genetics, *55*: 483–492. (p. 293).

KING, J. L. & JUKES, T. H. (1969). Non-Darwinian evolution. Science, N.Y., *164*: 788–798. (p. 296).

KINNEAR, N. B. (1906). On the mammals of Fair Isle, with a description of a new sub-species of *Mus sylvaticus*. Ann. Scot. nat. Hist., 65–68. (p. 201).

KIRK, R. L. (1968). *The Haptoglobin Groups in Man*. Basel: Karger. (p. 162).

KITCHING, J. A. & EBLING, F. J. (1967). Ecological studies at Lough Ine. In *Advances in Ecological Research*: 197–291. Cragg, J. B. (Ed.). London & New York: Academic. (p. 136).

KOEHN, R. K. (1969). Esterase heterogeneity: dynamics of a polymorphism. Science, N.Y., *163*: 943–944. (p. 304).

KOESTLER, A. (1971). *The Case of the Midwife Toad*. London: Hutchinson. (p. 21).

KREBS, C. J. (1964). The lemming cycle at Baker Lake, Northwest Territories, during 1959–62. Arctic Instit. North America, Tech. Paper no. 15. (p. 227).

KREBS, C. J., GAINES, M. S., KELLER, B. L., MYERS, J. H. & TAMARIN, R. H. (1973). Population cycles in small rodents. Science, N.Y., *179*: 35–41. (p. 227).

KÜCHEMANN, C. F., BOYCE, A. J. & HARRISON, G. A. (1967). A demographic and genetic study of a group of Oxfordshire villages. Hum. Biol., *39*: 251–276. (p. 107).

KÜCHEMANN, C. F., HARRISON, G. A., HIORNS, R. W. & CARRIVICK, P. J. (1974). Social class and marital distance in Oxford City. Ann. hum. Biol., *1*: 13–27. (p. 107).

KURTÉN, B. (1959). Rates of evolution in fossil mammals. Cold Spring Harb. Symp. quant. Biol., *24*: 205–215. (p. 215).

LACK, D. (1947). *Darwin's Finches*. Cambridge: University Press. (pp. 13, 198).

LACK, D. (1954). *The Natural Regulation of Animal Numbers*. Oxford: Clarendon. (p. 110).

LACK, D. (1966). *Population Studies of Birds*. Oxford: Clarendon. (pp. 110, 305).

LACK, D. (1968). *Ecological Adaptations for Breeding in Birds*. London: Methuen. (pp. 114, 229).

LAMOTTE, M. (1951). Recherches sur la structure génétique des populations naturelles de *Cepaea nemoralis* L. Biol. Bull. Suppl., France, *35*: 1–239. (p. 54).

LANGLET, O. (1959). A cline or not a cline – a question of Scots Pine. Sylvae genet., *8*: 13–22. (p. 31).

LARGEN, M. J. (1966). Some aspects of the biology of *Nucella lapillus* (L.). Ph.D. thesis, University of London. (p. 221).

LAWICK-GOODALL, J. (1971). *In the Shadow of Man: a study of Chimpanzees*. London: Collins. (p. 26).

LEDERBERG, J. (1966). Experimental genetics and human evolution. Am. Nat., *100*: 519–531. (p. 284).

LEE, B. T. O. & PARSONS, P. A. (1968). Selection, prediction and response. Biol. Rev., *43*: 139–174. (p. 183).

LEES, D. R. (1968). Genetic control of the melanic form *insularia* of the peppered moth *Biston betularia* (L.). Nature, Lond., *220*: 1249–1250. (p. 123).

LEES, D. R. (1971). The distribution of melanism in the pale brindled beauty moth, *Phigalia pedaria*, in Great Britain. In *Ecological Genetics and Evolution*: 175–201. Creed, E. R. (Ed.). Oxford: Blackwell. (pp. 127, 128).

LEES, D. R. & CREED, E. R. (1975). Industrial melanism in *Biston betularia*: the role of selective predation. J. Anim. Ecol., *44*: 67–83. (pp. 123, 124, 125, 128, 175).

LEES, D. R., CREED, E. R. & DUCKETT, J. G. (1973). Atmospheric pollution and industrial melanism. Heredity, *30*: 227–232. (p. 129).

LEOPOLD, A. S. (1944). The nature of heritable wildness in turkeys. Condor, *46*: 133–197. (p. 270).

LERNER, I. M. (1950). *Population Genetics and Animal Improvement*. Cambridge: University Press. (p. 159).

LERNER, I. M. (1954). *Genetic Homeostasis*. Edinburgh & London: Oliver & Boyd. (p. 273).

LEVENE, H. (1953). Genetic equilibrium when more than one ecological niche is available. Am. Nat., *87*: 311–313. (p. 299).

LEVINTON, J. (1973). Genetic variation in a gradient of environmental variability: marine Bivalvia (Mollusca). Science, N.Y., *180*: 75–76. (p. 303).

LEVINS, R. (1968). *Evolution in Changing Environments. Some theoretical explorations.* Princeton: University Press. (p. 302).

LEWIS, D. (1954). Comparative incompatibility in angiosperms and fungi. Adv. Genet., *6*: 235–285. (p. 65).

LEWIS, J. R. (1964). *The Ecology of Rocky Shores*. London: English Universities Press. (p. 136).

LEWONTIN, R. C. (1972). Testing the theory of natural selection. Nature, Lond., *236*: 181–182. (p. 13).

LEWONTIN, R. C. (1974). *The Genetic Basis of Evolutionary Change*. New York & London: Columbia University Press. (pp. 193, 195).

LEWONTIN, R. C. & HUBBY, J. L. (1966). A molecular approach to the study of genic heterozygosity in natural populations. II Amount of variation and degree of heterozygosity in natural populations of *Drosophila pseudoobscura*. Genetics, *54*: 595–609. (pp. 20, 287).

LI, C. C. (1955). *Population Genetics*. Chicago: University Press. (pp. 14, 167, 195).

LIDICKER, W. Z. (1973). Regulation of numbers in an island population of the California vole, a problem in community dynamics. Ecol. Monogr., *43*: 271–302. (p. 227).

LOCKLEY, R. M. (1942). *Shearwaters*. London: Dent. (p. 26).

LOCKLEY, R. M. (1943). *Dream Island Days: a Record of the Simple Life*. London: Witherby. (p. 234).

LOCKLEY, R. M. (1947). *Letters from Skokholm*. London: Dent. (p. 234).

LUDWIG, F. (1901). Variationsstatistische Probleme und Materialen. Biometrika, *1*: 11–29. (p. 31).

LUERS, H. (1953). Untersuchung zur Frage der Mutagenitat des Kontaktimsektizidf DDT an *Drosophila melanogaster*. Naturwissenshaften, *40*: 293. (p. 261).
LUND, M. (1968). Testing of poisons for control of rodents. Arsberetn St. Skadedyrlab. for 1967, 69–74. (p. 119).

MACALPINE, I. & HUNTER, R. (1968). *Porphyria – a Royal Malady*. London: British Medical Association. (p. 17).
MACARTHUR, R. H. (1960). On the relative abundance of species. Am. Nat., *94*: 25–36. (p. 300).
MACARTHUR, R. H. & WILSON, E. O. (1967). *The Theory of Island Biogeography*. Princeton: University Press. (pp. 56, 300).
MCFARQUAR, A. M. & ROBERTSON, F. W. (1963). The lack of evidence for coadaptation in crosses between geographical races of *Drosophila subobscura* Coll. Genet. Res., *4*: 104–131. (p. 195).
MCNEILLY, T. (1968). Evolution in closely adjacent plant populations. III *Agrostis tenuis* on a small copper mine. Heredity, *23*: 99–108. (p. 265).
MCPHEE, H. C. & WRIGHT, S. (1926). Mendelian analysis of the pure breeds of livestock. IV The British Dairy Shorthorns. J. Hered., *17*: 397–401. (p. 229).
MCWHIRTER, K. G. (1957). A further analysis of variability in *Maniola jurtina* L. Heredity, *11*: 359–371. (p. 296).
MCWHIRTER, K. G. (1969). Heritability of spot-number in Scillonian strains of the meadow brown butterfly (*Maniola jurtina*). Heredity, *24*: 314–318. (p. 217).
MCWHIRTER, K. G. & SCALI, V. (1965). Ecological bacteriology of the meadow brown butterfly. Heredity, *21*: 517–521. (p. 217).
MANLEY, B. F. J. (1975). A second look at some data on a cline. Heredity, *34*: 423–426. (p. 175).
MARCHANT, C. J. (1968). Evolution in *Spartina* (Gramineae). II Chromosomes, basic relationships and the problem of the *S.* × *townsendii* agg. J. Linn. Soc. (Bot.), *60*: 381–409. (p. 76).
MARSDEN-JONES, E. M. & TURRILL, W. B. (1938). Transplant experiments of the British Ecological Society at Potterne, Wiltshire. J. Ecol., *26*: 380–389. (p. 21).
MARSH, N. & ROTHSCHILD, M. (1974). Aposematic and cryptic Lepidoptera tested on the mouse. J. Zool., Lond., *174*: 89–122. (p. 134).
MARSHALL, D. R. & ALLARD, R. W. (1972). Maintenance of isozyme polymorphisms in natural populations of *Avena barbata*. Genetics, *66*: 393–399. (p. 304).
MATHER, K. (1943). Polygenic inheritance and natural selection. Biol. Rev., *18*: 32–64. (pp. 183, 292).
MATHER, K. (1953). The genetical structure of populations. Symp. Soc. exp. Biol., *7*: 66–95. (pp. 132, 292).
MATHER, K. (1955). Polymorphism as an outcome of disruptive selection. Evolution, Lancaster, Pa, *9*: 52–61. (p. 134).
MATHER, K. (1964). *Human Diversity*. Edinburgh & London: Oliver & Boyd. (p. 133).
MATHER, K. (1974). *Genetical Structure of Populations*. London: Chapman & Hall. (p. 292).
MATHER, K. & JINKS, J. L. (1971). *Biometrical Genetics*. London: Chapman & Hall. (pp. 14, 44, 159, 195).

MATTHEWS, L. H. (1952). *British Mammals*. London: Collins. (p. 201).
MAYR, E. (1954). Change in genetic environment and evolution. In *Evolution as a Process*: 157–180. Huxley, J., Hardy, A. C. & Ford, E. B. (Eds.). London: Allen & Unwin. (pp. 194, 220).
MAYR, E. (1959). Where are we? Cold Spring Harb. Symp. quant. Biol., *24*: 1–14. (pp. 28, 196).
MAYR, E. (1963). *Animal Species and Evolution*. London: Oxford. (pp. 28, 134).
MAYR, E. (1969). The biological meaning of species. Biol. J. Linn. Soc., *1*: 311–320. (p. 28).
MAYR, E. & AMADON, D. (1951). A classification of recent birds. Am. Mus. Novit., *1496*: 1–42. (p. 28).
MELLANBY, K. (1970). *Pesticides and Pollution*. London: Collins. (p. 251).
MEISE, W. (1928). Die Verbreitung der Aaskrähe (Formenkreis *Corvus corone* L.). J. Ornithol., *76*: 1–203. (p. 177).
MERRIFIELD, F. (1894). Temperature experiments in 1893 on several species of *Vanessa* and other Lepidoptera. Trans. ent. Soc., Lond., *1894*, 425–438. (p. 121).
MERRIL, C. R., GEIER, M. R. & PETRICCIANI, J. C. (1971). Bacterial virus gene expression in human cells. Nature, Lond., *233*: 398–400. (p. 284).
MILKMAN, R. D. (1967). Heterosis as a major cause of heterozygosity in nature. Genetics, *55*: 493–495. (p. 293).
MILLAIS, J. G. (1904). On a new British vole from the Orkney islands. Zoologist (4) *8*: 241–246. (p. 214).
MILLER, R. S. (1967). Pattern and process in competition. Adv. ecol. Res., *4*: 1–74. (p. 299).
MILNE, H. & ROBERTSON, F. W. (1965). Polymorphisms in egg albumen protein and behaviour in the eider duck. Nature, Lond., *205*: 367–369. (p. 226).
MONTAGU, I. G. S. (1922). On a further collection of mammals from the Inner Hebrides. Proc. zool. Soc., Lond., 929–941. (p. 201).
MOORE, H. B. (1936). The biology of *Purpura lapillus*. I Shell variation in relation to environment. J. mar. biol. Ass. U.K., *21*: 61–89. (p. 220).
MOORE, N. W. (1964). Intra- and interspecific competition among dragonflies (Odonata). J. Anim. Ecol., *33*: 49–71. (p. 301).
MUGGLETON, J., LONSDALE, D. & BENHAM, B. R. (1975). Melanism in *Adalia bipunctata* L. (Col., Coccinellidae) and its relationship to atmospheric pollution. J. appl. Ecol., *12*: 451–464. (p. 263).
MUIR, T. S. (1885). *Ecclesiological notes on some of the islands of Scotland*. Edinburgh: Douglas. (p. 237).
MÜLLER, H. J. (1950). Our load of mutations. Am. J. hum. Genet., *2*: 111–176. (pp. 25, 288).
MÜLLER, H. J. (1965). Means and aims in human genetic betterment. In *Control of Human Heredity and Environment*: 100–122. Sonneborn, T. M. (Ed.). London & New York: Macmillan. (p. 281).
MURRAY, J. (1972). *Genetic Diversity and Natural Selection*. Edinburgh: Oliver & Boyd. (p. 141).
MURTON, R. K. (1965). *The Wood-Pigeon*. London: Collins. (p. 113).
MURTON, R. K. (1971). *Man and Birds*. London: Collins. (p. 58).
MURTON, R. K., THEARLE, R. J. P. & COOMBS, C. F. B. (1974). Ecological studies of the feral pigeon *Columba livia* var. J. appl. Ecol., *11*: 841–854. (p. 110).

Murton, R. K., Westwood, N. J. & Thearle, R. J. P. (1973). Polymorphism and the evolution of a continuous breeding season in the pigeon, *Columba livia*. J. Reprod. Fert., suppl. 19: 563–577. (p. 109).

Myers, J. (1974). Genetic and social structure of feral house mouse populations on Grizzly Island, California. Ecology, *55* 747–759. (p. 246).

Nabours, R. K. (1929). The genetics of the Tettigidae (grouse locusts). Bibliogr. genet., *5*: 27–104. (p. 185).

Nabours, R. K., Larson, I. & Hartwig, N. (1933). Inheritance of color-patterns in the grouse locust *Acrydium arenosum* Burmeister (Tettigidae). Genetics, *18*: 159–171. (p. 185).

Neel, J. V. (1963). *Changing Perspectives on the Genetic Effects of Radiation*. Springfield: Thomas. (p. 258).

Nei, M., Maruyama, T. & Chakraborty, R. (1975). The bottleneck effect and genetic variability in populations. Evolution, Lancaster, Pa, *29*: 1–10. (p. 295).

O'Donald, P. (1967). On the evolution of dominance, over-dominance and balanced polymorphism. Proc. roy. Soc. B, *168*: 216–228. (p. 193).

O'Donald, P. (1968*a*). Natural selection by glow-worms in a population of *Cepaea nemoralis*. Nature, Lond., *217*: 194. (p. 149).

O'Donald, P. (1968*b*). Models of the evolution of dominance. Proc. roy. Soc. B, *171*: 127–143. (p. 193).

O'Donald, P. (1972). Natural selection of reproductive rates and breeding times and its effect on sexual selection. Am. Nat., *106*: 368–379. (p. 145).

O'Donald, P. (1973). A further analysis of Bumpus' data: the intensity of natural selection. Evolution, Lancaster, Pa, *27*: 398–404. (p. 132).

O'Donald, P. & Davis, P. E. (1959). The genetics of the colour phases of the Arctic Skua. Heredity, *13*: 481–486. (p. 142).

O'Donald, P. & Davis, J. W. F. (1975). Demography and selection in a population of Arctic Skuas. Heredity, *35*: 75–83. (p. 144).

O'Donald, P., Wedd, N. S. & Davis, J. W. F. (1974). Mating preferences and sexual selection in the Arctic Skua. Heredity, *33*: 1–16. (p. 145).

O'Reilly, R. A., Aggeler, P. M., Hoag, M. S., Leong, L. S. & Kropotkin, B. A. (1964). Hereditary transmission of exceptional resistance to coumarin anticoagulant drugs. New Engl. J. Med., *271*: 809–815. (p. 120).

Palmen, E. (1949). The *Diplopoda* of eastern Fennoscandia. Ann. zool. Soc. Vanamo, *13*: 1–54. (p. 97).

Park, T. (1962). Beetles, competition and populations. Science, N.Y., *138*: 1369–1375. (p. 305).

Parkin, D. T. (1972). Climatic selection in the land snail *Arianta arbustorum* in Derbyshire, England. Heredity, *28*: 49–56. (p. 184).

Parsons, P. A. (1971). Extreme-environment heterosis and genetic loads. Heredity, *26*: 479–483. (p. 296).

Parsons, P. A. & Bodmer, W. F. (1961). The evolution of overdominance: natural selection and heterozygote advantage. Nature, Lond., *190*: 7–12. (p. 183).

Passmore, J. (1974). *Man's Responsibility for Nature*. London: Duckworth. (p. 251).

Peacock, W. J. & Miklos, G. L. G. (1973). Meiotic drive in *Drosophila*. Adv. Genet., *17*: 361–409. (p. 228).

PENROSE, L. S. (1955). Evidence of heterosis in man. Proc. roy. Soc. B, *144*: 203–213. (pp. 65, 289).

PENROSE, L. S. (1959). *Outline of Human Genetics*. London: Heinemann. (p. 66).

PENROSE, L. S. (1961). Genetics of growth and development of the foetus. In *Recent Advances in Human Genetics*: 56–75. Penrose, L. S. (Ed.). London: Churchill. (p. 131).

PENROSE, L. S. (1963). *Biology of Mental Defect*, 3rd Edition. London: Sidgwick & Jackson. (p. 66).

PENROSE, L. S. & STERN, C. (1958). Reconsideration of the Lambert pedigree (Ichthyosis hystrix gravior). Ann. hum. Genet., *22*: 258–283. (p. 102).

PERRING, F. H. & SELL, P. D. (1968). *Critical Supplement to Atlas of the British Flora*. London & New York: Nelson.

PERRINS, C. M. (1965). Population fluctuations and clutch-size in the great tit. J. Anim. Ecol., *34*: 601–647. (p. 112).

PIANKA, E. R. (1970). On *r*- and *K*-selection. Am. Nat., *104*: 592–597. (p. 300).

POLANI, P. E. (1974). Chromosomal and other genetic influences on birth weight variation. Ciba Foundation Symp., no. 27: 127–164. (p. 131).

POLLARD, E., HOOPER, M. & MOORE, N. W. (1974). *Hedges*. London: Collins. (p. 267).

POPHAM, E. J. (1941). The variation in the colour of certain species of *Arctocorisa* (Hemiptera, Corixidae) and its significance. Proc. zool. Soc., Lond., *111*: 135–172. (p. 141).

POPHAM, E. J. (1942). Further experimental studies on the selective action of predators. Proc. zool. Soc., Lond., *112*: 105–117. (p. 142).

POPP, R. A. & BAILIFF, E. G. (1973). Sequence of amino acids in the major and minor β chains of the diffuse haemoglobin from BALB/c mice. Biochim. biophys. Acta, *303*: 61–67. (p. 246).

POWELL, J. R. (1971). Genetic polymorphisms in varied environments. Science, N.Y., *174*: 1035–1036. (p. 303).

PRAKASH, S. & LEWONTIN, R. C. (1968). A molecular approach to the study of genic heterozygosity in natural populations. III Direct evidence of coadaptation in gene arrangements of *Drosophila*. Proc. Nat. Acad. Sci., U.S.A., *59*: 398–405. (pp. 184, 291).

PROVINE, W. B. (1971). *The Origins of Theoretical Population Genetics*. Chicago & London: Chicago University Press. (p. 136).

PUNNETT, R. C. (1950). Early days of genetics. Heredity, *4*: 1–10. (p. 50).

PURDOM, C. E. (1963). *Genetic Effects of Radiations*. London: Newnes. (p. 254).

RAGGE, D. R. (1963). First record of the grasshopper *Stenobothrus stigmaticus* (Rambur) (Acrididae) in the British Isles, with other new distribution records and notes on the origin of the British Orthoptera. Entomologist, *96*: 211–217. (pp. 198, 199).

RAMEL, C. (1969). Genetic effects of mercury compounds. Hereditas, *61*: 208–254. (p. 256).

RAMSEY, P. (1970). *Fabricated Man. The Ethics of Genetic Control*. New Haven & London: Yale University Press. (p. 285).

RATSEY, S. (1973). The climate at and around Nettlecombe Court, Somerset. Fld Stud., *3*: 741–757. (p. 155).

RENNER, O. (1929). *Artbastarde bei Pflanzen*. Berlin: Borntraeger. (p. 291).

RICHARDSON, A. M. M. (1974). Differential climatic selection in natural populations of land snail *Cepaea nemoralis*. Nature, Lond., *247*: 572–573. (p. 154).

ROBERTS, D. F. & RAWLING, C. P. (1974). Secular trends in genetic structure: an isonymic analysis of Northumberland parish records. Ann. hum. Biol., *1*: 393–410. (p. 107).

ROTHENBUHLER, W. C. (1967). Genetic and evolutionary considerations of social behavior of honeybees and some related insects. In *Behavior-Genetic Analysis*: 61–111. Hirsch, J. (Ed.). New York: McGraw-Hill. (pp. 41, 42).

ROTHSCHILD, M. (1971). Speculations on mimicry with Henry Ford. In *Ecological Genetics and Evolution*: 202–233. Creed, R. (Ed.). Oxford: Blackwell. (p. 135).

ROTHSCHILD, M. (1972). Secondary plant substances and warning colouration in insects. In *Insect/Plant Relationships*: 59–83. Van Emden, H. F. (Ed.). Oxford: Blackwell. (p. 220).

ROTHSTEIN, R. I. (1973). The niche-variation model – is it valid? Am. Nat., *107*: 598–620. (p. 301).

ROUGHGARDEN, J. (1971). Density dependent natural selection. Ecology, *52*: 453–468. (p. 299).

ROWLANDS, I. W. (1964). Rare breeds of domesticated animals being preserved by the Zoological Society of London. Nature, Lond., *202*: 131–132. (p. 272).

RUSSELL, E. S. (1966). Lifespan and ageing pattern. In *Biology of the Laboratory Mouse*, 2nd Edition: 511–519. Green, E. L. (Ed.). New York: McGraw-Hill. (p. 242).

RUSSELL, E. S. & BERNSTEIN, S. E. (1966). Blood and blood formation. In *Biology of the Laboratory Mouse*, 2nd Edition: 351–372. Green, E. L. (Ed.). New York: McGraw-Hill. (p. 246).

RUSSELL, E. S. & MCFARLAND, E. C. (1974). Genetics of mouse hemoglobins. Ann. N.Y. Acad. Sci., *241*: 25–38. (p. 246).

RUSSELL, G. & MORRIS, P. (1970). Copper tolerance in the marine fouling alga *Ectocarpus siliculosus*. Nature, Lond., *228*: 288–289. (p. 267).

RYDER, M. L., LAND, R. B. & DITCHBURN, R. (1974). Coat colour inheritance in Soay, Orkney and Shetland sheep. J. Zool., Lond., *173*: 477–485. (p. 40).

SARGENT, T. D. (1971). Melanism in *Phigalia titea* (Cramer) (Lepidoptera: Geometridae). J. N.Y. ent. Soc., *79*: 122–129. (p. 127).

SAYERS, D. L. (1932). *Have His Carcase*. London: Gollancz. (p. 44).

SEARLE, A. G. (1949). Gene frequencies in London's cats. J. Genet., *49*: 214–220. (p. 51).

SEARLE, A. G. (1964). The gene-geography of cats. J. Cat Genet., *1*: 18–24. (p. 51).

SEARLE, A. G. (1974). Mutation induction in mice. Adv. rad. Biol., *4*: 131–207. (p. 255).

SELANDER, R. K., HUNT, W. G. & YANG, S. Y. (1969). Protein polymorphism and genic heterozygosity in two European subspecies of the house mouse. Evolution, Lancaster, Pa, *23*: 379–390. (p. 178).

SELANDER, R. K. & KAUFMAN, D. W. (1973*a*). Self-fertilization and genetic population structure in a colonizing land snail. Proc. Nat. Acad. Sci., U.S.A., *70*: 1186–1190. (p. 304).

SELANDER, R. K. & KAUFMAN, D. W. (1973*b*). Genic variability and strategies of adaptation in animals. Proc. Nat. Acad. Sci., U.S.A., *70*: 1875–1877. (pp. 20, 293, 302).

SELANDER, R. K., YANG, S. Y., LEWONTIN, R. C. & JOHNSON, W. E. (1970).

Genetic variation in the horseshoe crab (*Limulus polyphemus*), a phylogenetic "relic". Evolution, Lancaster, Pa, *24*: 402–414. (p. 303).

SEMENOFF, R. & ROBERTSON, F. W. (1968). A biochemical and ecological study of plasma esterase polymorphism in natural populations of the field vole, *Microtus agrestis*, L. Biochem. Genet., *1*: 205–227. (p. 227).

SHARPE, R. B. (1909). *A Handlist of the Genera and Species of Birds*. London: British Museum (Natural History). (p. 28).

SHELDON, R. W. & KERR, S. R. (1972). The population density of monsters in Loch Ness. Limnol. Oceanogr., *17*(5): 776–779. (p. 300).

SHEPPARD, P. M. (1951). Fluctuations in the selective value of certain phenotypes in the polymorphic land snail *Cepaea nemoralis* (L.). Heredity, *5*: 125–134. (p. 149).

SHEPPARD, P. M. (1959). Blood groups and natural selection. Br. med. Bull., *15*: 134–139. (p. 279).

SHEPPARD, P. M. (1975). *Natural Selection and Heredity*, 4th Edition. London: Hutchinson. (pp. 14, 294).

SHEPPARD, P. M. & FORD, E. B. (1966). Natural selection and the evolution of dominance. Heredity, *21*: 139–147. (p. 193).

SHORROCKS, B. (1975). The distribution and abundance of woodland species of British *Drosophila* (Diptera: Drosophilidae). J. Anim. Ecol., *44*: 851–864. (p. 195).

SHUGART, H. H. & BLAYLOCK, B. G. (1973). The niche-variation hypothesis: an experimental study with *Drosophila* populations. Am. Nat., *107*: 575–579. (p. 303).

SINSHEIMER, R. L. (1969). The prospect for designed genetic change. Am. Scient., *57*: 134–142. (p. 283).

SKELTON, M. (1969). Unpublished work, cited by Bradshaw (1971). (p. 267).

SLATKIN, M. (1973). Gene flow and selection in a cline. Genetics, *75*: 733–756. (pp. 173, 194).

SMITH, G. L. (1963). Studies in *Potentilla* L. New Phytol., *62*: 264–282; 283–300. (p. 95).

SMITH, G. R. & KOEHN, R. K. (1971). Phenetic and cladistic studies of biochemical and morphological characteristics of *Catostomus*. Syst. Zool., *20*: 282–297. (p. 304).

SMITH, J. M. (1962). Disruptive selection, polymorphism and sympatric speciation. Nature, Lond., *195*: 60–62. (p. 133).

SMITH, J. M. (1964). Group selection and kin selection. Nature, Lond., *201*: 1145–1147. (p. 230).

SMITH, J. M. (1966). Sympatric speciation. Am. Nat., *100*: 637–650. (pp. 134, 299).

SMITH, J. M. (1968). "Haldane's dilemma" and the rate of evolution. Nature, Lond., *219*: 1114–1116. (p. 293).

SMITH, J. M. (1970). The causes of polymorphism. In *Variation in Mammalian Populations*: 371–383. Berry, R. J. & Southern, H. N. (Eds.). London: Academic. (p. 277).

SMITH, R. H. (1974). Is the slow worm a Batesian mimic? Nature, Lond., *247*: 571–572. (p. 134).

SNAYDON, R. W. (1970). Rapid population differentiation in a mosaic environment. I The response of *Anthoxanthum odoratum* populations to soils. Evolution, Lancaster, Pa, *24*: 257–269. (p. 62).

SORSBY, A. (1958). Noah – an albino. Br. med. J., *ii*: 1587–1589. (p. 17).

SOULÉ, M. & YANG, S. Y. (1974). Genetic variation in side-blotched lizards on islands in the Gulf of California. Evolution, Lancaster, Pa, *27*: 593–600. (p. 304).

SOUTHERN, H. N. (1954). Mimicry in cuckoos' eggs. In *Evolution as a Process*: 219–232. Huxley, J. S., Hardy, A. C. & Ford, E. B. (Eds.). London: Allen & Unwin. (p. 116).

SOUTHERN, H. N. (1962). Survey of bridled guillemots, 1959–60. Proc. zool. Soc., Lond., *138*: 455–472. (pp. 168, 169).

SOUTHERN, H. N., CARRICK, R. & POTTER, W. G. (1965). The natural history of a population of guillemots (*Uria aalge* Pont.). J. Anim. Ecol., *34*: 649–665. (p. 168).

SOUTHWOOD, T. R. E. (1971). The role and measurement of migration in the population system of an insect pest. Trop. Sci., *13*: 275–278. (p. 302).

SOUTHWOOD, T. R. E., MAY, R. M., HASSELL, M. P. & CONWAY, G. R. (1974). Ecological strategies and population parameters. Am. Nat., *108*: 791–804. (pp. 301, 305).

SPIGHT, T. M. (1974). Sizes of population of a marine snail. Ecology, *55*: 712–729. (p. 222).

SPOONER, R. L., MAZUMDER, N. K., GRIFFIN, T. K., KINGWILL, R. G., WIJERATNE, W. V. S. & WILSON, C. D. (1973). Apparent heterozygote excess at the amylase I locus in cattle. Anim. Prod., *16*: 209–214. (p. 71).

STAIGER, H. (1954). Der chromosomendimorphism beim Prosobranchier *Purpura lapillus* in Beziehung zur Okologie der Art. Chromosoma, *6*: 419–478. (p. 78).

STAIGER, H. (1957). Genetical and morphological variation in *Purpura lapillus* with respect to local and regional differentiation of population groups. Ann. Biol., *33*: 251–258. (p. 79).

STEBBINGS, R. E. (1973). Size clines in the bat *Pipistrellus pipistrellus* related to climatic factors. Period. biol., *75*: 189–194. (p. 29).

STERN, C. (1970). Model estimates of the number of gene pairs in pigmentation variability of the Negro-Americans. Hum. Hered., *20*: 165–168. (p. 38).

SUMNER, F. B. (1932). Genetic, distributional, and evolutionary studies of the subspecies of deer mice (Peromyscus). Biblio. genet., *9*: 1–106. (p. 42).

SUOMALAINEN, E. (1950). Parthenogenesis in animals. Adv. Genet., *3*: 193–253. (p. 96).

SVED, J. A., REED, T. E. & BODMER, W. F. (1967). The number of balanced polymorphisms that can be maintained in a natural population. Genetics, *55*: 469–481. (p. 293).

SWYNNERTON, C. F. M. (1918). Rejection by birds of eggs unlike their own; with remarks on some of the cuckoo problems. Ibis, *6*: 127–154. (p. 115).

TATUM, E. L. (1965). Perspectives from physiological genetics. In *The Control of Human Heredity*: 20–34. Sonneborn, T. M. (Ed.). New York: Macmillan. (p. 283).

TAZIMA, Y., KADA, T. & MURAKAMI, A. (1975). Mutagenicity of nitrofuran derivatives, including furylfuramide, a food preservative. Mutation Res., *32*: 55–80. (p. 257).

TAYLOR, G. Rattray (1969). *The Biological Time Bomb*. London: Panther. (p. 285).

THODAY, J. M. (1961). The location of polygenes. Nature, Lond., *191*: 368–370. (pp. 183, 292).

THODAY, J. M. (1972). Disruptive selection. Proc. roy. Soc. B, *182*: 109–143. (pp. 134, 292).

THOMPSON, V. (1973). Spittlebug polymorphic for warning coloration. Nature, Lond., *242*: 126–128. (p. 219).

THOMPSON, W. A. & VERTINSKY, I. (1975). Bird flocking revisited: the case with polymorphic prey. J. Anim. Ecol., *44*: 755–765. (p. 73).

TODD, N. B., FAGEN, R. M. & FAGEN, K. (1975). Gene frequencies in Icelandic cats. Heredity, *35*: 172–183. (p. 51).

TUCK, L. M. (1960). *The Murres*. Monogr. nat. Pks Br., Canad. Wildl. Ser. No. 1. (p. 168).

TURESSON, G. (1922). The genotypical response of the plant species to the habitat. Hereditas, *3*: 211–350. (p. 286).

TURNER, J. R. G. (1968). On supergenes. II The estimation of gametic excess in natural populations. Genetica, *39*: 82–93. (p. 184).

TURNER, J. R. G. & WILLIAMSON, M. H. (1968). Population size, natural selection and the genetic load. Nature, Lond., *218*: 700. (p. 146).

TUTT, J. W. (1899). *A Natural History of the British Lepidoptera*. London: Swann Sonnenschein. (p. 121).

UCKO, P. J. & DIMBLEBY, G. W. (Eds.). (1969). *The Domestication and Exploitation of Plants and Animals*. London: Duckworth. (p. 268).

URSIN, E. (1952). Occurrence of voles, mice and rats (*Muridae*) in Denmark, with a special note on a zone of intergradation between two subspecies of the house mouse (*Mus musculus* L.). Vidensk. Meddr dank naturh. Foren., *114*: 217–244. (p. 178).

VAN VALEN, L. (1965). Morphological variation and width of ecological niche. Am. Nat., *99*: 377–390. (p. 301).

VAN VALEN, L. (1974). Molecular evolution as predicted by natural selection. J. molec. Evol., *3*: 89–101. (p. 298).

VARSHAVSKII, S. N. (1949). The age composition of the house mouse population (*Mus musculus* L.). Zool. Zh., *28*: 361–371 (in Russian). (p. 242).

VOGEL, F. & CHAKRAVARTTI, M. R. (1966). ABO blood groups and smallpox in a rural population of West Bengal and Bihar (India). Humangenetik, *3*: 166–180. (p. 279).

WADDINGTON, C. H. (1953*a*). Genetic assimilation of an acquired character. Evolution, Lancaster, Pa, *7*: 118–126. (p. 22).

WADDINGTON, C. H. (1953*b*). Epigenetics and evolution. Symp. Soc. exp. Biol., *7*: 186–199. (p. 208).

WADDINGTON, C. H. (1957). *The Strategy of the Genes*. London: Allen & Unwin. (p. 195).

WAHRMAN, J., GOITEIN, R. & NEVO, E. (1969). Mole rat *Spalax*: evolutionary significance of chromosome variation. Science, N.Y., *164*: 82–84. (p. 82).

WALLACE, B. (1956). Studies on irradiated populations of *Drosophila melanogaster*. J. Genet., *54*: 280–293. (p. 291).

WALLACE, B. (1963). Further data on the overdominance of induced mutations. Genetics, *48*: 633–651. (p. 291).

WALLACE, B. (1968). *Topics in Population Genetics*. New York: Norton. (p. 294).

WALLACE, B. (1970). *Genetic Load*. Englewood Cliffs: Prentice-Hall. (pp. 289, 294).

WALLACE, M. E. & MACSWINEY, F. J. (1975). Warfarin resistance and a new gene for obesity. Mouse News Letter, no. 53: 20. (p. 119).

WALLEY, K. A., KHAN, M. S. I. & BRADSHAW, A. D. (1974). The potential for evolution of heavy metal tolerance in plants. I Copper and zinc tolerance in *Agrostis tenuis*. Heredity, *32*: 309–319. (p. 265).

WALTON, J. R. (1966). Infectious drug resistance in *Escherichia coli* isolated from healthy farm animals. Lancet, *ii*: 1300–1302. (p. 262).

WARWICK, T. (1940). Field mice (*Apodemus*) from the Outer Hebrides, Scotland. J. Mammal., *21*: 247–251. (p. 201).

WATSON, A. (1970). Territorial and reproductive behaviour of red grouse. J. Reprod. Fert., suppl. *11*: 3–14. (p. 230).

WATSON, I. A. (1970). The utilization of wild species in the breeding of cultivated crops resistant to plant pathogens. In *Genetic Resources in Plants*: 441–457. Frankel, O. H. & Bennett, E. (Eds.). Oxford & Edinburgh: Blackwell. (p. 272).

WATSON, J. D. & CRICK, F. C. (1953). Genetical implication of the structure of deoxyribose nucleic acid. Nature, Lond., *171*: 964. (p. 33).

WATSON, J. D. (1968). *The Double Helix*. London: Weidenfeld & Nicolson. (p. 33).

WATSON, P. J. (1970). Evolution in closely adjacent plant populations. VII An entomophilous species, *Potentilla erecta*, in two contrasting habitats. Heredity, *24*: 407–422. (p. 229).

WATSON, P. J. & FYFE, J. L. (1975). *Potentilla erecta* in two contrasting habitats – a multivariate approach. Heredity, *34*: 417–422. (p. 229).

WEBER, W. (1950). Genetical studies on the skeleton of the mouse. III Skeletal variations in wild populations. J. Genet., *50*: 174–178. (p. 231).

WEINER, J. S. (1971). *Man's Natural History*. London: Weidenfeld & Nicolson. (pp. 275, 276).

WELDON, W. F. R. (1901). A first study of natural selection in *Clausilia laminata* (Montagu). Biometrika, *1*: 109–124. (p. 136).

WEST, D. A. (1964). Polymorphism in the isopod *Sphaeroma rugicauda*. Evolution, Lancaster, Pa, *18*: 671–684. (p. 139).

WHEELER, L. L. & SELANDER, R. K. (1972). Genetic variation in populations of the house mouse, *Mus musculus*, in the Hawaiian islands. Studies in Genetics, VII. Univ. Texas Publs. No. 7213: 269–296. (p. 244).

WHITE, D. F. B. (1876). On melanochroism and leucochroism. Entomologist's mon. Mag., *13*: 172–179. (p. 120).

WHITTAKER, R. H., LEVIN, S. A. & ROOT, R. B. (1973). Niche, habitat, and ecotope. Am. Nat., *107*: 321–338. (p. 299).

WHITTEN, M. J. & FOSTER, G. G. (1975). Genetical methods of pest control. A. Rev. Ent., *20*: 461–476. (p. 261).

WICKLER, W. (1968). *Mimicry in Plants and Animals*. London: Weidenfeld & Nicolson. (p. 135).

WILLIAMS, G. C. (1966). *Adaptation and Natural Selection*. Princeton: University Press. (p. 229).

WILLIAMSON, K. (1958). Population and breeding environment of the St. Kilda and Fair Isle wrens. Br. Birds, *51*: 369–393. (p. 53).

WILLIAMSON, K. (1959). Changes of mating within a colony of Arctic Skuas. Bird Study, *6*: 51–60. (p. 144).

WILLIAMSON, K. (1965). *Fair Isle and Its Birds*. Edinburgh & London: Oliver & Boyd. (p. 142).

WILLIAMSON, M. H. (1958). Selection, controlling factors, and polymorphism. Am. Nat., *92*: 329–335. (p. 72).

WILSON, E. O. & BOSSERT, W. H. (1971). *Primer of Population Biology*. Stamford, Conn.: Sinauer. (p. 14).

WOLDA, H. (1963). Natural populations of the polymorphic landsnail *Cepaea nemoralis* (L.). Archs. neerl. Zool., *15*: 381–471. (p. 154).

WRIGHT, S. (1931). Evolution in Mendelian populations. Genetics, *16*: 97–159. (pp. 166, 195).

WRIGHT, S. (1949). Adaptation and selection. In *Genetics, Paleontology and Evolution*: 365–389. Jepsen, J. L., Simpson, G. G. & Mayr, E. (Eds.). Princeton: University Press. (p. 167).

WRIGHT, S. (1952). The genetics of quantitative variability. In *Quantitative Inheritance*: 5–41. London: HMSO. (p. 193).

WRIGHT, S. (1965*a*). The interpretation of population structure by *F*-statistics with special regard to systems of mating. Evolution, Lancaster, Pa, *19*: 395–420. (pp. 193, 229).

WRIGHT, S. (1965*b*). Factor interaction and linkage in evolution. Proc. roy. Soc. B, *162*: 80–104. (p. 194).

WRIGHT, S. & DOBZHANSKY, T. (1946). Genetics of natural populations. XII Experimental reproduction of some of the changes caused by natural selection in certain populations of *Drosophila pseudoobscura*. Genetics, *31*: 125–156. (p. 295).

WU, L., BRADSHAW, A. D. & THURMAN, D. A. (1975). The potential for evolution of heavy metal tolerance in plants. III The rapid evolution of copper tolerance in *Agrostis stolonifera*. Heredity, *34*: 165–187. (p. 266).

WYNNE-EDWARDS, V. C. (1962). *Animal Dispersion in Relation to Social Behaviour*. Edinburgh & London: Oliver & Boyd. (pp. 113, 230, 305).

WYNNE-EDWARDS, V. C. (1963). Intergroup selection in the evolution of social systems. Nature, Lond., *200*: 623–626. (p. 230).

WYNNE-EDWARDS, V. C. (1974). Society versus the individual in animal evolution. Presidential address to Section D, British Association for the Advancement of Science. (p. 230).

ZIMMERMAN, K. (1959). Uber eine Kreuzung von Unterarten der Feldmaus *Microtus arvalis*. Zool. Jb. (Syst.), *87*: 1–12. (p. 214).

ZOHARY, D. (1970). Centers of diversity and centers of origin. In *Genetic Resources in Plants*: 33–42. Frankel, O. H. & Bennett, E. (Eds.). Oxford: Blackwell. (p. 268).

ZOHARY, D. & IMBER, D. (1963). Genetic dimorphism in fruit types in *Aegilops speltoides*. Heredity, *18*: 223–231. (p. 273).

SPECIES INDEX

PLANTS

INVERTEBRATA

VERTEBRATA

GENERAL INDEX

(For names of authors, see Bibliography)